Radiometric Dating Errors

Radiometric Dating Errors

A rebuttal of Brent Dalrymple's Book "The Age Of The Earth"

Advantage BOOKS

Paul Nethercott

Radiometric Dating Error by Paul Nethercott

ISBN: 978-1-59755-667-5

Published by: ADVANTAGE BOOKS™ -Longwood, FL - www.advbookstore.com

Library of Congress Catalog Number: 2021951295

AUTHOR: Nethercott, Paul
TITLE***: Radiometric Dating Error***
PUBLISHER: Advantage Books, Longwood, FL 2021
ISBN: (print): 9781597556675
CATEGORIES: Science & Mathematics: Earth Sciences, Science & Math: Earth Sciences - Geology

First Printing: November 2021
21 22 23 24 25 26 27   10 9 8 7 6 5 4 3 2 1

Table of Contents

Introduction

How reliable is radiometric dating? We are repeatedly told that it proves the Earth to be billions of years old. If radiometric dating is reliable than it should not contradict the evolutionary model. According to the Big Bang theory the age of the Universe is 10 to 15 billion years.[1] Standard evolutionist publications give the age of the universe as 13.75 billion years. Standard evolutionist geology views the Earth as being 4.5 billion years old. Here are some quotes from popular text: "The age of the Earth is 4.54 ± 0.05 **billion** years." [2, 3, 4] "The Solar System, formed between 4.53 and 4.58 billion years ago." [5] The Milky way galaxy is 8 to 10 billion years old. [6]

If we run the isotopic ratios give in standard geology magazines through the computer program Isoplot [7] we find that the Uranium/Thorium/Lead isotopic ratios in the rocks disagree radically other dating methods. The U/Th/Pb ratios give ages older than the evolutionist age of the Earth, Solar System, Galaxy and Universe. How can Earth rocks be dated as being older than the Big Bang? Here are examples of isotopic ratios taken from several articles in major geology magazines which give absurd dates. There is a wealth of more material on the Internet. [8, 9] This book analyses Dalrymple's claims [10-13] In detail.

The Bible's View On The Age of the Earth
One thing that makes the Bible unique compared to other religious texts is its view on the age and origin of the Earth. In the fourth commandment [Exodus 20:8-11] God told the children of Israel that because He created the Earth in six days and rested on the seventh day they should work six days and rest on the seventh day. Nobody would ever think that he commanded them to work for millions of years and rest millions of years. In the creation week [Genesis 1:1-31] each day is called evening and morning [1:5, 1:8, 1:13, 1:19, 1:23, 1:31]. In the Bible evening and morning is always a 24 hour day. In Exodus 31:13-18 God told the children of Israel to keep the Sabbath because it is the same one in the creation week. Since human lifespan is only 80 years we could never keep a Sabbath millions of years long. A literal week [168 hours long] is what the Bible teaches

In the book of Isaiah [41:4, 64:4] the prophet clearly believed that humans have been on the planet since the beginning. The prophet Zechariah [12:1] tells us that God made Adam when He made the foundations of the Earth. Hebrews 1:10 tells us that He made the foundations of the Earth in the beginning. The time interval between the beginning [Genesis 1:1] and God creating Adam and Eve is five days or 120 hours.

In Matthew 19:4-6 and Mark 10:6-9 Jesus said that God made humans at the beginning [Genesis 1:1]. According to the Bible the age of the Earth is the age of mankind. According to the theory of evolution the Earth existed 4.5 billion years before the arrival of humans. In Luke 1:70 the Bible sates that God's holy prophets have been since the world began. If you accept the Earth is 4.5 billion years old that would mean humans have been here that long which is impossible. The Bible cannot be reconciled with an old Earth. Out of 6,000 years of Earth history man has been here for 99.9996804%. If we say that humans arrived 100 thousand years ago and you accept the Earth is 4.5 billion years old then man has been absent for 99.9999574% of Earth history.

The book of Luke [11:50, 51] tells us that God's prophets have been since the foundations of the world. Hebrews 1:10 tells us that this is the beginning. The age of the Earth is therefore the age of mankind. In Acts 3:26 the Bible sates that God's holy prophets have been since the world began.

Jesus said in John 8:44 that Satan tempted Adam and Eve in the beginning. The Bible states that there were 21 generations from Adam to Abraham [Luke 3:34-38]. From the creation week to the building of the pyramids [Genesis 12:15] could only be a few thousand years. Unless we say the Bible is complete rubbish we have to accept that the beginning is only a few thousand years ago. The apostle Paul tells us in Hebrews 4:3, 4 that all of God's creative work was finished at the foundations of the world. Hebrews 1:10 tells us that this is the beginning. This leaves no room for creating anything after the beginning. All of His creative work was done in the creation week. Nothing before or after.

Hebrews 9:26 tells us that sinners have been in existence since the foundations of the world. Since the foundations of the world were made at the beginning this means the age of the Earth is the age of mankind. The Bible tells us that there were no sinners before Adam and Eve [Romans 5:12, 1 Corinthians 15:22]. The creation of the Earth and the arrival of humans overlapped each other. There is no gap of millions of years. The last book of the Bible tells us that Jesus was the "Lamb slain from the foundation of the world" Revelation 13:8, 17:8. Since the foundation of the world were made in the beginning [Hebrews 1:10] this puts the age of the Earth and the entrance of sin at the same time. If the beginning were billions of years ago you would have sin before Adam and Eve.

1 **http://en.wikipedia.org/wiki/Age_of_the_universe**
2 **http://en.wikipedia.org/wiki/Age_of_the_Earth**
3 **http://sp.lyellcollection.org/content/190/1/205**
The age of the Earth, G. Brent Dalrymple, Geological Society, London, Special Publications, January 1, 2001, Volume 190, Pages 205-221
4 **The age of the earth, Gérard Manhes, Earth and Planetary Science Letters, Volume 47, Issue 3, May 1980, Pages 370–382**
5 **https://en.wikipedia.org/wiki/Formation_and_evolution_of_the_Solar_System**
6 **https://en.wikipedia.org/wiki/Galaxy_formation_and_evolution**
7 http://www.creationismonline.com/Isoplot/Isoplot.html
https://www.bgc.org/isoplot
8 https://www.icr.org/rate
9 https://creation.com/topics
10 **Ancient Earth, Ancient Skies, The Age Of Earth And Its Cosmic Surroundings, G. Brent Dalrymple, Stanford University Press, Stanford, California, 2004 https://archive.org/download/B-001-001-783/B-001-001-783.pdf**
11 **How Old is the Earth? A Response to "Scientific" Creationism, by G. Brent Dalrymple, 2006 http://www.talkorigins.org/faqs/dalrymple/contents.html**
12 **The Age of the Earth, G. Brent Dalrymple, Stanford University Press; 1st edition (February 1, 1994)**
13 **G. Brent Dalrymple, http://www.blc.arizona.edu/courses/schaffer/449/Geology/Dalrymple%20Geol%20Time.pdf**

Chapter 1

The ^{40}Argon/^{39}Argon Dating Method

207Pb/206Pb and 40Argon/39Argon ages from Southwest Montana

These rocks from North America were dated in 2002 using both [1] ^{40}Argon/^{39}Argon and Lead-Lead dating methods. Again the no dates beside the ^{207}Pb/^{206}Pb ratios. If we add dates we soon see why. The first table in his article has dates [2] using the ^{40}Ar–^{39}Ar dating method. The third table [2] has the ^{207}Pb/^{206}Pb ratios.

Table 1

Sample Name	Argon/Argon Max Age	Argon/Argon Min Age	Pb Dating Max Age	Pb Dating Min Age
RRCR2	1,818	1,695	4,471	1,895
RRSW1	1,806	1,740	5,011	4,032
HLM2	1,853	1,620	4,522	1,848
TRMR2	1,729	1,199	5,049	2,644

If we use the computer program Isoplot [3] and calculate the ages of the ^{207}Pb/^{206}Pb ratios we see why not dates have been put beside them. The Argon-Argon and Lead-Lead dating methods are extremely discordant. The author's use of data is very selective. Dates that agree are added and those that do not are omitted. This happens over and over in geology magazines. We can see from the table below that many dates are older than the evolutionist view of the age of Earth. How can such an absurdity be possible? How can the Earth be older than itself?

Table 2

Sample Name	Million Years	Age Category
RRSW1	5,005	Older Than The Solar System
RRSW1	5,011	Older Than The Solar System
RRSW1	4,939	Older Than Earth
TRMR2	5,015	Older Than The Solar System
TRMR2	5,049	Older Than The Solar System

^{207}Pb/^{206}Pb Dates

Shocked Meteorites: Argon-40/Argon-39

Dated in 1997 by scientists [4] from Germany and France, these meteorite samples gave astounding results also. Many dates were older than the evolutionist age of the Solar System, older than the evolutionist age of the galaxy and older than the Big Bang. Most age results that were hundreds or thousands of percent discordant.

Table 3

Sample Name	Maximum Million Years	Minimum Million Years	Difference Million Years	Percent Difference
A. Rose City (H5/S6) host rock	4,766	193	4,573	2,469
B. Rose City (H5/S6) melt	4,529	2,126	2,403	213
C. Rose City (H5/S6) host rock #1	3,876	231	3,645	1,678

D. Rose City (H5/S6) host rock #2	3,259	293	2,966	1,112
E. Travis County (H5/S4) whole rock	3,614	295	3,319	1,225
F. Yanzhuang (H6/S6) host rock	5,598	65	5,533	8,612
G. Yanzhuang (H6/S6) melt fragment	10,217	1,902	8,315	537
H. Yanzhuang (H6/S6) melt vein	7,016	1,314	5,702	534
I. Alfianello (L6/S5) whole rock	3,470	968	2,502	358
J. Bluff (L6/S6) host rock	13,348	506	12,842	2,638
K. Bluff (L6/S6) melt	3,773	554	3,219	681
L. Mbale (L5-6) whole rock	3,531	466	3,065	758
M. McKinney (L4/S4-5) whole rock	1,821	499	1,322	365
N. Ness County (L6/S6) host rock #I	5,052	987	4,065	512
O. Ness County (L6/S6) host rock #2	6,668	1,997	4,671	334
P. Paranaiba (L6/S6) host mk #I	3,332	453	2,879	736
Q. Paranaiba (L6/s6) host rock #2	5,593	3,110	2,483	180
R. Taiban (L5/S6) host rock	2,845	492	2,353	578
S. Taiban (L5/S6) melt	1,435	156	1,279	920
T. Walters (L6/S4) host rock	3,452	1,592	1,860	217
U. Walters (L6/S4) melt	4,074	2,026	2,048	201
V. Beeler (LU/S4) host rock #I	6,466	798	5,668	810
W. Beeler (LL6/S4) host rock #2	6,609	1,491	5,118	443
X. ALHA 8101 1 (eucrite) clast	3,818	375	3,443	1,018
Y. ALHA 8101 1 (eucrite) melt	2,827	244	2,583	1,159

Blue squares are dates over 4.5 billion years old.

Argon Diffusion Properties

Dating done in 1980 of various meteorites gave many discordant values.[13] Six were dated as older than the evolutionist view of the age of the Solar System.

Table 4

Meteor's Name	Maximum Billion Years	Minimum Billion Years	Percentage Difference
Wellman	5.2	3.737	139%
Wickenburg	3.005	0.568	529%
Shaw	5.15	4.17	123%
Louisville	5.5	0.51	1,078%
Arapahoe	9.71	0.89	1,091%
Farmington	3.7	0.511	724%
Lubbock	9.4	0.12	7,833%
Orvinio	8.78	0.764	1,149%

U-Th-Pb Dating of Abee E4 Meteorite

This dating was done in 1982 by scientists from the NASA, Johnson Space Center, Houston Texas and the U.S. Geological Survey, Denver, Colorado.[7] The two table below [Table 5, 6] are a summary of Argon dating done on different meteorite samples.[8] Both sample record dates older than the evolutionist age of the solar system. The original article has undated $^{207}Pb/^{206}Pb$ ratios. If we run the through Isoplot [7] we find the ratios [9, 10] give the results in tables 7 and 8. All are much older than the evolutionist age of the solar system.

Table 5

Abee clast 2, 2, 05		
Maximum Age	7,200	Million Years
Minimum Age	3,990	Million Years
Average Age	4,640	Million Years
Age Difference	3,210	Million Years
Difference	180%	Percent
Standard Deviation	840	Million Years

Table 6

Abee clast 3, 3, 06		
Maximum Age	8,900	Million Years
Minimum Age	3,580	Million Years
Average Age	4,610	Million Years
Age Difference	5,320	Million Years
Difference	248%	Percent
Standard Deviation	1,360	Million Years

Table 7

Meteorite Name	Pb-206/207 Ratio	Pb-206/207 Age
Abee 1	1.0992	5,370
	1.0945	5,364
	1.0947	5,364
	1.0330	5,283
Abee 2	1.1000	5,371
	1.0966	5,367
	0.8958	5,082
Abee 3	1.0976	5,368
	1.0967	5,367
	1.0708	5,333

Table 8

Meteorite Name	Pb-207/206 Ratio	Pb-207/206 Age
Abee 1	1.0993	5,370
	1.1005	5,372
	1.0994	5,370
Abee 2	1.1005	5,372
	1.0991	5,370
Abee 3	1.0999	5,371
	1.0993	5,370
Indarch	1.1005	5,372
St. Sauveur	0.7015	4,734
Canyon Diablo	1.1060	5,379

The original article has undated ^{232}Thorium and ^{238}Uranium ratios. If we run these through Isoplot [7] we find the ratios [9, 10] give the results in table 9. All these dates are between 18 and 93 billion years old. Much older than the evolutionist age of the universe.

Table 9

206Pb/238U Million Years	208Pb/232Th Million Years	207Pb/206Pb Million Years
27,806	86,711	5,370
26,605	84,996	5,364
27,370	86,616	5,364
23,272	85,323	5,283
28,051	85,725	5,371
27,476	83,944	5,367
18,801	93,166	5,082
28,127	82,811	5,368
26,517	81,174	5,367
22,143	75,483	5,333

Argon-39/Argon-40 Ages

These samples were dated in 2003 by scientists from the NASA Johnson Space Center, Houston, Texas, and the Lockheed-Martin Corporation, Houston, Texas.[11] The Monahans chondrite and halite was dated in 2001 as being over eight billion years old. [12]

Table 10

Maximum Age	8,058	Million Years
Minimum Age	3,899	Million Years
Average Age	4,474	Million Years
Age Difference	4,159	Million Years
Difference	206%	Percent

40-Argon/39-Argon Ages of Allende

Scientist from the Max-Planck-Institute, Heidelberg, Germany, dated these samples in 1980. [13] Seven samples were dated as being over five billion years old. [14] The data in table 11 contains fifteen dates [13] over 4.6 billion years old and six that are over five billion years old. The data in table 12 contains eight dates [14] over 4.6 billion years old and twenty that are over five billion years old.

Table 11

Number	Mineral	K/Ar	Ar/Ar
1	Matrix	3.8	
2	Whole	4.43	4.57
		4.62	
3	Monosomatic	4.63	
4	Barred	4.26	4.56
5	Fine	4.53	
6	Granular	4.51	
7	Granular	4.56	
8	Black	4.52	
		4.33	4.47
9	Fine	4.44	4.55
10	Fine	4.82	
11	Iquffy	4.49	4.5
12	Coarse	4.2	4.52
13	Coarse	4.4	4.53
14	Gray	4.23	4.48
15	White	5.12	4.98
16	White		5.43
17	Coarse	5.08	4.92
18	Coarse		5.37
19	Coarse	4.92	4.68
20	Coarse	5.54	5.27
21	Coarse	5.26	4.62
23	Coarse	4.57	

Table 12

Sample Name	Maximum Million Years	Minimum Million Years	Difference Million Years	Percentage Difference
Sample 01	4,455	2,452	2,003	181%
Sample 02	5,067	3,027	2,040	167%
Sample 03	4,919	4,092	827	120%
Sample 04	4,939	4,363	576	113%
Sample 05	4,691	2,248	2,443	208%
Sample 06	4,943	4,102	841	120%
Sample 07	4,835	4,166	669	116%
Sample 08	4,776	4,207	569	113%
Sample 09	5,004	3,682	1,322	135%
Sample 10	4,505	1,871	2,634	240%
Sample 11	4,707	3,631	1,076	129%
Sample 12	5,641	4,330	1,311	130%
Sample 13	4,549	4,396	153	103%
Sample 19	5,590	4,110	1,480	136%
Sample 20	5,812	4,367	1,445	133%
Sample 21	5,784	4,256	1,528	135%
Sample 23	7,460	3,967	3,493	188%

Ar-39/Ar-40 Dating of IAB Iron Meteorites

In 1979 this dating was carried out by the Department of Physics, University of California, Berkeley. [15] One of the meteorites was dated at almost ten billion years old. [16] I will use the following Argon/Argon dating formula [17] listed in Brent Dalrymple's book:

$$T = 1.1804 \times 10^{9} Log_{E}\left(J\left[\frac{{}^{40}Ar}{{}^{39}Ar} \right] + 1 \right)$$

Where T is the age in years and J is the special constant. If we run a list of eighty Argon 40/39 ratios listed [18] in Niemeyer's article through Microsoft Excel, we get eighty dates. The J value [18] is listed in the article as 0.03754. Twenty-six dates [32%] are over 4.6 billion years old. Twenty-one dates [26%] are over 5 billion years old. Thirteen dates [16%] are impossible future ages. The dates vary from negative 2.42 billion to positive 9.59 billion years old. There is a 12-billion-year range of dates. In the table below we can see the comparison between the so called "Model Age" [16] and dates calculated from the eighty ratios [18].

Table 13

Meteor Sample	Max Age Billion Years	Min Age Billion Years	Model Age Billion Years	Range Billion Years
Landes	6.01	-2.42	4.55	8.43
Copiapo	5.89	-1.16	4.47	7.05
Woodbine	9.59	0.48	4.61	9.11
Mundrabilla silicate	6.70	-0.57	4.59	7.27
Unetched	6.01	-1.04	4.54	7.05
Etched	6.98	0.09	4.57	6.89
Mundrabilla troilite	4.22	-0.62	9.5	4.84

40Ar-39Ar Studies of Whole Rock Nakhlites

These whole rock nakhiltes were dated in 2004 by scientists from the Lunar and Planetary Laboratory, University of Arizona, Tucson, Arizona.[19]

Table 14

Table Number	Maximum Million Years	Minimum Million Years	Difference Million Years	Difference Percent
Table 1	1,405	262	1,143	536%
Table 2	1,409	199	1,210	708%
Table 3	1,425	761	664	187%

$^{40}Ar/^{39}Ar$ Dating Of Desert Meteorites

Dated in 2005 by scientists [20] from Germany and Russia, these meteorite samples gave astounding results. Many dates were older than the evolutionist age of the Solar System. [21]

Table 16

Sample Name	Million Years
Table A1. Dhofar 007 whole rock.	7,632
	6,033
	5,498
Table A2. Dhofar 007 plagioclase.	7,582
	7,011
	4,753
	4,741
Table A3. Dhofar 300 whole rock.	9,015
	8,485
	5,516
	5,137
Table A5. Dhofar 300 pyroxene	8,957
	6,064
	5,656
	4,998
	4,720
Table A5. Dhofar 300 plagioclase.	9,680
	5,793
	5,721
	5,395
	5,237
	5,035
	4,788

References

1 207Pb–206Pb and 40Ar–39Ar ages from SW Montana, Precambrian Research, 2002, Volume 117, Pages 119 - 143
2 Reference 1, Page 128, 133
3 http://www.creationismonline.com/Isoplot/Isoplot.html
https://www.bgc.org/isoplot
Excel formulas
=AgePb6U8 (Pb206/U238), Returns the 206Pb/238U age. Input is the radiogenic 206Pb/238U ratio.
=AgePb76 (Pb207/Pb206), Returns the 207Pb/206Pb age. Input is the radiogenic 207Pb/206Pb ratio.
=AgePb7U5 (Pb207/U235), Returns the 207Pb/235U age. Input is the radiogenic 207Pb/235U ratio.
=AgePb8Th2 (Pb208/Th232), Returns the 208Pb/232Th age. Input is the radiogenic 208Pb/232Th ratio.
4 Joachim Kunz, Shocked meteorites: Argon-40-Argon-39, Meteoritics And Planetary Science, 1997, Volume 32, Pages 647 - 670
5 Reference 4, Pages 664 to 670
6 D. D. Bogard, Ar Diffusion Properties, Meteorites, Geochemica Et Cosmochemica Acta, 1980, Volume 44, Pages 1667 - 1682
7 Reference 6, Pages 1670, 1671
8 D. D. Bogard, U-Th-Pb dating of Abee E4 Meteorite, Earth and Planetary Science Letters, 1983, Volume 62, Pages 132 – 146
9 Reference 8, Page 134, 135
10 Reference 8, Page 139
11 Reference 8, Page 142
12 D. D. Bogard, The Monahans chondrite and halite, Meteoritics And Planetary Science, 2001, Volume 36, Pages 107 - 122
13 Reference 12, Pages 120-122
14 Elmar K. Jessberger, 40-Ar/39-Ar Ages of Allende, Icarus, 1980, Volume 42, pages 384
15 Reference 14, Pages 390 – 403
16 Sidney Niemeyer, Ar-39/Ar-40 dating of IAB iron meteorites, Geochemica Et Cosmochemica Acta, 1979, Volume 43, Pages 1829 - 1840
17 Reference 16, Page 1834
18 Reference 16, Page 1830-1831
19 Meteoritics & Planetary Science 39, Nr 5, 755–766 (2004)
20 Meteoritics & Planetary Science, 2005, Volume 40, Number 9/10, Pages 1433–1454
21 Reference 20, Pages 1452 – 1454

Chapter 2

The Mythology Of Modern Dating

Does Radiometric dating agree with the Geological Column?

U-Th-Pb Dating of Hydrothermal ore Deposits

These rocks from Hubei Province in China were dated in 2008 by scientist from the University of Hong Kong, using the 206Pb/238U and 208Pb/232Th age dating methods. [1] According to the article the true age of the rock formation is between 100 million years and 140 million years old: "Both the quartz diorite intrusion and ore bodies, yield weighted mean 206Pb/238U ages of 136.0±1.5 Ma and 120.6±2.3 Ma (2σ), respectively, in agreement within analytical uncertainty to their 208Pb/232Th ages. In situ analysis of epidote-enclosed hydrothermal titanite in thin sections of a skarn ore sample yields a mean 206Pb/238U age of 135.9±1.3 Ma and 208Pb/232Th age of 138.2±4.5 Ma, whereas titanite in calcite from a calcite-dominated vein cross-cutting the skarn ore body has consistent 206Pb/238U and 208Pb/232Th ages of ca. 121 Ma." [1] The article contains two tables [2] with Uranium/Thorium/Lead ratios that have no dates beside them. If we put the tables into Microsoft Excel and use the computer program Isoplot [3] we can calculate dates from the undated isotopic ratios. There is a 12,616-million-year range between the youngest and oldest dates.

Table 1	207Pb/206Pb	207Pb/235U	206Pb/238U	208Pb/232Th
Average	2,492	165	131	961
Maximum	4,398	392	142	12,721
Minimum	676	105	118	115

Isotope Evolution in the HIMU

These rocks from St. Helena Island in the Atlantic Ocean were dated in 2014 by scientist from The University of Tokyo, using the Pb–Sr–Nd–Hf–He isotopic data together with 40Ar/39Ar and K/Ar age dating methods. [4] According to the article the true age of the rock formation is between 8 million years and 12 million years old: "Although isotopic variations are small in the St. Helena lavas (20.6–21.0 for 206Pb/204Pb) between 12 and 8 Ma, the younger lavas have more HIMU-like isotopic compositions than the older lavas." [4] The article contains tables [5] with Lead 207/206 ratios that have no dates beside them. If we put the tables into Microsoft Excel and use the computer program Isoplot we can calculate dates from the undated isotopic ratios. There is a 4,800-million-year range between the so-called true age and oldest dates.

Table 2	Age (Ma)
Average	4,849
Maximum	4,856
Minimum	4,839

U-Th-Pb Dating

These rocks from the Southern Alpine Domain, Italy were dated in 2013 by scientist from the University of Bern, Switzerland, using the 206Pb/238U and 208Pb/232Th age dating methods. [6] According to the article the true age of the rock formation is between 10 million years and 420 million years old: "SHRIMP analyses of Tara allanite yielded a weighted mean 208Pb/232Th age of 414.9 ± 3.3 Ma (2σ; n = 26), and a mean 206Pb/238U age of 419.3 ± 7.7 Ma (2σ; n = 23). LA-ICP-MS single-spot mean 208Pb/232Th data yielded an age of 417.5 ± 1.4 Ma." [6] The article contains tables [7] with Uranium/Thorium/Lead ratios that have no dates beside them. If we put the tables into Microsoft Excel and use the computer program Isoplot we can calculate dates from the undated isotopic ratios. Out of the 276 dates there is a 15,347-million-year range between the youngest and oldest dates. Thirty-one dates [11%] are over 10000 million years old. Ninety-one dates [33%] are over 9000 million years old. Fifteen dates [5%] are impossible future ages. Two hundred and twenty-three dates [81%] are over 500 million years old. Two hundred and thirty-eight dates [86%] are too old or too young.

Table 3, Tara Allanite

206Pb/238U Age	207Pb/235U Age	208Pb/232Th Age	206Pb/238U Age	207Pb/235U Age	208Pb/232Th Age
3,481	-84	10,377	3,263	567	10,293
3,232	281	10,195	3,244	565	10,207
3,574	-141	10,159	3,317	600	10,232
3,496	477	9,974	3,173	575	10,060
3,224	545	10,097	3,252	588	10,085
3,070	449	9,863	3,082	559	9,826
2,973	530	9,999	3,082	558	10,024
3,306	159	9,838	3,126	566	9,838
3,717	-271	9,688	3,102	489	9,626
3,432	-4,491	9,875	3,002	473	9,751
3,351	-255	10,007	3,164	554	9,994
3,717	545	10,377	3,317	600	10,293
2,973	-4,491	9,688	3,002	473	9,626

Table 4, AVC Allanite

206Pb/238U Age	207Pb/235U Age	208Pb/232Th Age	206Pb/238U Age	207Pb/235U Age	208Pb/232Th Age
2,867	320	10,207	3,209	545	10,268
3,451	-141	10,134	3,165	547	10,110
3,038	319	9,900	3,271	564	10,011
2,854	196	9,826	3,189	551	9,900
3,130	147	9,788	3,098	535	9,813
3,014	683	9,801	3,126	540	9,826
3,142	137	9,701	3,154	544	9,713
3,228	-115	9,738	3,046	526	9,713
4,188	-2,251	9,563	3,130	541	9,487
3,515	489	9,335	3,094	535	9,297
3,243	-22	9,799	3,148	543	9,814
4,188	683	10,207	3,271	564	10,268
2,854	-2,251	9,335	3,046	526	9,297

Table 5, Bona Allanite

206Pb/238U Age	207Pb/235U Age	208Pb/232Th Age	206Pb/238U Age	207Pb/235U Age	208Pb/232Th Age
2,256	-26	9,676	2,588	386	9,888
2,053	-47	9,788	2,674	383	9,912
4,910	7	9,937	2,622	417	9,776
2,020	94	9,475	2,631	419	9,663
2,377	25	9,475	2,699	430	9,713
1,647	80	9,220	2,238	356	9,475
2,178	11	9,713	2,601	414	9,863
1,954	492	9,424	2,545	405	9,813
2,549	-88	9,130	2,695	424	9,437
3,604	416	9,550	2,665	420	9,399
2,555	97	9,539	2,596	405	9,694
4,910	492	9,937	2,699	430	9,912
1,647	-88	9,130	2,238	356	9,399

Table 6, Plešovice Zircon

206Pb/238U Age	207Pb/235U Age	208Pb/232Th Age
3,428	559	10,856
3,302	550	10,558
3,317	539	10,510
3,255	534	10,183
3,201	526	10,293
3,110	521	9,999
3,421	555	10,606
3,371	555	10,522
3,379	549	10,353
3,371	533	10,220
3,316	542	10,410
3,428	559	10,856
3,110	521	9,999

The Unique Achondrite Ibitira

These basaltic meteorite from Brazil were dated in 2013 by scientist from the Australian National University, using the Uranium/Lead age dating methods. [8] According to the article the true age of the meteorite is 4,555 million years old: "This value results in corrections of 1.1 Ma for Pb–Pb dates calculated using the previously assumed invariant 238U/235U value of 137.88. Using the determined 238U/235U value, the 7 most radiogenic Pb isotopic analyses for acid-leached pyroxene-rich and whole rock fractions yield an isochron Pb–Pb age of 4556.75 ± 0.57 Ma, in excellent agreement with the results of Mn–Cr chronology which give the ages of 4557.4 ± 2.5 Ma and 4555.9 ± 3.2 Ma using the U-corrected Pb–Pb age" [8] The article contains a table [9] with Uranium/Thorium/Lead ratios that have no dates beside them. If we put the tables into Microsoft Excel and use the computer program Isoplot we can calculate dates from the undated isotopic ratios. There is an 8,078-million-year range between the youngest and oldest dates. Forty-six dates are over 5 billion years old.

Table 7

207Pb/235U	207Pb/235U	206Pb/238U	206Pb/238U	207Pb/206Pb	207Pb/206Pb
5,131	4,765	6,676	5,264	4,672	4,557
5,639	4,750	8,949	5,249	4,635	4,557
5,610	4,750	8,739	5,215	4,631	4,556
5,608	4,749	8,722	5,214	4,614	4,556
5,566	4,746	8,627	5,209	4,608	4,556
5,438	4,697	7,901	5,037	4,607	4,556
5,353	4,642	7,539	4,830	4,604	4,556
5,285	4,635	7,419	4,821	4,589	4,556
5,228	4,620	7,083	4,761	4,588	4,556
5,081	4,617	6,425	4,726	4,584	4,556
5,056	4,606	6,383	4,698	4,584	4,555
5,044	4,603	6,377	4,693	4,582	4,554
5,030	4,597	6,303	4,638	4,582	4,553
5,010	4,572	6,182	4,604	4,573	4,553
4,970	4,522	6,060	4,495	4,571	4,553
4,955	4,486	5,955	4,259	4,564	4,552
4,906	4,392	5,781	4,077	4,563	4,552
4,857	4,357	5,586	3,839	4,563	4,552
4,853	4,227	5,586	3,539	4,561	4,551
4,851	4,172	5,499	3,315	4,560	4,546
4,831	4,060	5,452	3,084	4,559	4,539
4,812	3,943	5,436	2,669	4,558	4,538
4,805	3,874	5,407	2,582	4,558	4,536
4,792	3,811	5,399	2,519	4,558	4,533
4,779	2,526	5,377	871	4,557	4,527
4,775		5,305		4,557	4,523

Pb Isotopic Analysis

These rocks from Iceland were dated in 2003 by scientist from the University of Iowa, using the Lead/Lead age dating methods. [10] The article does not give a true age of the rock formation. The article contains five tables [11] with Lead 207/206 ratios that have no dates beside them. If we put the tables into Microsoft Excel and use the computer program Isoplot we can calculate dates from the undated isotopic ratios. Most of them are older than the evolutionist age of the Earth. One hundred and ninety-three dates [97%] are over 4.9 billion years old. One hundred and eleven dates [56%] are over 5 billion years old.

Table 8

207Pb/206Pb Age (Ma)	207Pb/206Pb Age (Ma)	207Pb/206Pb Age (Ma)	207Pb/206Pb Age (Ma)	207Pb/206Pb Age (Ma)	207Pb/206Pb Age (Ma)	207Pb/206Pb Age (Ma)
5,538	5,112	5,112	5,042	4,980	4,976	4,968
5,538	5,112	5,112	5,042	4,980	4,976	4,968
5,538	5,112	5,112	5,042	4,980	4,976	4,968
5,537	5,112	5,112	5,035	4,980	4,975	4,968
5,537	5,112	5,112	5,005	4,980	4,975	4,968
5,536	5,112	5,105	5,005	4,980	4,970	4,968
5,534	5,112	5,105	5,005	4,979	4,970	4,942
5,534	5,112	5,104	5,005	4,979	4,970	4,941
5,533	5,112	5,104	5,005	4,979	4,970	4,941
5,533	5,112	5,104	5,005	4,979	4,970	4,941
5,523	5,112	5,104	5,005	4,979	4,970	4,941
5,523	5,112	5,104	5,005	4,979	4,970	4,938
5,523	5,112	5,104	5,005	4,979	4,969	4,883
5,494	5,112	5,104	5,005	4,979	4,969	4,141
5,494	5,112	5,104	5,005	4,979	4,969	4,141
5,459	5,112	5,104	5,005	4,979	4,969	4,141
5,459	5,112	5,104	5,005	4,979	4,969	4,140
5,451	5,112	5,103	5,005	4,979	4,969	4,140
5,450	5,112	5,101	5,005	4,979	4,969	
5,450	5,112	5,101	5,005	4,979	4,969	
5,449	5,112	5,101	4,985	4,979	4,969	
5,449	5,112	5,101	4,985	4,978	4,968	
5,449	5,112	5,101	4,985	4,976	4,968	
5,448	5,112	5,101	4,985	4,976	4,968	
5,448	5,112	5,100	4,985	4,976	4,968	
5,448	5,112	5,070	4,985	4,976	4,968	
5,448	5,112	5,043	4,980	4,976	4,968	
5,113	5,112	5,043	4,980	4,976	4,968	
5,112	5,112	5,043	4,980	4,976	4,968	
5,112	5,112	5,043	4,980	4,976	4,968	

U–Th–Pb Systematics of Allanite

These rocks from the Archaean Fiskenaesset anorthosite complex, western Greenland were dated in 2012 by scientist from the Isotope Geosciences Laboratory, British Geological Survey, using the Uranium/Thorium/Lead age dating methods. [12] According to the article the true age of the rock formation is between 5 million years and 4 billion years old: "The four shards analysed for Th isotopic composition yield apparent 208Pb/232Th ages between 5 and 4133 Ma." [13] There is a huge spread of date ranges throughout the entire article. The article contains a table [14] with Uranium/Thorium/Lead ratios that have no dates beside them. If we put the tables into Microsoft Excel and use the computer program Isoplot we can calculate dates from the undated isotopic ratios. There is an 11,083-million-year range between the youngest and oldest dates. Such a huge spread of assumed ages is meaningless.

Table 9

	207Pb/206Pb	207Pb/235U	206Pb/238U	208Pb/232Th	206Pb/238U	207Pb/206Pb
Average	1,474	1,298	1,277	2,023	1,438	3,346
Maximum	2,707	2,744	2,850	10,619	3,125	4,901
Minimum	-464	14	23	6	87	1,754

The Paleo-Tethyan Mian-Lueyang

These rocks from the Qinling Mountains, central China were dated in 2001 by scientist from the University of California, San Diego, using the Lead, Neodymium and Strontium age dating methods. [15] According to the article the true age of the rock formation is between 340 million years and 350 million years old: "This indicates that a portion of the modern Indian MORB mantle isotopic domain could have been in existence for at least 350 Ma." [15] The article contains tables [16] with Uranium/Thorium/Lead ratios that have no dates beside them. If we put the tables into Microsoft Excel and use the computer program Isoplot we can calculate dates from the undated isotopic ratios. There is an 80,000-million-year range between the youngest and oldest dates. Thirty-six dates [100%] are over 5 billion years old. Twenty-one dates [58%] are over 10 billion years old. Sixteen dates [44%] are over 20 billion years old.

Table 10

207Pb/206Pb	208Pb/232Th	206Pb/238U
5,111	86,040	27,626
5,110	81,479	26,654
5,108	76,575	24,689
5,104	76,345	24,029
5,100	68,008	22,082
5,095	65,466	18,973
5,092	64,670	18,904
5,091	63,308	17,076
5,090	60,967	16,070
5,089	34,712	10,127
5,087	32,366	8,551
5,077		
5,067		
5,066		

The Homestake Gold Deposit

These rocks from Black Hills, South Dakota, USA were dated in 2008 by scientist from the University of Copenhagen, using the Lead 207/206 age dating methods. [17] According to the article the true age of the rock formation is between 1,300 million years and 2,900 million years old. [18] "Lead stepwise leaching (PbSL) data for monazite bearing garnet separated from a sample of Homestake Iron Formation has yielded an isochron age of 1746±10Ma (2; MSWD= 0.42), which represents a maximum age for both the isoclinal folding and subsequent gold mineralization." [17] The article contains tables [19] with Uranium/Thorium/Lead ratios that have no dates beside them. If we put the tables into Microsoft Excel and use the computer program Isoplot we can calculate dates from the undated isotopic ratios. There is a 3,342-million-year range between the youngest and oldest dates. Out of the 139 dates 129 [93%] are over 4,000 million years old. One hundred and eight dates [78%] are older that the evolutionist age of the Earth. Sixty-seven dates [48%] are over 5 billion years old

Table 11

207Pb/206Pb	207Pb/206Pb	207Pb/206Pb	207Pb/206Pb	207Pb/206Pb
5,185	5,133	5,090	5,026	4,936
5,184	5,132	5,090	5,025	4,929
5,183	5,132	5,087	5,020	4,928
5,183	5,131	5,071	5,018	4,928
5,175	5,125	5,070	5,012	4,926
5,169	5,124	5,068	5,010	4,923
5,163	5,120	5,068	5,002	4,908
5,162	5,120	5,067	4,999	4,883
5,159	5,119	5,067	4,999	4,880
5,156	5,119	5,064	4,994	4,874
5,149	5,117	5,060	4,977	4,858
5,147	5,117	5,058	4,973	4,857
5,146	5,116	5,058	4,972	4,831
5,146	5,113	5,057	4,965	4,830
5,145	5,106	5,056	4,957	4,807
5,144	5,104	5,053	4,957	4,803
5,143	5,101	5,050	4,955	4,796
5,142	5,101	5,044	4,951	4,796
5,141	5,093	5,038	4,949	4,792
5,138	5,090	5,027	4,941	4,791

The Paleoproterozoic Huronian Supergroup

These rocks from the Huronian Supergroup, Canada were dated in 1999 by scientist from the State University of New York, using the Uranium/Lead age dating methods. [20] According to the article the true age of the rock formation is between 2,100 million years and 2,200 million years old: "Lower Huronian (McKim, Pecors) samples align along 207Pb:204Pb–206Pb:204Pb slopes equivalent to 2,170 Ma and 2,212 Ma, respectively. These ages are at the minimum age limit on sedimentation and within uncertainty of the Nipissing Diabase (2,219 Ma)." [20] The article contains a table [21] with Lead 207/206 ratios that have no dates beside them. If we put the tables into Microsoft Excel and use the computer program Isoplot we can calculate dates from the undated isotopic ratios. There is complete disagreement between the so called true [Model] age and the isotope ratio age.

Table 12

	Average	Maximum	Minimum	Model (Max)	Model (Min)
McKim	4,569	5,012	3,753	3,000	2,840
Pecors	4,700	4,918	4,452	2,930	2,930
Gowganda	4,391	4,716	3,992	3,000	2,840
Gordon	3,845	4,445	3,028	2,760	2,550

Angrite Sahara 99555

These meteorites found in the Sahara Desert was dated in 2008 by scientist from The Australian National University, using the Uranium/Lead age dating methods. [22] According to the article the true age of the meteorite is 4,564 million years old: "The Pb–Pb age of SAH of 4566.18 ± 0.14 Ma, reported by Baker et al., differs from the Pb–Pb age of D'Orbigny, another basaltic angrite, of 4564.42 ± 0.12 Ma" [22] The article contains a table [23] with Uranium/Lead ratios that have no dates beside them. If we put the tables into Microsoft Excel and use the computer program Isoplot we can calculate dates from the undated isotopic ratios. There is a 932-million-year range between the youngest and oldest dates. Seven dates are over 5 billion years old.

Table 13

	207Pb/235U	206Pb/238U	Pb 207/206	Pb 207/206
Average	4,686	4,976	4,565	4,565
Maximum	4,758	5,224	4,567	4,568
Minimum	4,479	4,292	4,563	4,564

Mantle Xenoliths from Namibia

These rocks from Proterozoic Rehoboth Terrane, Namibia were dated in 2012 by scientist from the Goethe-University in Frankfurt, using the Rhenium/Osmium age dating methods. [24] According to the article the true age of the rock formation is 2,100 million years old and the recent Kimberlite intrusion is 70 million years old: "The Proterozoic (1.8 to 1.6 Ga) Rehoboth Terrane is separated from the Archaean Kaapvaal craton to its east by the 2.1 to 1.75 Ga Kheis–Maghondi belt and bordered to its west by the 1.35 to 1.0 Ga Namaqua–Natal belt and was intruded by the kimberlites of the Gibeon field around 70 Ma ago." [25] The article contains two tables [26] with Osmium 187/188 ratios that have no dates beside them. If we put the tables into Microsoft Excel and use the standard dating formulas [27-30] we can calculate dates from the undated isotopic ratios.

$$t = \frac{1.04 - ({}^{187}Os / {}^{186}Os)}{0.050768} \qquad \textbf{1}$$

In the above formula, t = billions of years. The same date can be calculated from the Osmium 187/188 ratios. If we use another formula [31] we can convert the Osmium 187/188 ratio to the Osmium 187/186 ratio.

$$\frac{{}^{187}Os}{{}^{186}Os} \times 0.12035 = \frac{{}^{187}Os}{{}^{188}Os} \qquad \textbf{2}$$

$$\frac{{}^{187}Os}{{}^{186}Os} = \frac{({}^{187}Os \div {}^{188}Os)}{0.12035} \qquad \textbf{3}$$

$$t = \frac{1.04 - \left(\frac{({}^{187}Os \div {}^{188}Os)}{0.12035}\right)}{0.050768} \qquad \textbf{4}$$

Table 14. There is a 75,000-million-year range between the youngest and oldest dates.

Sample A	Sample B	Sample A	Sample B	Model Age	Model Age
Million Years	Million Years	Million Years	Million Years	Million Years	Million Years
-431		1,369	1,418	1,780	
-5,702		1,434	1,434	1,870	-149,740
-361		1,402	1,434		
-12,248		1,074	976	1,550	-2,050
-448		1,353	1,353	1,730	
-43,345		2,204	2,105	2,520	-1,570
-337		1,778	1,729		
-28,615		-8,615	-8,533		
-333		583	583	1,100	
-12,248		1,320	1,385	1,750	
-194	-192	1,091	976	1,700	-590
-153	-153	1,205	1,189		
-232	-232	1,991	2,040	2,280	
-443	-376	2,253	2,253	2,460	
-307	20,425	2,040	1,942	2,360	-1,870
-309	20,358	2,089	2,089		
-353	-291	2,220	2,236	2,440	
-314	-422	2,515	2,515	2,710	5,760
-305	-286	2,024	1,909	2,340	-1,960
-153	-153	2,253	2,220		
-443	-422	-4,114	-4,048		
-44,982		992	894	1,640	-530
-34,556	20,011	-317	-268	390	
-34,474	-34,474	-4,801	-4,736		
-34,622	-34,900	-8,909	-8,828		
-2,958	20,469				
-2,975	20,435				
-2,975	20,376				

Neo-Tethyan Ophiolite in SW Turkey

These rocks from Southwest Turkey were dated in 2010 by scientist from the Karadeniz Technical University, Turkey, using the Rhenium/Osmium age dating methods. [32] According to the article the true age of the rock formation is between 250 million years and 1,000 million years old: "The Re–Os isotope systematics of the Muğla peridotites gives model age clusters of ~250 Ma, ~400 Ma and ~750 Ma that may record major tectonic events associated with the geodynamic evolution of the Neo-Tethyan, Rheic, and Proto-Tethyan oceans, respectively. Furthermore, >1000 Ma model ages can be interpreted as a result of an ancient melting event before the Proto-Tethys evolution." [32] The article contains a table [33] with Osmium 187/188 ratios that have no dates beside them. If we put the tables into Microsoft Excel and use standard dating formulas [27-31] we can calculate dates from the undated isotopic ratios. There is a 178,000-million-year range between the youngest and oldest dates.

Table 15

187Os/188Os	Depletion Age	Model Age
Million Years	Million Years	Million Years
420	Age	Age
-175,916		
-644	Future	Future
-713	Future	Future
-93	274	376
718	1,014	1,014
2,822	2,892	2,892
403	728	728
578	747	750
-376	14	14
61	252	252

Central Asian Orogenic Belt

These rocks from northeast China were dated in 2010 by scientist from the Chinese Academy of Sciences, using the Rhenium/Osmium age dating methods. [34] According to the article the true age of the rock formation is between 1,900 million years and 2,100 million years old: "The unradiogenic 187Os/188Os ratios of the refractory harzburgites give Re depletion ages (TRD) of 1.9–2.1 Ga." [34] The article contains one table [35] with Osmium 187/188 ratios that have no dates beside them and another with calculated ages. If we put the table into Microsoft Excel and use the standard dating formulas we can calculate dates from the undated isotopic ratios. There is a 23,920-million-year range between the youngest and oldest calculated dates listed in the article. There is a 25,000-million-year range between the youngest and oldest calculated dates.

Table 16

Sample	Os/Os (Ma)	TDM (Ma)	TDM (Ma)	TRD (Ma)	TDA (Ma)
KL3-24	1,161	1,610	10,620		760
KL3-26	135	740	-2,490	2,310	-30
KL3-27-1	-543	160	-20	1,820	690
KL3-27-2	-760	-30	-40		700
KL3-28	97	710	1,010	1,430	14,650
KL3-30	300	880	-3,830		740
KL3-31	503	1,050	-1,780		
KL3-38	580	1,120	6,520	1,150	-30
KL3-40	1,732	2,090	2,170		
KL3-41	-1,104	-330	10		
08KL-01	565	1,110	-200	1,520	470
08KL-02	1,497	1,890	2,540		
08KL-03	-24	600	-1,170	4,350	-520
08KL-04	929	1,420	2,640	3,470	630
08KL-05	765	1,280	1,380	490	-9,270
08KL-07	1,511	1,910	-640	570	440

08KL-09	-397	280	-620	2,310	390
08KL-10	-271	390	-150	5,910	430
08KL-11	477	1,030	-100	820	540
08KL-12	-76	560	690	1,390	590
08KL-13	693	1,220	1,420	1,280	1,280

Hebi, North China Craton

These rocks from Hebi, North China Craton were dated in 2012 by scientist from the Chinese Academy of Sciences, using the Rhenium/Osmium and Rubidium/Strontium age dating methods. [36] According to the article the true age of the rock formation is between 1,800 million years and 3,000 million years old: "Their bulk rock 187Os/188Os ratios give TRD ages varying from Paleoproterozoic to Neoarchean (1.8–2.6 Ga), which is slightly younger than the previous reported TRD ages of two sulfide grains (2.5 and 3.0 Ga) in Hebi mantle xenoliths." [37] The article contains a table [38] with Osmium 187/188 ratios that have no dates beside them. If we put the tables into Microsoft Excel and use the standard dating formulas we can calculate dates from the undated isotopic ratios. The dates I calculated agree with the article. The same table contains a list of calculated dates and there is an 11,500-million-year range between the youngest and oldest of those dates.

Table 17

187Os/188Os	TDM	MA
1,794	2,100	2,300
2,372	2,600	2,900
1,930	2,300	3,400
1,901	2,200	-3,100
1,356	1,800	8,400
1,791	2,100	2,500
1,961	2,300	3,000
1,953	2,300	4,600

Re-Os Isotopic Results

These rocks from South China were dated in 2011 by scientist from the Chinese Academy of Sciences, using the Rhenium/Osmium age dating methods. [39] According to the article the true age of the rock formation is between 1,800 million years and 2,200 million years old: "A correlation between 187Os/188Os and Al_2O_3 exists among the Ningyuan mantle xenoliths, which if interpreted as an isochron analog yields a model age of ~2.2 Ga. This age is older than the Re depletion age (TRD) of the harzburgite (~1.8 Ga), which represents a minimum age of melt depletion." [39] The article contains a table [40] with Osmium 187/188 ratios that have no dates beside them. If we put the tables into Microsoft Excel and use the standard dating formulas we can calculate dates from the undated isotopic ratios. There is a 2,042-million-year range between the youngest and oldest dates. The dates do not even overlap the assumed model age. The same table contains a list of calculated dates and there is a 10,640-million-year range between the youngest and oldest of those dates.

Table 18

187Os/188Os	MA	TRD
655	2,050	1,260
-57	3,280	720
-93	4,630	700
236	2,500	940

151	10,430	910
-80	2,340	690
-675	-210	250
-333	2,140	480
43	2,130	780
583	2,590	1,220
-423	1,580	400
1,367	2,320	1,820
-219	3,240	580

Lithospheric Mantle Evolution

These rocks from the Atherton Volcanic Province in north Queensland were dated in 2010 by scientist from the Macquarie University, using the Rhenium/Osmium age dating methods. [41] According to the article the true age of the rock formation is between 350 million years and 2,200 million years old: "Collision and accretion processes have probably initiated a melt-extraction event followed by cratonic lithosphere stabilisation at ~2.2 Ga (TMA model age). Metasomatism of the mantle lithosphere most likely involved infiltration of asthenospheric melts/fluids during lithospheric thinning and rifting beneath the Chudleigh Province at ~1.82 Ga, 0.81 Ga and 0.35 Ga (TRD Rhenium-depletion model ages), beneath the Atherton Province at ~1.75 Ga and 0.44 Ga (TRD), and during suturing at ~1.23 Ga (TRD)." [41] The article contains a table [42] with Osmium 187/188 ratios that have no dates beside them as well as two columns [TMA (Ma), TRD (Ma)] with calculated ages beside them. If we put the tables into Microsoft Excel and use the standard dating formulas, we can calculate dates [Column 2] from the undated isotopic ratios. The assumed model age [350-2,200 Ma] only allows and 1,850-million-year age range.

As far as the dates that I calculated, only Sapphire Hill falls within the accepted range. With Sapphire Hill [Table 20] there is a 20,690-million-year range between the youngest and oldest dates. With Mount Quincan [Table 21] there is a 28,130-million-year range between the youngest and oldest dates.

Table 19

Lucie's Crater	187Os/188Os	TMA (Ma)	TRD (Ma)
Maximum	583	1,510	870
Minimum	-2,281	-1,820	-1,750
Difference	2,864	3,330	2,620
Batchelor's Crater	187Os/188Os	TMA (Ma)	TRD (Ma)
Maximum	3,398	4,840	3,360
Minimum	-1,512	2,200	-1,040
Difference	4,910	2,640	4,400
Sapphire Hill	187Os/188Os	TMA (Ma)	TRD (Ma)
Maximum	2,024	8,620	2,160
Minimum	469	-12,070	760
Difference	1,555	20,690	1,400
Mount Quincan	187Os/188Os	TMA (Ma)	TRD (Ma)
Maximum	1,614	3,420	1,800
Minimum	-1,413	-24,710	-950
Difference	3,028	28,130	2,750

The Age of Lithospheric Mantle

These rocks from the Mongolian Orogenic Belt in north China were dated in 2002 by scientist from the Chinese Academy of Sciences, using the Rhenium/Osmium age dating methods. [43] According to the article the true age of the rock formation is between 600 million years and 2,800 million years old: "Two cratonic blocks, the NCC and the XMOB, with crustal residence ages of about 2500–2800 and 600–1000 Ma, respectively, overlie SCLM with both Proterozoic and Phanerozoic model ages." [44] The article contains a table [45] with Osmium 187/188 ratios that have no dates beside them as well as two columns [TRD (Ma), TMA (Ma)] with calculated ages beside them. If we put the table into Microsoft Excel and use the standard dating formulas we can calculate dates [Column 2] from the undated isotopic ratios. The assumed model age [600-2,800 Ma] only allows and 2,200-million-year age range. As far as the dates that I calculated, five of the seven rock formations have impossible negative or future ages. As far as the dates listed in the magazine article, three of the seven rock formations have impossibly old ages.

Table 20

Aobaoshan	$^{187}Os/^{188}Os$	TRD (Ma)	TMA (Ma)
Maximum	-178	110	12,330
Minimum	-268	30	1,720
Difference	90	80	10,610
Bolishan	$^{187}Os/^{188}Os$	TRD (Ma)	TMA (Ma)
Maximum	1,151	1,310	4,520
Minimum	-610	260	200
Difference	1,761	1,050	4,320
Bobotushan	$^{187}Os/^{188}Os$	TRD (Ma)	TMA (Ma)
Maximum	1,050	1,220	2,790
Minimum	272	520	610
Difference	777	700	2,180
Wangqing	$^{187}Os/^{188}Os$	TRD (Ma)	TMA (Ma)
Maximum	1,174	1,330	15,410
Minimum	-317	10	150
Difference	1,491	1,320	15,260
Longquan	$^{187}Os/^{188}Os$	TRD (Ma)	TMA (Ma)
Maximum	983	1,160	1,240
Minimum	-263	30	50
Difference	1,246	1,130	1,190
Dayishan	$^{187}Os/^{188}Os$	TRD (Ma)	TMA (Ma)
Maximum	1,048	1,220	1,420
Minimum	-611	280	430
Difference	1,660	940	990
Dalongwan	$^{187}Os/^{188}Os$	TRD (Ma)	TMA (Ma)
Maximum	1,074	1,240	3,000
Minimum	371	610	1,360
Difference	704	630	1,640

Late Cenozoic Arctic Ocean

These rocks samples dredged from the Arctic Ocean bottom from northern Canada or Queen Elizabeth Island region were dated in 1996 by scientist from the University of Wisconsin, [46] using the stratigraphy age dating methods. [47] According to the article the true age of the rock formation is between 0.5 million years and 5.1 million years old. [47] The article contains a table [48] with Lead 207/206 ratios that have no dates beside them. If we put the tables into Microsoft Excel and use the computer program Isoplot we can calculate dates from the undated isotopic ratios. There is a 5,234-million-year range between the oldest age and stratigraphy age.

Table 21

207Pb/206Pb	207Pb/206Pb	207Pb/206Pb	207Pb/206Pb
4,986	4,991	5,008	4,970
4,967	4,981	4,984	4,973
4,975	4,985	4,987	4,978
4,971	5,239	4,986	4,974
4,972	4,995	4,988	4,971
4,980	4,987	4,987	4,967
4,980	4,977	4,988	4,995
4,991	4,972	4,965	4,972
4,990	4,971	4,960	4,960
4,984	4,989	4,963	4,990
4,981	4,984	4,969	4,982

French Massif Central

These rocks from the French Alps were dated in 2005 by scientist from the University of London, using the Uranium/Thorium/Lead age dating methods. [49] According to the article the true age of the rock formation is between 300 million years and 380 million years old: "Lu–Hf isotopic data for these clinopyroxenes plot close to a 360 Ma reference 'isochron' and individually the clinopyroxenes yield depleted mantle Hf model ages between 299 and 376 Ma." [50] The article contains two tables [51] with Lead 206/207/208 ratios and one table [51] with 238U/204Pb and 232Th/204Pb ratios that have no dates beside them. If we put the tables into Microsoft Excel and use the computer program Isoplot we can calculate dates from the undated isotopic ratios. We can combine the lead ratios of tables one and two with the 238U/204Pb and 232Th/204Pb ratios of table three to get Uranium/Thorium ages as well. There is a 20,496-million-year range between the youngest and oldest dates.

Table 22

206Pb/238U	208Pb/232Th	207Pb/206Pb	207Pb/206Pb	207Pb/206Pb
15,943	21,165	5,008	5,014	4,948
13,385	17,660	4,988	4,873	4,947
7,502	15,590	4,939	4,871	4,947
6,684	9,160	4,934	4,898	4,936
5,772	9,061	4,926	4,939	4,936
4,225	6,269	4,867	4,961	4,930
3,394	5,931	4,867	4,930	4,928
1,531	5,670	4,858	4,933	4,932
1,426	5,262	4,868	4,932	4,932
1,401	5,247	5,068	4,933	4,963

1,162	5,042	5,070	4,934	4,965
669	4,445	5,066	4,935	4,945
	3,203	5,063	4,945	4,943
		5,045	4,912	5,041
		4,998	4,912	5,000
		4,997	4,953	5,001
		4,999	4,953	5,031

Evolution of Mauna Kea Lavas

These volcanic rocks from Hawaii were dated in 2002 by scientist from the Max-Planck-Institute for Chemistry, using the Lead/Lead age dating methods. [52] According to the article the true age of the rock formation is between 125 thousand years and 550 thousand years old. [53] The article contains a table [57] with Lead 207/206 ratios that have no dates beside them. If we put the tables into Microsoft Excel and use the computer program Isoplot we can calculate dates from the undated isotopic ratios. There is a 5,060-million-year range between the 'true age' and oldest dates I calculated from the Lead ratios.

Table 23

Model Age Million Years	207Pb/206Pb Million Years	Age Ratio	Model Age Million Years	207Pb/206Pb Million Years	Age Ratio
0.125	5,006	40,047	0.498	4,989	10,019
0.232	4,991	21,514	0.501	4,989	9,959
0.244	4,991	20,454	0.503	4,990	9,920
0.255	4,990	19,569	0.503	4,990	9,920
0.313	4,987	15,934	0.503	4,990	9,920
0.342	4,981	14,565	0.505	4,990	9,881
0.342	4,981	14,566	0.509	4,985	9,794
0.349	4,982	14,275	0.510	4,983	9,771
0.366	4,982	13,613	0.515	4,985	9,680
0.369	4,983	13,505	0.515	4,986	9,681
0.392	4,984	12,714	0.520	4,985	9,587
0.400	4,984	12,459	0.523	4,985	9,532
0.404	4,992	12,356	0.525	4,985	9,496
0.411	4,989	12,139	0.526	4,982	9,471
0.425	4,987	11,735	0.529	4,976	9,407
0.425	4,987	11,735	0.529	4,976	9,407
0.429	4,980	11,609	0.529	4,976	9,407
0.442	4,982	11,271	0.530	4,986	9,407
0.452	4,986	11,032	0.530	4,986	9,407
0.454	4,981	10,972	0.532	4,989	9,378
0.462	4,980	10,779	0.532	4,989	9,378
0.462	4,990	10,800	0.536	4,981	9,293
0.462	4,990	10,800	0.538	4,986	9,267
0.467	4,977	10,658	0.538	4,986	9,267

0.467	4,977	10,658	0.541	4,983	9,210
0.468	4,977	10,635	0.541	4,978	9,201
0.471	4,985	10,583	0.545	4,985	9,148
0.479	4,983	10,402	0.547	4,978	9,100
0.484	4,986	10,302	0.550	4,978	9,051
0.484	4,986	10,303	0.550	4,986	9,065
0.489	4,980	10,183	0.550	4,986	9,066
0.494	4,985	10,091	0.552	4,986	9,033
0.496	4,989	10,059	0.552	4,986	9,033

U-Th-Pb Geochronology

These rocks from the Kola Peninsula in Russia were dated in 2011 by scientist from the Russian Geological Research Institute (St. Petersburg, Russian Federation), using the 206Pb/238U and Lead 207/206 age dating methods. [54] According to the article the true age of the rock formation is between 370 million years and 380 million years old: "The batch calculations of baddeleyite data show a concordant age of 379.1±3.7 Ma, and a weighted mean 206Pb/238U age of 376.5±4.3 Ma." [54] The article contains a table [55] with model ages between 342 and 396 million years old. The article contains a table [59] with Uranium/Thorium/Lead ratios that have no dates beside them and ratios that have calculated dates beside them. If we put the tables into Microsoft Excel and use the computer program Isoplot we can calculate dates from the undated isotopic ratios. The table has dates beside the 208Pb/232Th ratios but no dates beside the 238U/206Pb and 207Pb/206Pb ratios. There is a 4,871-million-year range between the youngest and oldest dates. The 238U/206Pb and 207Pb/206Pb differ completely with the 208Pb/232Th age so the author deliberately did not put dates beside them.

Table 24

	208Pb/232Th	238U/206Pb	207Pb/206Pb
Average	396	1,054	3,381
Maximum	521	5,140	4,741
Minimum	306	269	1,318

Diamond Facies Pyroxenites

These rocks from the Beni Bousera Peridotite Massif, North Morocco were dated in 1992 by scientist from the University of Leeds in England, using the Rubidium/Strontium and Neodymium/Samarium age dating methods. [56] According to the article the true age of the rock formation is anywhere between 4 million years and 20,158 million years old. [57] The author admits that no coherent dates could be obtained: "The absence of coherent isochronous relationships in the Beni Bousera peridotites combined with their Sr-Nd isotope variability imply a multistage evolution. A complex, multistage evolution is also indicated by highly variable model ages." [58] The article contains a table [59] with Lead 207/206 ratios that have no dates beside them. If we put the tables into Microsoft Excel and use the computer program Isoplot we can calculate dates from the undated isotopic ratios. The dates are a 500 million years older than the evolutionist age of the Earth.

Table 25

207Pb/206Pb	207Pb/206Pb	Sm/Nd	Sm/Nd	Rb/Sr
4,905	4,993	4	1,170	221
4,911	4,993	36	1,416	300
4,918	4,994	234	1,747	1,148
4,957	4,995	353	2,006	1,155
4,958	4,998	355	2,341	1,163

4,965	4,999	646	3,705	1,174
4,972	5,013	898	3,978	1,455
4,992	5,018	948	4,318	1,735
4,993	5,022	960	4,906	1,923
		1,002	5,324	2,040
		1,062	10,042	2,052
		1,152	20,158	2,261
				2,294
				3,794

Indian Ocean Seamount Province

These rocks from the Christmas Island Seamount Province in the northeast Indian Ocean were dated in 2011 by scientist from the University of Sydney, using the 40Ar/39Ar age, Rubidium/Strontium, Neodymium/Samarium, Lutetium/Hafnium and high-precision Lead isotope analyses age dating methods. [60] According to the article the true age of the rock formation is between 50 million years and 140 million years old: "The ages of the seamounts and the underlying crust decrease from east to west: from Argo Basin Province (AP, 136 Myr; underlying crust 154-134 Myr) to Eastern Wharton Basin Province (EWP, 115-94 Myr; crust 120-105 Myr from SE to NW) to Vening-Meinesz Province (VMP, 95-64 Myr; crust 100-78 Myr from SE to NW) to Cocos-Keeling Province (CKP, 56-47 Myr; crust 67-61 Myr from S to N." [60] The article contains tables [61] with Rubidium/Strontium, Neodymium/Samarium, Lutetium/Hafnium and Uranium/Thorium/Lead ratios that have no dates beside them. If we put the tables into Microsoft Excel and use the computer program Isoplot we can calculate dates from the 189 undated isotopic ratios. There is an 11,314-million-year range between the youngest and oldest dates. In table 27 we can see that of the 189 Uranium/Thorium/Lead dates 188 [99.47%] are over 1 billion years old.

Table 26

Table 27	207Pb/206Pb	208Pb/232Th	206Pb/238U	176Lu/177Hf	87Rb/86Sr	147Sm/144Nd
Average	5,015	7,740	5,191	76	68	70
Maximum	5,025	11,317	5,191	142	136	136
Minimum	4,921	1,943	890	4	4	4

Table 27

Number Of Dates	Age Range Dates	Percentage Of All Dates
188	Dates Over 1 Billion Years Old	99.47%
184	Dates Over 2 Billion Years Old	97.35%
170	Dates Over 3 Billion Years Old	89.95%
157	Dates Over 4 Billion Years Old	83.07%
111	Dates Over 5 Billion Years Old	58.73%
59	Dates Over 6 Billion Years Old	31.22%
43	Dates Over 7 Billion Years Old	22.75%
36	Dates Over 8 Billion Years Old	19.05%
31	Dates Over 9 Billion Years Old	16.40%
25	Dates Over 10 Billion Years Old	13.23%

A Pb Isotope Investigation

These rocks from the Kola Peninsula in Russia were dated in 2013 by scientist from the *Massachusetts Institute of Technology*, using the Uranium/Thorium/Lead age dating methods. [62] According to the article the true age of the rock formation is between 370 million years and 380 million years old: "Our most precise ages of 373 ± 32 Ma, based on a 206Pb/204Pb vs. 238U/204Pb isochron for all the samples and 376 ± 13 Ma, based on a 208Pb/204Pb vs. 232Th/204Pb isochron determination for an urtite with a particularly high Th/Pb lie within the generally accepted range of the earlier age measurements." [62] The article contains a table [63] with Uranium/Thorium/Lead ratios that have no dates beside them. If we put the table into Microsoft Excel and use the computer program Isoplot we can calculate dates from the undated isotopic ratios. There is an 11,788-million-year range between the youngest and oldest dates.

Table 28	Average	Maximum	Minimum
207Pb/206Pb	4,891	5,011	4,798
206Pb/238U	3,534	7,462	1,959
208Pb/232Th	6,203	13,590	1,802

Conclusion

Evolutionists Schmitz and Bowring claim that Uranium/Lead dating is 99% accurate. [64] Looking at some of the dating it is obvious that precision is much lacking. The Bible believer who accepts the creation account literally has no problem with such unreliable dating methods. Much of the data used in this dating method is selectively taken to suit and ignores data to the contrary.

Yuri Amelin states in the journal Elements that radiometric dating is extremely accurate: "However, four 238U/235U-corrected CAI dates reported recently (Amelin et al. 2010; Connelly et al. 2012) show excellent agreement, with a total range for the ages of only 0.2 million years – from 4567.18 ± 0.50 Ma to 4567.38 ± 0.31 Ma." [65-67] To come within 0.2 million years out of 4,567.18 million years means an accuracy of 99.99562%. Looking at some of the dating it is obvious that precision is much lacking. The Bible believer who accepts the creation account literally has no problem with such unreliable dating methods. Much of the data in radiometric dating is selectively taken to suit and ignores data to the contrary.

Prominent evolutionist Brent Dalrymple states: "Several events in the formation of the Solar System can be dated with considerable precision." [68] Looking at some of the dating it is obvious that precision is much lacking. He then goes on: "Biblical chronologies are historically important, but their credibility began to erode in the eighteenth and nineteenth centuries when it became apparent to some that it would be more profitable to seek a realistic age for the Earth through observation of nature than through a literal interpretation of parables." [69] The Bible believer who accepts the creation account literally has no problem with such unreliable dating methods. Much of the data in Dalrymple's book is selectively taken to suit and ignores data to the contrary.

The Geological Column

Eon	Era	Period	Began	Finished
			Million Years Ago	Million Years Ago
Phanerozoic	Cenozoic	Quaternary	3	0
		Neogene	23	3
		Paleogene	65	3
	Mesozoic	Cretaceous	146	65
		Jurassic	201	145
		Triassic	252	201
	Paleozoic	Permian	299	52
		Carboniferous	359	299
		Devonian	419	359
		Silurian	443	419

		Ordovician	485	443
		Cambrian	541	485
Proterozoic	Neoproterozoic	Ediacaran	635	541
		Cryogenian	850	635
		Tonian	1,000	850
	Mesoproterozoic	Stenian	1,200	1,000
		Ectasian	1,400	1,200
		Calymmian	1,600	1,400
	Paleoproterozoic	Statherian	1,800	1,600
		Orosirian	2,050	1,800
		Rhyacian	2,300	2,050
		Siderian	2,500	2,300
Formation	Of The	Earth	4,500	4,300
Formation	Of The	Solar System	4,600	4,500
Formation	Of The	Galaxy	11,000	10,000
Formation	Of The	Universe	13,500	13,500

References

1 Chemical Geology, Volume 270 (2010) Page 56, U–Th–Pb Dating of Hydrothermal ore Deposits
2 Reference 1, page 63, 65
3 http://www.creationismonline.com/Isoplot/Isoplot.html
https://www.bgc.org/isoplot
4 http://dx.doi.org/10.1016/j.gca.2014.03.016,
Geochimica et Cosmochimica Acta, 2014, Page 1, Isotope evolution in the HIMU
5 Reference 4, page 5, 6, 7
6 Chemical Geology, Volume 371 (2014) Pages 46–47, U–Th–Pb dating
7 Reference 6, page 54, 57
8 Geochimica et Cosmochimica Acta, Volume 132 (2014) Pages 259–273, The Unique Achondrite Ibitira
9 Reference 8, page 262, 263
10 Chemical Geology, Volume 211 (2004) Pages 275, Pb isotopic analysis
11 Reference 10, pages 283, 286-288, 291, 294
12 Geochimica et Cosmochimica Acta, Volume 135 (2014) Pages 1, 3, U–Th–Pb systematics of allanite
13 Reference 12, page 8
14 Reference 12, page 6, 7
15 Earth and Planetary Science Letters, Volume 198 (2002) Pages 323, The Paleo-Tethyan Mian-Lueyang
16 Reference 15, page 328
17 Precambrian Research, Volume 172 (2009) Pages 1, The Homestake Gold Deposit
18 Reference 17, page 4
19 Reference 17, page 11, 12, 13
20 Precambrian Research, Volume 102 (2000) Pages 263–278, The Paleoproterozoic Huronian Supergroup
21 Reference 20, page 269
22 Geochimica et Cosmochimica Acta, Volume 72 (2008) Pages 4874, Angrite Sahara 99555
23 Reference 22, page 4876
24 Lithos, Volume 184–187 (2014) Pages 478, Mantle Xenoliths from Namibia
25 Reference 24, page 479
26 Reference 24, page 481, 483

27 Principles of Isotope Geology, Second Edition, By Gunter Faure, Published By John Wiley And Sons, New York, 1986. Pages 269.
28 Isotopes in the Earth Sciences, By H.G. Attendorn, R. Bowen, Chapman And Hall Publishers, London, 1994. Page 289 http://books.google.com.au/books?id=k90iAnFereYC&printsec=frontcover
29 Introduction to Geochemistry: Principles and Applications, Page 241, By Kula C. Misra, Wiley-Blackwell Publishers, 2012 http://books.google.com.au/books?id=ukOpssF7zrIC&printsec=frontcover
30 Radioactive and Stable Isotope Geology, Issue 3, By H. G. Attendorn, Robert Bowen, Page 298, Chapman and Hall Publishers, London, 1997 http://books.google.com.au/books?id=-bzb_XU7OdAC&printsec=frontcover
31 http://www.geo.cornell.edu/geology/classes/Geo656/656notes03/656%2003Lecture11.pdf
32 Lithos, Volume 132-133 (2012) Pages 50
33 Reference 32, page 62
34 Lithos, Volume 126 (2011) Pages 233, Central Asian Orogenic Belt
35 Reference 34, page 241, 242
36 Chemical Geology, Volume 328 (2012) Pages 123, Hebi, North China Craton
37 Reference 36, page 132
38 Reference 36, page 131
39 Chemical Geology, Volume 291 (2012) Pages 186, Re–Os Isotopic Results
40 Reference 39, page 194
41 Lithos, Volume 125 (2011) Pages 405, Lithospheric Mantle Evolution
42 Reference 41, page 417
43 Chemical Geology, Volume 196 (2003) Pages 107, The age of Lithospheric Mantle
44 Reference 43, page 127
45 Reference 43, page 116
46 Geochimica et Cosmochimica Acta, 1997, Volume 61, Number 19, Pages 4181-4200, Late Cenozoic Arctic Ocean
47 Reference 46, page 4185
48 Reference 46, page 4186, 4187
49 Geochimica et Cosmochimica Acta 71 (2007) 1290, French Massif Central
50 Reference 49, page 1299
51 Reference 49, page 1294, 1295, 1300
52 Geochemistry, Geophysics and Geosystems, May 2003, Volume 4, Number 5, Page 1, Evolution of Mauna Kea Lavas, doi:10.1029/2002GC000339
53 Reference 52, page 3, 4
54 Gondwana Research, Volume 21 (2012) Pages 728, U–Th–Pb geochronology
55 Reference 54, page 732
56 Journal of Petrology, 1993, Volume 34, Part 1, pages 125, Diamond Facies Pyroxenites
57 Reference 56, page 144, 145
58 Reference 56, page 155
59 Reference 56, page 146
60 Nature Geoscience, Volume 4, 2011, Pages 883, Indian Ocean Seamount Province
61 http://www.nature.com/ngeo/journal/v4/n12/extref/ngeo1331-s2.xls
62 Doklady Earth Sciences, 2014, Volume 454, Part 1, Pages 25, A Pb Isotope Investigation
63 Reference 62, page 27
64 An assessment of high-precision U-Pb geochronology. Geochimica et Cosmochimica Acta, 2001, Volume 65, Pages 2571-2587
65 Dating the Oldest Rocks in the Solar System, Elements, 2013, Volume 9, Pages 39-44
66 Amelin, Earth and Planetary Science Letters, 2010, Volume 300, Pages 343-350
67 Connelly, Science, 2012, Volume 338, Pages 651-655
68 The Age Of The Earth, By G. Brent Dalrymple, 1991, Stanford University Press, Stanford, California, Page 10.
69 Reference 68, Page 23

Chapter 3

Rocks With Future Dates

Norwegian Caledonides: An Isotopic Investigation

These rocks from Norway were dated [1] in 2009 using the Rubidium/Strontium and Neodymium/Samarium method. The rock samples gave ages [2] between -31 billion and 76 billion years old! Since the Earth exists in the present how can rocks have formed in the future? How can a rock be 60 billion years older than the Big Bang explosion?

"Re/Os model ages determined by LA-ICPMS from Fe–Ni sulfides (primarily pentlandite) scatter across the entire history of the Earth, and a few give meaningless future ages or ages older than the Earth." [3]

"The model ages show enormous scatter both within and between bodies and range from meaningless future dates to equally meaningless dates older than the Earth." [4]

Of all the samples 20 are older than the Earth, 8 are older than the Galaxy, 7 are older than the Universe and 19 have negative ages. [2] There is a 96,557 million year spread of dates between the youngest [-31,071 Ma] and the oldest [76,523 Ma] ages.

Table 1 [4]

CHUR	CHUR	DM
Age (Ma)	Age (Ma)	Age (Ma)
-20,034	4,849	-31,071
-7,491	4,960	-2,394
-6,102	6,510	-2,104
-2,184	7,049	-546
-1,220	7,816	-249
-1,201	9,751	-90
-1,038	9,789	4,580
-836	13,959	5,525
-685	17,711	5,788
-655	18,704	6,598
-483	38,213	7,031
-200	76,523	7,710
-137	76,523	40,285
4,527		64,577

Shaded ages are negative/future.

Multi-stage Origin of Roberts Victor Eclogites

These rocks from South Africa were dated [5] in 2011 using the Rubidium/Strontium and Neodymium/Samarium method. The rock samples gave ages [6] between -22 billion and 20 billion years old! Since the Earth exists in the present how can rocks have formed in the future? How can a rock be 5 billion years older than the Big Bang explosion? The author admits that the dates are impossible: "Type I eclogites show wide variations in model ages, from negative values to values much larger than the age of Earth. Sr model ages of Type I samples are all negative. Nd TCHUR ranges from -22.4 to 6.6 Ga, and Nd TDM from -2.3 to 8.1 Ga. Most of the Hf data give future ages; RV07-03, -18 and HRV247 give reasonable model ages, but the model ages of RV07-16 are older than Earth itself." [6]

Table 2[6]

87Rb/86Sr	147Sm/144Nd	147Sm/144Nd	176Lu/177Hf	176Lu/177Hf
-1,580	1,120	1,620	4,410	4,490
-2,640	6,630	4,240	-1,110	-3,430
-1,630	1,120	1,600	-7,150	19,870
-1,610	2,820	2,830	-2,340	-12,340
-2,510	2,270	2,540	80	-140
-1,640	1,700	2,090	-10,020	19,310
-1,540	2,380	2,600		
-1,560	2,410	2,600	-670	-1,900
-1,550	-9,900	8,100	-470	-1,350
-1,350	-480	-1,510	6,570	5,410
-1,570	-22,420	6,270	1,210	2,020
-1,410	-730	-2,290	-1,570	-5,510
-1,720	1,700	2,100	-7,000	18,170
-1,650	5,070	3,710	-4,770	
1,290	1,690	80	1,410	
-2,040	2,990	2,920	-1,970	-11,440
	-140	1,000	980	1,900
2,770	2,960	2,970	1,600	1,370
-1,130	6,360	7,660	1,570	1,500
3,260	2,710	2,710	2,150	2,130
	2,850	2,850	2,350	2,320
2,070			1,560	1,470

Blue ages are over 4.5 billion years old.

There is a 42,290 million year spread of dates between the youngest [-22,420 Ma] and the oldest [19,870 Ma] ages.

Re-Os Systematics of Mantle Xenoliths

These rocks from Tanzania were dated[7] in 1999 using the Rubidium/Strontium and Neodymium/Samarium method. The rock samples gave ages[8] between 4 billion years old to seven future ages! Since the Earth exists in the present how can rocks have formed in the future? The author admits this in two different places:

"Corresponding to Re depletion (TRD) model ages of 2.8 Ga to the future, respectively"[8]

"Collectively, the deep samples have more radiogenic Os isotopic compositions, corresponding to TRD ages that range from 1 Ga to the future."[9]

Re/Os Isotopes of Sulfides

These rocks from eastern China were dated[10] in 2006 using the Rhenium/Osmium method. The rock samples gave ages[11] between 40 billion to -87 billion years! Since the Earth exists in the present how can rocks have formed in the future? How can a rock be 70 billion years older than the Big Bang explosion? The author admits this major problem in four different places:

"Widespread Mesozoic magmatism in the Cathaysia block may be represented by abundant mantle sulfides with mildly superchondritic187Os/188Os and 'future' model ages."[12]

"Many of the peridotites studied here contain several generations of sulfides, spanning from Archean to 'future' model ages." [13]

"Samples with higher Re/Os may give 'future' ages, or ages older than Earth." [13]

"However, TMA calculations may yield both future ages and ages older than the Earth, because Re may be added to, or removed from, a xenolith by processes in the mantle and in the host basalt." [14]

In table 3 we can see the minimum ages, and in table 4 the maximum ages. There is 127-billion-year difference between the oldest [39 billion years] and the youngest [-87 billion years]. If the universe is only 13 billion years old how can there be such a wide range of ages?

Table 3 [11]

187Re/188Os	187Re/188Os	187Re/188Os	187Re/188Os	187Re/188Os	187Re/188Os
-87,817	-3,053	-1,917	-837	-191	6,054
-47,693	-3,031	-1,916	-763	-183	6,054
-27,938	-3,011	-1,908	-763	-181	6,088
-16,952	-2,939	-1,908	-732	-168	6,106
-15,940	-2,924	-1,870	-725	-158	6,398
-12,854	-2,915	-1,867	-712	-154	6,428
-10,838	-2,902	-1,860	-690	-142	6,437
-10,501	-2,882	-1,860	-649	-118	6,437
-7,384	-2,830	-1,841	-611	-114	6,470
-7,124	-2,824	-1,835	-608	-110	6,519
-6,574	-2,822	-1,829	-599	-98	6,736
-6,558	-2,814	-1,798	-595	-97	6,743
-6,398	-2,786	-1,644	-589	-55	7,395
-6,203	-2,741	-1,641	-527	-49	7,441
-6,138	-2,729	-1,567	-515	-42	8,008
-5,956	-2,552	-1,519	-497	-31	8,044
-5,922	-2,543	-1,432	-496	-30	8,862
-5,892	-2,386	-1,426	-453	-22	8,889
-5,038	-2,366	-1,374	-407	-19	9,449
-5,000	-2,214	-1,321	-344	-18	9,464
-4,084	-2,191	-1,107	-344	-15	10,382
-4,010	-2,137	-1,098	-316	-10	10,701
-3,773	-2,121	-1,001	-315	-7	10,736
-3,752	-2,061	-992	-310	-2	18,606
-3,504	-2,023	-916	-286	5,318	20,073
-3,503	-2,004	-852	-283	5,700	22,664
-3,473	-1,979	-840	-279	5,709	24,677
-3,434	-1,953	-840	-231	5,977	34,329
-3,070	-1,922	-837	-191	6,001	39,229

There is a 127,046 million year spread of dates between the youngest [Negative] and the oldest [Positive] ages. The values in table 4 are taken from figure 4 in Xisheng Xu's article. [16] There is 16-billion-year difference between the oldest [9.464 billion years] and the youngest [-6.74 billion years]. If the universe is only 13 billion years old how can there be such a wide range of ages?

Table 4[13]

187Re/188Os	187Re/188Os	187Re/188Os	187Re/188Os	187Re/188Os
Cathaysia	Cathaysia	Yangtze	Sino-Korean	Xing-Meng
-6,574	-1,870	-3,752	-2,824	-2,061
-5,922	-1,860	-3,504	-2,214	-1,953
-5,038	-1,641	-2,822	-2,137	-1,829
-4,084	-1,426	-2,786	-916	-496
-3,473	-1,374	-2,729	-589	-344
-3,070	-1,107	-1,922	-527	-191
-3,053	6,054	-1,908	-344	-191
-2,939	6,398	-1,519	5,709	6,054
-2,915	6,437	5,287	6,437	7,395
-2,543	6,743	5,977		
-2,366	8,008	8,889		
-2,023	9,464			
-1,908				

Lu-Hf Geochronology

These granulite xenoliths from the Kilboume Hole, New Mexico, [15] have been dated in 1997 using the Lu-Hf isotope system. The author admits that impossible dates have been generated: "The Nd isotope model ages presented in Table 3 are generally negative for the garnet granulites. A future age, or one that is older than the actual differentiation event, represents a rotation of a sample's apparent Nd isotope evolution curve, caused by increasing the Sm/Nd ratio at some time in the past." [16]

The values in table 5 contain numerous negative ages. [17] One sample (CKH63) has dates that vary from -3,297 to 2,478 million years old. That means a 5.7-billion-year difference. Earth rocks can only be 4.5 billion years old so how can there be such a wide variation?

Table 5

Sample	Nd	Nd	Hf	Hf
CKH39	-194	-1,051	7	-18
CKH64	209	-514	14	-5
CKH63	-659	-3,297	1,706	2,478
CKH58	1,137	1,649	1,152	1,719

Table 6

Sample	206Pb/238U	207Pb/235U	207Pb/206Pb
63a	426	611	1371
63d	317	490	1410
63e	98	161	1238
63j	430	622	1402
63g	136	242	1457
63b	319	483	1362
63c	425	624	1429

The Uranium/Lead dates [18] listed in table 6 shows that there is major discordance between various methods. Sample 63e has a 1,260% difference in ages. The author's choice of 'true age' is arbitrary.

Isotopic Disequilibrium

These mineral samples from Mono Lake, California and Seram, Indonesia [19] have been dated in 1998 using the Rb/Sr and Pb/U isotope systems. These mineral samples from Mono Lake, California are supposed to be 11.9 million years old: "The HIGH glasses are all less radiogenic than the source granite at 11.9 Ma. Within the HIGH glasses there is a general positive correlation between 87Sr/86Sr (11.9 Ma) and Rb/Sr." [20] If we run the isotopic ratios [21] listed in table 2 in the article through Isoplot [22] we get dates from 3,913 to 11,500 million years old! That means they are between 328 and 966 times too old!

The mineral samples from Seram, Indonesia are supposed to be 5.5 million years old: "The most precise muscovite and biotite Ar/Ar ages obtained from the complex 5.90 Ma and 5.51 Ma, respectively." [26] If we run the isotopic ratios listed in table 4 [23] in the article through Isoplot we get dates from 4,980 to 11,660 million years old! That means they are between 906 and 2,120 times too old! The Rb/Sr dates listed in table 5 [24] in the article have a range of 95 million years.

"In contrast, the plagioclase from the leucosome and the three matrix samples from the melanosome of BK 21B yield 'future ages' from -11 and -15 Ma." [25] There is a 11,674 million year spread of dates between the youngest [-14.7 Ma] and the oldest [11,660 Ma] ages.

Table 7 [21, 23, 24]

Table 2	Table 2	Table 4	Table 4	Table 5
206Pb/238U	207Pb/206Pb	206Pb/238U	207Pb/206Pb	87Rb/86Sr
5,902	3,914	4,493	4,982	-14.7
5,976	3,914	10,822	4,985	-13.3
6,403	3,913	9,728	4,984	-11
6,157	3,913	11,216	4,980	4.79
7,801	3,914	10,980	4,982	12
8,006	3,913	11,660	4,982	31.4
8,320	3,919	7,133	4,981	32.2
8,522	3,916	10,168	4,982	33.9
8,726	3,917	10,235	5,041	44
8,368	3,920	8,167	5,031	65.2
11,501	3,920			79

Multiple Metasomatic Events

These mineral samples from the Labait volcano, north-central Tanzania [26] have been dated in 2008 using the Rb/Sr and Sm/Nd isotope systems. The author admits that the dates give several negative ages:

"These deeper more fertile peridotites yield younger Re/Os ages (1 Ga to future ages) and represent either mixtures of ancient lithosphere with the underlying asthenosphere or recent additions to the base of the lithosphere." [27] If we use mathematical formulas [28] given in standard geology text, we can arrive at ages from the Rb/Sr and Nd/Sm ratios. The formula for Rb/Sr age is given as:

$$t = \frac{2.303}{\lambda} \log\left(\frac{(87Sr/86Sr) - (87Sr/86Sr)_0}{(87Rb/86Sr)} + 1 \right) \quad [1]$$

Where t equals the age in years. λ equals the decay constant. (87Sr/86Sr) = the current isotopic ratio. $(87Sr/86Sr)_0$ = the initial isotopic ratio. (87Rb/86Sr) = the current isotopic ratio. The same is true for the formula below.

$$t = \frac{2.303}{\lambda}\log\left(\frac{(143Nd/144Nd)-(143Nd/144Nd)_0}{(147Sm/144Nd)}+1\right) \quad [2]$$

There is a 4,205 million year spread [29] of dates between the youngest [-2,192 Ma] and the oldest [2,013 Ma] ages.

Table 8 [29]

87Rb/86Sr	147Sm/144Nd	Model Age, Nd/Sm
0	0.628	-2,192
0	0	-1,115
0	0.495	-573
0	1.016	-59
0	0	181
1.495	1.623	416
0	1.070	639
1.504	1.644	664
0.400	1.129	669
0	2.188	676
0	0	698
0	1.909	698
0	1.807	1,564
1.457	0	1,760
0	0	2,013

Re-Os Evidence

These mineral samples from central eastern China, [27] have been dated in 2006 using the Re/Os isotope systems. The author admits that the dates give several negative ages: "Ages (-6,900 to 7,330 Ma) of the Raobozhai peridotites vary widely from geologically meaningless to future ages." [28] The dating gave four impossible future ages. [29] According to Re/Os isochron diagrams [30] for Xugou peridotites, the formation is 2,000 million years old. There is a 7,330 million year spread of dates between the youngest [Negative] and the oldest [7,330 Ma] ages.

Table 9 [30]

Sample Number	187Re/188Os Age (Ma)	187Re/188Os Age (Ma)
NSDZ01(FC)	600	2,000
NSDZ02(FC)	900	2,100
NSDZ02(FC)	1,200	2,200
04XG14(FC)	1,400	1,900
04XG26(FC)	1,400	2,000
04XG26(EM)	1,500	2,200
ZK4-13U	1,600	4,200
ZK4-56L	1,700	Negative
ZK84-10U	1,800	Negative
ZK84-48L	1,800	2,100

ZK84-58L	1,800	1,600
ZK84-69L	1,800	1,500
ZK90-5U	1,900	Negative
ZK90-7U	2,000	7,330
ZK92-12U	Negative	3,300

Central Asian Orogenic Belt

These mineral samples from north-eastern China, [34] have been dated in 2010 using the Re/Os isotope systems. According to Re/Os isochron dates [35] the formation is 2,000 million years old. The author admits that the dates give several negative ages: "Other samples give TMA either older than the age of the Earth or a future age, suggesting a disturbance of the Re–Os isotope system in these samples." [36] There is a 18,480 million year spread [37] of dates between the youngest [-3,830 Ma] and the oldest [14,650 Ma] ages.

Table 10

Sample	147Sm/144Nd	176Lu/177Hf	187Re/188Os	187Re/188Os
KL-3-24	820	-30	-330	-3,830
KL-3-26	1,150	-30	-30	-2,490
KL3-27-1	1,280	390	160	-1,780
KL3-27-2	1,390	430	280	-1,170
KL3-28	1,430	440	390	-640
KL3-30	1,520	470	560	-620
KL3-38	1,820	540	600	-200
08KL-01	2,310	590	1,030	-20
08KL-03	2,310	630	1,050	10
08KL-04	3,470	690	1,110	690
08KL-05	4,350	700	1,120	1,010
08KL-07	5,910	740	1,220	1,380
08KL-09		760	1,280	1,420
08KL-10		1,280	1,420	2,170
08KL-11		14,650	1,890	2,640
08KL-12			1,910	6,520
08KL-13			2,090	10,620

The Mamonia Complex, Cyprus

These mineral samples from Mamonia complex, Cyprus, [38] have been dated in 2008 using the Re/Os isotope systems. According to Re/Os isochron dates [38] the formation is from three age clusters at 250 Ma, 600–800 Ma and 1,000 Ma. Four [39] of the thirty dates had future ages. This is a serious issue of having so many impossible dates: "The minimum ages of the Mamonia spinel peridotites varies from negative (future age) to 1150 Ma." [40]

Table 11

187Os/188Os	187Re/188Os	187Re/188Os
-1,217	Future Age	Future Age
-284	255	662
-317	271	633
-137	472	732
-104	483	710
-22	541	843
534	1,150	1,550
-153	449	679
174	740	1,080
92	473	1,090
-333	448	637
-88	480	545
-759	Future Age	Future Age
-235	209	248
-170	540	562
-1,348		
-55		
387		

"The calculation of the ages of the melting event (depletion in Re) gives inconclusive results varying from future ages to >1000 Ma." [41]

A Paleozoic Convergent Plate

These mineral samples from Austria, [42] have been dated in 2004 using the Re/Os isotope systems. Even though the Earth is supposed to be only 4.5 billion years old some dates are twice as old: "Rhenium-Osmium model ages range between future ages and 9.1 Ga." [43] If we enter the isotopic ratios [44] into Microsoft Excel and use the standard mathematical formula [45] we find that the dates are between 100 and 2,500 percent in error.

$$t = \frac{2.303}{\lambda}\log\left(\frac{(187Os/188Os)-(187Os/188Os)_0}{(187\mathrm{Re}/188Os)}+1\right)$$

$$\lambda = \frac{0.693}{h}$$

h = half-life 41.6 billion years

t = the rocks age in years

Table 12

Billion Years 187Os/188Os	Billion Years 187Re/188Os	Age Difference	Age Ratio Percentage
352	-4,300	4,652	1,221
376	-500	876	133
349	1,900	1,551	545

352	8,800	8,448	2,500
356	9,100	8,744	2,559
357	6,200	5,843	1,739
350	400	50	114
354	1,300	946	368
350	1,200	850	343
355	3,300	2,945	930
350	1,100	750	314
351	2,100	1,749	598
350	4,300	3,950	1,230

There is a 13,400 million year spread [44] of dates between the youngest [-4,300 Ma] and the oldest [9,100 Ma] ages.

Northern Canadian Cordillera Xenoliths

These mineral samples from Northern Canada, [46] have been dated in 1999 using the Re/Os isotope systems. According to Re/Os isochron dates [47] the formation's true age is 1.64 billion years old. Of the forty-one dates, fifteen [37%] were negative ages. [48] Many of the dates were impossible future ages: "The decoupling of 187Re/188Os and 187Os/188Os observed in the Canadian Cordillera xenolith data also affects the calculation of Os model ages, and leads to 'future' ages or ages older than the Earth." [49] If we enter the Osmium isotopic ratios [4] into Microsoft Excel and use the standard mathematical formula [48] we find that the dates have a 6,500 million year range between the youngest [-1,983 Ma] and the oldest [4,500 Ma] ages.

Table 13

187Re/188Os	187Re/188Os	187Os/188Os
	80	-530
Future	460	-84
	Future	-924
	500	-40
Future	430	-122
Future	100	-513
Future	520	-24
320	Future	-1,053
Future	480	-62
	1,110	685
Future	220	-366
	520	-24
550	350	-215
2,400	250	-340
4,500	630	109
Future	920	459
Future	Future	-721
540	Future	-1,983
680	640	120
610	510	-26
Future	330	-237

840	750	251
		37
Future	Future	-939
		-973

Xenoliths From Yangyuan and Fansi

These mineral samples from North China Craton, [49] have been dated in 2007 using the Re/Os isotope systems. According to Re/Os isochron dates [49] the formation's true age is 2.6 billion years old. Many of the dates were impossible future ages: "Nd model ages range from future ages to older than that of the Earth." [50] If we enter the Osmium isotopic ratios [51] into Microsoft Excel and use the standard mathematical formula [45] we find that the dates have a 13,500 million year range between the youngest [-10,800 Ma] and the oldest [2,690 Ma] ages.

Table 14

187Re/188Os	187Re/188Os	187Os/188Os
-500	1,180	682
	2,140	2,089
1,000	-530	-1,201
9,700	2,110	1,729
-200		
2,600	1,710	1,271
-10,800	2,630	2,318
	1,430	960
500		
2,000	1,450	976
2,800	330	-268
-3,500	1,520	1,058
2,200	1,210	714
	2,690	2,384
	-310	-972
	420	-170
	1,680	1,238
	1,720	1,287
	1,420	943

Formation of the North Atlantic Craton

These mineral samples from west Greenland, [52] have been dated in 2010 using the Re/Os isotope systems. According to Re/Os isochron dates [52] the formation's true age is 2.0 to 3.0 billion years old. Many of the dates were impossible future ages: "The WG-NAC peridotites, unsurprisingly, yield a substantial number of TMa model ages that are older than the earliest solids in the solar system or Earth (16%) or result in future ages (15%). This means that a third of the samples investigated here do not provide realistic TMa mantle melting ages. Os isotope data acquired by laser ablation measurements of sulphides in peridotites typically lack precise Re/Os data, and also yield a high proportion of samples with extremely scattered and unrealistic TMa mantle melting ages that range from future ages to those exceeding the age of the Earth." [53]

"These Os isotope systematics yield equally diverse TRD model ages ranging from Paleoarchean in individual samples to future ages." [53] There is a 34,000 million year spread of dates [54] between the youngest [-14,258 Ma] and the oldest [19,831 Ma] ages. The data in table 15 correspond to tables 1, 2 and 3 in the original article.

Table 15

	187Os/188Os	187Re/188Os	187Re/188Os	187Re/188Os
Table 1	Age	Model Age	Re Age	Eruption Age
Maximum	2,842	5,872	2,886	3,523
Minimum	56	-3,214	546	717
Difference	2,786	9,086	2,340	2,806
Table 2				
Maximum	19,831	6,994	2,571	2,609
Minimum	-106	-14,258	408	460
Difference	19,936	21,252	2,163	2,149
Table 3				
Maximum	2,942	6,950	2,968	3,096
Minimum	-112	442	402	416
Difference	3,054	6,508	2,566	2,680

In Situ Measurement of Re-Os Isotopes

These mineral samples from the Siberian and Slave Cratons, and the Massif Central, France, [55] have been dated in 2010 using the Re/Os isotope systems. According to Re/Os isochron dates [56] the formation's true age is 2.3 to 3.6 billion years old. Many of the dates were impossible future ages: "Therefore, both TRD and TMA yield unrealistic ages (future or unreasonably old, respectively)." [57]

Table 16

Billion Years	Billion Years
-1.89	-7.12
-1.3	-3.54
-1.2	-1.99
3.52	7.69
5.41	14.81

If we look at table 16 we see the bottom row has the difference between the oldest and youngest dates [58] in the original article.

Conclusion

Prominent evolutionist Brent Dalrymple states: "Several events in the formation of the Solar System can be dated with considerable precision." [59] Looking at some of the dating it is obvious that precision is much lacking. He then goes on: "Biblical chronologies are historically important, but their credibility began to erode in the eighteenth and nineteenth centuries when it became apparent to some that it would be more profitable to seek a realistic age for the Earth through observation of nature than through a literal interpretation of parables." [60] The Bible believer who accepts the creation account literally has no problem with such unreliable dating methods. Much of the data in Dalrymple's book is selectively taken to suit and ignores data to the contrary.

References

1 Norwegian Caledonides: An isotopic investigation, Lithos, Volume 117, 2010, Pages 1–19
2 Reference 1, Pages 6, 7
3 Reference 2, Pages 7
4 Reference 2, Pages 11
5 Multi-stage origin of Roberts Victor Eclogites, Lithos, Volume 142-143, 2012, Pages 161–181
6 Reference 5, Page 169
7 Re-Os Systematics of Mantle Xenoliths, Geochimica et Cosmochimica Acta, Volume 63, Number 7/8, Pages 1203–1217, 1999
8 Reference 7, Page 1206
9 Reference 7, Page 1213
10 Re/Os Isotopes of Sulphides, Lithos, Volume 102, 2008, Pages 43-64
11 Reference 10, Pages 46-50
12 Reference 10, Pages 43
13 Reference 10, Pages 52
14 Reference 10, Pages 53
15 Lu-Hf Geochronology, Chemical Geology, Volume 142, 1997, Pages 63-78
16 Reference 15, Page 73
17 Reference 15, Page 70
18 Reference 15, Page 71
19 Isotopic Disequilibrium, Chemical Geology, Volume 162, 2000, Pages 169-191
20 Reference 19, Page 175
21 Reference 19, Page 174
22 http://www.creationismonline.com/Isoplot/Isoplot.html
https://www.bgc.org/isoplot
23 Reference 19, Page 177
24 Reference 19, Page 179
25 Reference 19, Page 180
26 Multiple Metasomatic Events, Lithos, Volume 112-S, 2009, Pages 896-912
27 Reference 26, Page 897
28 Radioactive and Stable Isotope Geology, By H. G. Attendon, Chapman And Hall Publishers, 1997. Page 73 [Rb/Sr], 195 [K/Ar], 295 [Re/OS], 305 [Nd/Sm].
29 Reference 26, Page 910
30 Re–Os Evidence, Chemical Geology, Volume 236, 2007, Pages 323-338
31 Reference 30, Page 334
32 Reference 30, Page 331
33 Reference 30, Page 332
34 Central Asian Orogenic Belt, Lithos, Volume 126, 2011, Pages 233-247
35 Reference 34, Page 233
36 Reference 34, Page 241
37 Reference 34, Page 241, 242
38 The Mamonia Complex in Cyprus, Chemical Geology, Volume 248, 2008, Pages 195
39 Reference 38, Page 198
40 Reference 38, Page 208
41 Reference 38, Page 209
42 A Paleozoic Convergent Plate, Chemical Geology, Volume 208, 2004, Pages 141-156
43 Reference 42, Page 150
44 Reference 42, Page 146
45 Principles of Isotope Geology, Second Edition, By Gunter Faure, Published By John Wiley And Sons, New York, 1986, Page 266.
46 Northern Canadian Cordillera Xenoliths, Geochimica et Cosmochimica Acta, Volume 64, 2000, Number 17,

Pages 3061-3071
47 Reference 46, Page 3067
48 Reference 46, Page 3064
49 Xenoliths from Yangyuan and Fansi, Lithos, Volume 102, 2008, Page 25
50 Reference 49, Page 29
51 Reference 49, Page 37
52 Formation of the North Atlantic Craton, Chemical Geology, Volume 276, 2010, Pages 166-187
53 Reference 52, Page 181
54 Reference 52, Page 170-174
55 In situ Measurement of Re-Os Isotopes, Geochimica et Cosmochimica Acta, 2002, Volume 66, Number 6, Pages 1037-1050
56 Reference 55, Page 1045
57 Reference 55, Page 1047
58 Reference 55, Page 1046
59 The Age Of The Earth, By G. Brent Dalrymple, 1991, Stanford University Press, Stanford, California, Page 10.
60 Reference 59, Page 23

Chapter 4

Impossible Radiometric Dates

Evolution Beneath the Kaapvaal Craton

These rocks from South Africa were dated [1] in 2004 using the Rhenium/Osmium dating method. The rock samples gave ages [2] between -279 and 79 billion years old! There is a 358,000-million-year [2] spread of dates between the youngest [Negative] and the oldest [Positive] ages. Of the 374 dates, 92 [25%] are negative. The author admits in several places that many ages are impossibly old or young:

"In some cases these define plausible ages (Fig. 8a) but in most the 'ages' are greater than the age of the Earth (Fig. 8b), and all of these correlations are regarded as mixing lines." [3]

"Both types of high-Fe samples have high proportions of sulphides with young to negative TRD ages." [4]

"Negative model ages are meaningless numbers, and are plotted at increments of .0.1 Ga to illustrate the relative abundance of sulphides."[4]

Table 1

Average	-5	3
Maximum	5	79
Minimum	-279	-124

Table 2

Age Type	Amount	Percent
Negative Ages	92	24.59
Older Than The Earth	35	9.35
Older Than The Galaxy	11	2.94
Older Than The Universe	8	2.13

Central Asian Orogenic Belt

These rocks from Northern China were dated [5] in 2010 using the Rhenium/Osmium dating method. The rock samples in table 2 in the article gave ages [6] between -9 and 14 billion years old! There is a 14,450 million year spread of dates between the youngest [Negative] and the oldest [Positive] ages. The rock samples in table 3 in the article gave ages [7] between -3.8 and 10.6 billion years old! There is a 23,920 million year spread of dates between the youngest [Negative] and the oldest [Positive] ages. The author admits in several places that many ages are impossibly old or young:

"Whereas two samples give model ages close to, or even greater than, the age of the Earth." [8]

"Other samples give TMA either older than the age of the Earth or a future age, suggesting a disturbance of the Re–Os isotope system in these samples." [6]

"Thirteen Keluo mantle xenoliths yield impossible TMA model ages, i.e., negative or greater than the Earth's age, reflecting the modification of Re/Os ratios shortly before, during or since basalt entrainment." [9]

Table 3

	187Re/188Os Billion Years	187Re/188Os Billion Years
Average	0.94	0.86
Maximum	2.09	10.62
Minimum	-0.33	-3.83

Table 4

	147Sm/144Nd Billion Years	176Lu/177Hf Billion Years
Average	2.06	0.73
Maximum	5.91	14.65
Minimum	0.49	-9.27

If we use the Rhenium/Osmium dating formula shown in Gunter Faure's book [10] and enter a set of isotopic ratios listed in the original online article [11] we find the rock formation is less than 500 thousand years old.

$$t = \frac{2.303}{\lambda}\log\left(\frac{(187Os/188Os) - (187Os/188Os)_0}{(187\,\mathrm{Re}/188Os)} + 1\right)$$

$$\lambda = \frac{0.693}{h}$$

h = half life, 41.6 billion years

t = the rocks age in years

Norwegian Caledonides

These rocks from western Norway were dated [12] in 2009 using the Samarium/Neodymium dating method. The rock samples in the article gave ages [13] between -64 and 76 billion years old! There is a 141,100 million year spread of dates between the youngest [Negative] and the oldest [Positive] ages. The author admits in several places that many ages are impossibly old or young:

"Re–Os model ages determined by LA-ICPMS from Fe–Ni sulfides (primarily pentlandite) scatter across the entire history of the Earth, and a few give meaningless future ages or ages older than the Earth." [14]

"Table 2 lists model ages based on primitive (CHUR) and depleted (DM) mantle models. The model ages show enormous scatter both within and between bodies and range from meaningless future dates to equally meaningless dates older than the Earth." [15]

"These filters eliminate most of the negative dates and leave only three apparent ages older than the Earth." [15]

Table 5

	Million Years	Million Years
Average	4,510	1,400
Maximum	76,523	40,384
Minimum	-7,491	-64,577

Re-Os Isotopes of Sulfides

These rocks from eastern China were dated [16] in 2007 using the Rhenium/Osmium dating method. The rock samples in the article gave ages [17] between -47 and 39 billion years old! There is an 86,900 million year spread of dates between the youngest [Negative] and the oldest [Positive] ages. Out of the 348 dates, 72 (21%) were negative and 19 (5%) were older than the evolutionist age of the Earth. The author admits in several places that many ages are impossibly old or young:

"Re/Os versus TMA and TRD model ages, showing how samples with higher Re/Os may give 'future' ages, or ages older than Earth." [18]

"Many of the peridotites studied here contain several generations of sulfides, spanning from Archean to 'future' model ages." [18]

"However, TMA calculations may yield both future ages and ages older than the Earth, because Re may be added to, or removed from, a xenolith by processes in the mantle and in the host basalt." [19]

"A plot of TRD model ages that includes the "future" ages required by sulfides with super chondritic 187Os/188Os shows a marked peak at -180 Ma for the samples from the Cathaysia block." [20]

Table 6

	Million Years	Million Years
Average	462	1,369
Maximum	4,461	39,229
Minimum	-6,558	-47,693

Archean Man Shield, West Africa

These rocks from Sierra Leone were dated [21] in 2001 using the Rhenium/Osmium and Uranium/Lead dating method. The Uranium/Lead dating system gave an average age [22] of 2.5 billion years. The Rhenium/Osmium dating system gave an average age [23] of 8 billion years. The rock samples in the article gave ages [23] between 1.2 and 77 billion years old! There is a 76,000 million year spread of dates between the youngest [Negative] and the oldest [Positive] ages. The author admits in several places that many ages are impossibly old or young:

"For the high MgO samples, more than half of the Re/ Os model ages are older than the age of the Earth, indicating they either experienced recent Re loss or gain of radiogenic Os." [24]

"Five out of 13 of the low MgO samples also have Re/Os model ages older than the Earth." [24]

Table 7

Statistics	Re/Os	206Pb/238U	207Pb/235U	207Pb/206Pb
Average	8,092	2,367	2,649	2,910
Maximum	77,160	3,185	3,412	3,562
Minimum	1,390	1,204	1,873	2,743

Lithospheric Mantle Evolution

These rocks from north Queensland were dated [25] in 2010 using the Rhenium/Osmium dating method. The rock samples in the article gave ages [26] between -24 and 8.6 billion years old! There is a 33,330 million year spread of dates between the youngest [Negative] and the oldest [Positive] ages. Out of the 54 dates, 13 (24%) were negative and two were older than the evolutionist age of the Earth. The author admits that many ages are impossibly old or young: "Sulfides deposited from fluids with variable Re/Os have Os-isotope compositions that either plot in the field with $\gamma Os>0$ and Re/Os> CHUR, and with negative TRD and TMA ages or they plot in the field with $\gamma Os>0$ and Re/Os> CHUR, and with negative TMA and positive TRD ages." [27]

Table 8

	Billion Years	Billion Years
Average	-0.44	0.93
Maximum	8.62	3.36
Minimum	-24.71	-1.75

Upper Crust in North-East Australia

These rocks from north Queensland were dated [28] in 2010 using the Rhenium/Osmium dating method. The rock samples in the article gave ages [29] between -3.2 and 9.7 billion years old! There is a 12,950 million year spread of dates between the youngest [Negative] and the oldest [Positive] ages. Out of the 31 dates, 6 (20%) were negative and one was older than the evolutionist age of the Earth. The author admits that many ages are impossibly old or young: "Some garnet-rich granulites from the McBride Province yielded negative Hf and Nd model ages, whereas the Mt Quincan granulite yields model ages both older than the Earth and negative; these are not useful and are rejected." [30]

Table 9

Average	2.01	1.50
Maximum	9.73	3.97
Minimum	-0.80	-3.22

The Kaapvaal Cratonic Lithospheric Mantle

These rocks from South Africa were dated [31] in 2006 using the Samarium/Neodymium and Lutetium/Hafnium dating methods. The rock samples in the first table [Table 10] in the article gave ages [32] between -67 and 30 billion years old! There is a 97,790 million year spread of dates between the youngest [Negative] and the oldest [Positive] ages. Out of the 57 dates, 17 (30%) were negative and four were older than the evolutionist age of the Earth. The author admits that many ages are impossibly old or young:

"The large difference in Sm/Nd, but the relatively similar Nd isotope compositions of the garnet and cpx from the same sample result in generally young two-point cpx garnet Sm/Nd 'ages' for the Kimberley samples ranging from negative to 202 Ma." [33]

"Evidence that complete equilibration was not achieved in many of the samples comes from the observation that tie-lines connecting the garnet and Sm/Nd data for seven samples provide ages younger than the time of kimberlite eruption, including a number of samples that give negative ages." [34] "Negative Sm/Nd garnet ages are not uncommon for peridotite xenoliths and were first described in samples from Kimberley." [34]

Table 10

Minimum	Maximum
-67.49	4.85
-8.15	25.46
-2	30.3

If we put the Samarium/Neodymium and Lutetium/Hafnium ratios in first table [32] in the article into Microsoft Excel and use the dating formulas [35, 36] listed in Gunter Faure's book we find that the average age is just 100 million years! The spread of dates is not 100 billion years but just 100 million years!

$$t = \frac{2.303}{\lambda}\log\left(\frac{(143Nd/144Nd)-(143Nd/144Nd)_0}{(144Sm/147Nd)}+1\right)$$

h = half-life, 106 billion years

$$t = \frac{2.303}{\lambda}\log\left(\frac{(176Hf/177Hf)-(176Hf/177Hf)_0}{(176Lu/177Hf)}+1\right)$$

h = half-life, 37.3 billion years

Table 11

Billion Years
0.6
12.2
14.5
21.8
34.6

If we look at the dates in table eleven [37] there is a **34,000** million year spread of dates between the youngest [Negative] and the oldest [Positive] ages. If we look at the dates in table twelve [34] there is a **99,908** million year spread of dates between the youngest [Negative] and the oldest [Positive] ages.

Table 12

Statistical Data	Billion Years Sm-Nd	Billion Years Lu-Hf
Minimum	-2,247	-2,377
Maximum	96,661	1,995
Difference	98,908	4,372

In Situ Analysis of Sulphides

These rocks from South Australia and France were dated [38] in 2001 using the Rhenium/Osmium dating methods. The rock samples in the second table in the article gave ages [39] between -17 and 34 billion years old! With the South Australian rocks, there is a 51,000 million year spread of dates between the youngest [Negative] and the oldest [Positive] ages. The author admits that many ages are impossibly old or young:

"It is obviously not the case here, given that TMA model ages for some sulphides or samples are unrealistic, giving future ages or ages older than 4.5 Ga." [39]

"Interstitial sulphides in GRM-2 yield future TRD ages and unrealistic TMA ages, again indicating that the Os isotopic composition is not related to time-integrated in situ Re decay." [40]

Table 13

Billion Years	Billion Years
-17.4	4.35
-9.5	5.2
-7.06	8.3
-2.35	8.8
-0.3	34

South Australian rocks

Table 14

Billion Years	Billion Years
-32	3.11
-2.08	3.93
-1.79	6.7
-1.43	7.4
-1.42	16

French rocks

With the French rocks, [41] there is a 48,000 million year spread of dates between the youngest [Negative] and the oldest [Positive] ages.

Southern African Peridotite Xenoliths

These rocks from South Africa were dated [42] in 1988 using several dating methods. If we insert the isotopic ratios listed one table [43] we find that the Rubidium/Strontium ratios give ages between 83 and between 1,100 million years old. If we insert the Lead/Lead ratios listed in the same table, we find the rock is between 4,700 and 5,000 million years old. If we insert the Osmium ratios listed in another table [44] and use the dating formula shown in Gunter Faure's book [45] we find the rock is between -3,300 and 13,500 million years old. There is a **16,000** million year spread of dates between the youngest [Negative] and the oldest [Positive] ages.

$$t = \frac{1.04 - (^{187}Os/^{186}Os)}{0.050768}$$

In the above formula, t = billions of years.

Table 15

Dating Summary	Age 87Rb/86Sr	Age 187Os/186Os	Age Neodymium	Age 207Pb/206Pb
Maximum	1,100	13,551	1,630	5,064
Minimum	83	-3,309	520	4,700
Difference	1,017	16,860	1,110	364

Xenoliths from Kimberley, South Africa

These rocks from South Africa were dated [46] in 2007 using the Rhenium/Osmium dating method. The rock samples in the article gave ages [47] between -117,980 and 143,830 million years old! With the rocks, there is a 261,810 million year spread of dates between the youngest [Negative] and the oldest [Positive] ages. The author admits that many ages are impossibly old or young:

"The very old Re–Os model age of websterite DJ0217 of 7 Ga testifies to a more complex history for this sample." [48]

"The olivines from these samples also provide negative Re–Os model ages suggesting recent modification of their Re–Os systematics." [49]

"On a Re–Os isochron diagram, the whole-rock—olivine tie-line for DJ0259 corresponds to an age of 5.2 Ga. This unrealistic age coupled with the radiogenic Os, but near chondritic Re/Os ratio of the olivine suggests that the olivine in this dunite was either added recently, or interacted extensively with modern mantle melts, for example the host kimberlite." [49]

Table 16

Mineral	Average	Maximum	Minimum	Difference
Dunite	970	3,250	-3,470	6,720
Dunite	1,918	14,580	-15,020	29,600
Wehrlite	2,375	3,190	900	3,100
Wehrlite	3,096	21,670	-11,150	32,820
Websterite	-19,150	3,050	-117,980	121,030
Websterite	24,503	143,830	450	143,380

Conclusion

Yuri Amelin states in the journal Elements that radiometric dating is extremely accurate: "However, four 238U/235U-corrected CAI dates reported recently (Amelin et al. 2010; Connelly et al. 2012) show excellent agreement, with a total range for the ages of only 0.2 million years – from 4567.18 ± 0.50 Ma to 4567.38 ± 0.31 Ma." [50-52]

To come within 0.2 million years out of 4567.18 million years means an accuracy of 99.99562%. Looking at some of the dating it is obvious that precision is much lacking. The Bible believer who accepts the creation account literally has no problem with such unreliable dating methods. Much of the data in radiometric dating is selectively taken to suit and ignores data to the contrary.

References

1 Evolution Beneath the Kaapvaal Craton, Chemical Geology, Volume 208, 2004, Pages 89-118
2 Reference 2, pages 101-105
3 Reference 2, pages 107
4 Reference 2, pages 110
5 Central Asian Orogenic Belt, Lithos, Volume 126, 2011, Pages 233-247
6 Reference 5, page 241
7 Reference 5, page 242
8 Reference 5, page 239
9 Reference 5, page 244
10 Principles of Isotope Geology, Second Edition, By Gunter Faure, Published By John Wiley And Sons, New York, 1986, Page 266.
11 http://www.sciencedirect.com/science/article/pii/S0024493711002179
12 Norwegian Caledonides: An isotopic investigation, Lithos, Volume 117, 2010, Pages 1-19
13 Reference 12, pages 6, 7
14 Reference 12, page 7
15 Reference 12, page 11
16 Re/Os Isotopes of Sulfides, Lithos, Volume 102, 2008, Pages 43-64
17 Reference 16, pages 46-50
18 Reference 16, page 52
19 Reference 16, page 53
20 Reference 16, page 61
21 Archean Man Shield, West Africa, Precambrian Research, Volume 118, 2002, Pages 267-283
22 Reference 21, pages 273, 274
23 Reference 21, page 277
24 Reference 21, page 276
25 Lithospheric Mantle Evolution, Lithos, Volume 125, 2011, Pages 405-422
26 Reference 25, page 417
27 Reference 25, page 415

28 Upper Crust in North-East Australia, International Journal Earth Science, 2012, Volume 101, Pages 1091-1109
29 Reference 28, Pages 1099, 1101
30 Reference 28, Pages 1098
31 The Kaapvaal Cratonic Lithospheric Mantle, Journal Of Petrology, 2007, Volume 48, Number 3, Pages 589-625
32 Reference 31, pages 600-601
33 Reference 31, pages 609
34 Reference 31, pages 612
35 Reference 10, pages 205
36 Reference 10, pages 252
37 Reference 38, pages 610
38 In Situ Analysis of Sulphides, Earth and Planetary Science Letters, 2002, Volume 203, Pages 651-663
39 Reference 38, page 654
40 Reference 38, page 659
41 Reference 38, page 655
42 Southern African Peridotite Xenoliths, Geochimica et Cosmochimica Acta, 1989, Volume 53, Pages 1583-1595
43 Reference 42, page 1587
44 Reference 42, page 1588
45 Reference 10, page 269
46 Xenoliths from Kimberley, South Africa, Geochimica et Cosmochimica Acta, 2008, Volume 72, Pages 5722-5756
47 Reference 46, page 5737
48 Reference 46, page 5743
49 Reference 46, page 5744
50 Dating the Oldest Rocks in the Solar System, Elements, 2013, Volume 9, Pages 39-44
51 Amelin, Earth and Planetary Science Letters, 2010, Volume 300, Pages 343-350
52 Connelly, Science, 2012, Volume 338, Pages 651-655

Chapter 5

Concordia Isochron Dating

Age and Mineralogy of Supergene Uranium

Theses rocks from the Bohemian Massif, Southeast Germany [1] were dated in 2010 using the Uranium-Lead dating method. The table in the essay has three columns of isotopic ratios, $^{206}Pb/^{238}U$, $^{207}Pb/^{235}U$ and $^{207}Pb/^{206}Pb$. You will notice in Table 4 the original article [2] that there are dates besides the $^{206}Pb/^{238}U$ and $^{207}Pb/^{235}U$ ratios but no dates beside the $^{207}Pb/^{206}Pb$ ratios. The first two sets of ratios and dates agree with each other between 94 and 101 percent accuracy. If we use the computer program Isoplot [3] and calculate the ages of the $^{207}Pb/^{206}Pb$ ratios we see why not dates have been put beside them. In **Table 1** we can see that many dates are negative. That is logically impossible. How can the rock have formed millions of years in the future?

Table 1

Sample Name	Pb-206/207 Negative Ages	Sample Name	Pb-206/207 Negative Ages
A30	-29	A06	-29
A35	-8	A10	-45
A04	-18	A11	-83
A07	-8	A12	-23
A10	-8	A13	-133
A11	-13	A17	-116
A18	-8	A19	-72
A19	-18	A21	-2
A20	-8	A26	-34
		A27	-13
		A29	-45
		A39	-8
		A40	3
		A41	-50

In **Table 2** we can see that the 207Pb/206Pb dates are between 1,000 to 21,000 percent discordant when compared to the two Uranium-Lead dating methods. Here is just one of many times where geology journals use selective evidence to try and prove evolution. If the third column or ratios were dated and added to the essay you can see how silly it would look.

Table 2

Sample Name	Difference Percent	Sample Name	Difference Percent
A26	1,087	A01	1,006
A29	1,192	A16	1,073
A25	1,202	A32	1,891
A41	1,338	A31	2,067
A07	1,964	A30	3,070
A19	2,385	A29	3,539
A10	2,389	A33	10,452
A22	2,551	A36	16,112
A18	3,126		

A30	3,129		
A24	3,360		
A09	3,612		
A13	4,616		
A05	4,881		
A06	4,982		
A11	5,350		
A25	5,479		
A08	5,628		
A42	6,215		
A04	6,551		
A22	7,031		
A43	10,253		
A17	10,673		
A21	15,256		
A20	21,500		

207Pb-206Pb and 40Ar-39Ar ages from SW Montana

These rocks from North America were dated in 2002 using both [4] ^{40}Argon/^{39}Argon and Lead-Lead dating methods. Again the no dates beside the ^{207}Pb/^{206}Pb ratios. If we add dates we soon see why. The first table in his article has dates [5] using the ^{40}Ar–^{39}Ar dating method. The third table [6] has the ^{207}Pb/^{206}Pb ratios.

Table 3

Sample Name	K-Ar Dating Max Age	K-Ar Dating Min Age	Pb Dating Max Age	Pb Dating Min Age
RRCR2	1,818	1,695	4,471	1,895
RRSW1	1,806	1,740	5,011	4,032
HLM2	1,853	1,620	4,522	1,848
TRMR2	1,729	1,199	5,049	2,644

If we use the computer program Isoplot and calculate the ages of the ^{207}Pb/^{206}Pb ratios we see why not dates have been put beside them. The Potassium-Argon and Lead-Lead dating methods are extremely discordant. The author's use of data is very selective. Dates that agree are added and those that do not are omitted. This happens over and over in geology magazines. We can see from the table below that many dates are older than the evolutionist view of the age of Earth. How can such an absurdity be possible? How can the Earth be older than itself?

Table 4

Sample Name	Million Years	Age Category
RRSW1	5,005	Older Than The Solar System
RRSW1	5,011	Older Than The Solar System
RRSW1	4,939	Older Than Earth
TRMR2	5,015	Older Than The Solar System
TRMR2	5,049	Older Than The Solar System

^{207}Pb/^{206}Pb Dates

Uranium-Thorium-Lead Dating

This dating [7] was done in 1999 on meteorite samples by the Department of Earth and Planetary Sciences, Hiroshima University in Japan. Below we can see the isotopic ratios take from Table 2 in the original article. [8] Using the computer program Isoplot we calculate the ages of the $^{207}Pb/^{206}Pb$ ratios we see why not dates have been put beside them.

Table 5

Pb-207 Pb-206	Million Years	Age Category
0.889	5,071	Older Than Solar System
0.916	5,114	Older Than Solar System
0.876	5,051	Older Than Solar System
0.869	5,039	Older Than Solar System
0.922	5,123	Older Than Solar System
0.867	5,036	Older Than Solar System

5,051 to 5,123 million years old.

Diagram 1

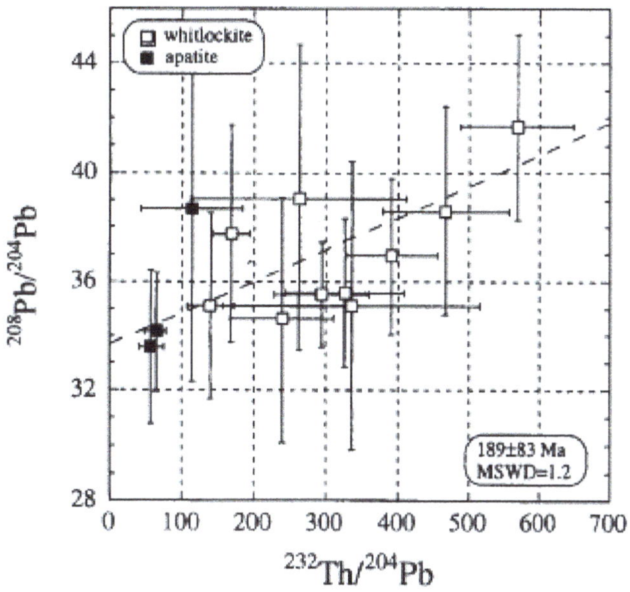

According to the Iscohron [1, 2 and 3] diagrams in the article [9] the meteorites are only supposed to be 200 million years old! This means that the dates are 4,800 million years in error. The ratio of the so called "true" age versus the $^{207}Pb/^{206}Pb$ age is 25 to 1. The author deliberately chose not to put the dates beside the isotopic ratios because they would show how utterly ridiculous the whole system is. According to the Iscohron diagram in the article, the maximum error level is only 83 million years. The error level is 4934 years if we compare it to the $^{207}Pb/^{206}Pb$ age. This means the error level is 59 times in error.

Pb–Pb Dating Of Chondrules

The meteorite samples [10] were dates in 2009 by scientists form the Geological Museum, University of Copenhagen and The University of Texas at Austin. If we use Isoplot and run some of the $^{207}Pb/^{206}Pb$ ratios given in the article [11] through Microsoft Excel, we see that many of the ratios produce ages over 5 billion years old. Below we can see a Concordia diagram taken from the article [12] that shows the age of the rocks to be 4,565 million years old. As you can see the diagram claims that the error margins are only 810,000 years! If we add the $^{207}Pb/^{206}Pb$ ratios dates, we can see that the diagram is out by 550 million years. That means the error margin given in the diagram is 677 times to short!

Diagram 2

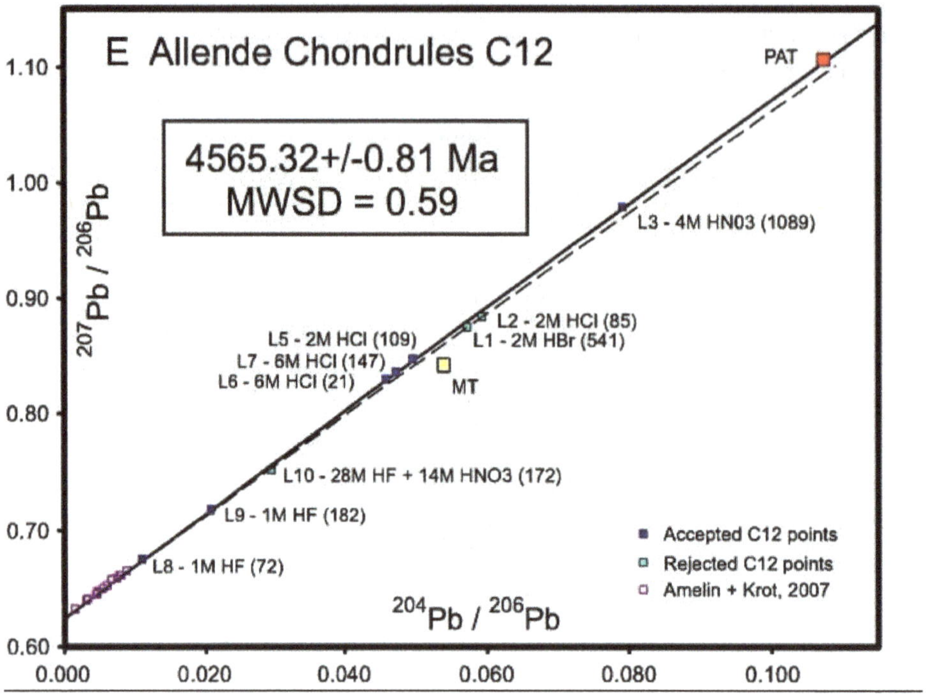

Table 6

Sample Number	Age Million Years	Age Category
C2-L1	5,194	Older Than Solar System
C2-L2	5,190	Older Than Solar System
C2-L3	5,089	Older Than Solar System
C2-L6	5,020	Older Than Solar System
C4	5,174	Older Than Solar System
C4-L6	5,013	Older Than Solar System
C4-L7	5,094	Older Than Solar System
C4-L8	5,051	Older Than Solar System
C7	5,091	Older Than Solar System
C7-L7	5,032	Older Than Solar System
C7-L8	5,021	Older Than Solar System
C12-10	5,050	Older Than Solar System
C12-L2	5,063	Older Than Solar System
C12-L3	5,206	Older Than Solar System
C12-L5	5,002	Older Than Solar System

5,002 to 5,206 million years old.

Pb-Pb Dating Constraints

This dating [13] was done in 2007 on meteorite samples by the Washington State University, Department of Geology. We can see from table seven which data in my essay the data was obtained from in Audrey Bouvier's essay.

Table 7

Her Essay	My Essay
Table 2, Page 1587	Table 8
Table 3, Page 1588	Table 9
Table 4, Page 1589	Table 10
Table 5, Page 1590	Table 11
Table 6, Page 1590	Table 12

One of the concordia diagrams [14] in the article gives the following data:

Chondrules: 4565.5 ± 1.2 Ma
Pyroxenes: 4564.3 ± 0.8 Ma
Phosphates: 4562.7 ± 0.7 Ma

We are told that the date of 4,565 million years old is only one million years in error at the maximum. If run some of the $^{207}Pb/^{206}Pb$ ratios given in the article through Isoplot, we see that many of the ratios produce ages over 5 billion years old. The oldest is 5,379 million years. The error margin given in the article is 814 times in error.

Table 8

Sample Name	Age Million Years	Age Category
Allende, Whole-rock-R0	5,334	Older Than Solar System
CV3, L0	5,325	Older Than Solar System
MNHN, L1	5,250	Older Than Solar System
MNHN, L2	5,258	Older Than Solar System
MNHN, L1	5,296	Older Than Solar System
MNHN, L2	5,029	Older Than Solar System
UCLA, L1	5,244	Older Than Solar System
UCLA, L1	5,244	Older Than Solar System
UCLA, L1	5,245	Older Than Solar System
UCLA, Olivine-R0	5,344	Older Than Solar System
UCLA, L0	5,336	Older Than Solar System
Murchison, Whole-rock-R0	5,333	Older Than Solar System
CM2, L0	5,321	Older Than Solar System
CM2, CAI-R0-Murch	5,238	Older Than Solar System
CM2, L0	5,267	Older Than Solar System
ENSL, Blanke	5,016	Older Than Solar System
Canyon-Diablo, Troilitef	5,379	Older Than Solar System

5,016 to 5,379 million years old.

Table 9

Pb-206/Pb-207 Ratio	Age Million Years	Age Category
0.86665	5,035	Older Than Solar System
0.84518	5,000	Older Than Solar System
0.86306	5,030	Older Than Solar System
0.84983	5,008	Older Than Solar System
0.96359	5,185	Older Than Solar System

0.98081	5,210	Older Than Solar System
0.91120	5,106	Older Than Solar System
1.09068	5,359	Older Than Solar System
0.87958	5,056	Older Than Solar System
0.96906	5,193	Older Than Solar System

5,000 to 5,359 million years old.

Table 10

Pb-206/Pb-207 Ratio	Age Million Years	Age Category
0.85705	5,020	Older Than Solar System
0.85871	5,022	Older Than Solar System
0.85888	5,023	Older Than Solar System
0.85681	5,019	Older Than Solar System

5,019 to 5,023 million years old.

Table 11

Pb-206/Pb-207 Ratio	Age Million Years	Age Category
0.90695	5,100	Older Than Solar System
0.86255	5,029	Older Than Solar System
0.85613	5,018	Older Than Solar System
0.86644	5,035	Older Than Solar System
0.92835	5,133	Older Than Solar System
0.91990	5,120	Older Than Solar System
0.92542	5,128	Older Than Solar System
0.90807	5,101	Older Than Solar System
0.90861	5,102	Older Than Solar System

5,018 to 5,133 million years old.

Table 12

Pb-206/Pb-207 Ratio	Age Million Years	Age Category
0.88990	5,073	Older Than Solar System
0.87125	5,043	Older Than Solar System
0.89581	5,082	Older Than Solar System
0.89269	5,077	Older Than Solar System
0.85401	5,015	Older Than Solar System
0.89561	5,082	Older Than Solar System
0.98433	5,215	Older Than Solar System
0.92618	5,129	Older Than Solar System
0.99857	5,235	Older Than Solar System
0.95025	5,166	Older Than Solar System
1.01559	5,259	Older Than Solar System

5,015 to 5,259 million years old.

U-Th-Pb Dating of Hydrothermal ore Deposits

This dating [15] was done in 2010 on rocks from eastern China. If we look at one of the tables [16] in the original essay we see four columns of isotopic data $^{207}Pb/^{206}Pb$, $^{207}Pb/^{235}U$, $^{206}Pb/^{238}U$ and $^{208}Pb/^{232}Th$. Three have dates beside them but here are no dates beside the $^{207}Pb/^{206}Pb$ ratios. If we run the $^{207}Pb/^{206}Pb$ ratios through Isoplot we soon see why there are no dates beside them. According to the Concordia diagrams in the essay [17] the rocks are supposed to be 137 million years old with an average age of 120 million years.

Table 13

Sample Name	Maximum Age	Minimum Age	Average Age
TLS01	2,508	272	943
TLS02	346	8	254
S38	1,682	-294	354
S38	2,508	-139	899
S39	440	-325	94

$^{207}Pb/^{206}Pb$ dates.

Table 14

Sample Name	Maximum Age	Minimum Age	Difference Age	Percentage Difference	Age Category
S38-1-a1	12,721	136	12,585	9,253%	Older Than Galaxy
S38-3-a1	7,663	136	7,527	5,534%	Older Than Solar System
S38-3-a2	11,457	44	11,413	25,938%	Older Than Galaxy
S38-3-a3	7,175	130	7,045	5,419%	Older Than Solar System

Some of the dates listed in the article [16] are older than the age of the Solar System and Galaxy! The author offers an explanation: "Due to the very low Th contents in the calcite-hosted titanite, no meaningful 208Pb/232Th ages were obtained." [18]

U-Th-Pb Dating of Yucca Mountain, Nevada

This dating was done [19] in 2008 by the U.S. Geological Survey office in Denver, Colorado. You will notice in Table 1 the original article [20] that there are no dates beside the $^{207}Pb/^{206}Pb$, U/Pb and Th/Pb ratios. If we use the computer program Isoplot and calculate the ages of the $^{207}Pb/^{206}Pb$ ratios we see why not dates have been put beside them.

Table 15

Sample Name	208Pb/232Th Age	Pb 207-206 Age	206Pb/238U Age	Sample Name	208Pb/232Th Age	Pb 207-206 Age	206Pb/238U Age
1939Pb1-Cc	35	4,999		2074Pb2-Op	22	472	145
2055Pb6-Cc		4,837		2089APbl-Ccl			1
2055Pb7-Cc1		4,959		2089APb1-Cc2a			8,010
2055Pb7-Cc2		4,949	1,404	2089APb1-Cc2b			6,416
2055Pb10-Cc		4,883	5,612	2089APb1-Cc2c			7,556
2055Pb11-Cc		1,384	1,044	2089APb1-Cc3	25		6,372
2055Pb11-Op	42		408	2089APb2-Cc	14	3,365	4,511
2055Pb12-Cc		282	1	2089APb3-Mn	75	4,989	15,434
2055Pb12-Op	355,962	-332	830	2092Pb1-Cc	74,588	1,956	
2057-Pb1-Cc	445	2,258	6	2098Pb3-Cc	3,853	5,806	
2057-Pb2-Cc	454	3,818	1,711	2155Pb1-Cc	4,184	6,349	
2057-Pb2-Op	58		437	2177Pb1-Cc	74,641	4,017	
2059Pb4-Cc	1,839	2,759	1	2177Pb1-Mn	2,291	5,453	

2062Pb1-Cc		-445	2,173	2231Pb1-Cc	7,787		
2062Pb2-Cc	28	4,943	13,503	2233Pb1-Ch2	22,868	7,933	
2062Pb3-Mn	489	5,023	1,978	2233Pb2-Ch2	4,831	7,583	
2065Pb4-Cc	208	3,092		2233Pb4-Ch	7,649	6,821	
2074Pb1-Cc2			696	2247Pb1-Cc1	74,706	3,052	
2074Pb1-Cc3	938	4,122	776	2247Pb1-Op	2,291	5,449	
2074Pb2-Cc1		2,639	176	2293Pb1-Cc	75,628	5,706	
2074Pb2-Cc2	1,557	3,884	185				

The dates are between 1 and 366,000 million years old. Table 3 in the original article [21] has dates older than the universe and extreme discordance with up to 2 million percent. The average discordance is 212,000 percent!

40Ar/39Ar and U-Th-Pb Dating

This meteorite sample [22] was dated in 1983 by Donald Bogard from the Johnson Space Center, Houston Texas. If we look in Table 5 [23] in the original article, we see that there are dates beside the $^{207}Pb/^{208}Pb$ ratios. If we run the $^{207}Pb/^{206}Pb$ ratios through Isoplot we see that they uniformly differ with the $^{207}Pb/^{208}Pb$ dates given in the essay. The author's choice to drop these dates and only have dates beside the $^{207}Pb/^{208}Pb$ ratios is just an arbitrary choice.

Table 16

Age Pb-207/208	Age Pb-207/206	Age Category
4,560	5,370	Older Than Solar System
4,720	5,364	Older Than Solar System
4,560	5,364	Older Than Solar System
4,450	5,283	Older Than Solar System
4,700	5,371	Older Than Solar System
4,540	5,367	Older Than Solar System
4,410	5,082	Older Than Solar System
4,560	5,368	Older Than Solar System
4,700	5,367	Older Than Solar System
4,500	5,333	Older Than Solar System

Isotopic Lead Investigations

These meteorite samples were dated in 1975 by the Department of Geological Sciences, University of California, Santa Barbara, California. [24] From Table 2 in the original article we can calculate the $^{207}Pb/^{206}Pb$ ratios [25] and then we run them through Isoplot. The ages are consistently older than the age of the Solar System.

Table 17

Sample Name	Pb 206/207 Ages	Age Category
7-1	5,175	Older Than Solar System
7-2	5,300	Older Than Solar System
7-3	5,287	Older Than Solar System
7-4	5,346	Older Than Solar System
4-1	5,337	Older Than Solar System
W-2	5,342	Older Than Solar System
Allende-1	5,297	Older Than Solar System
Allende-2	5,326	Older Than Solar System
Allende	5,262	Older Than Solar System

9-1	5,324	Older Than Solar System
M-2	5,322	Older Than Solar System
9-3	5,339	Older Than Solar System
9-4	5,334	Older Than Solar System
ChL-1 (IC)	5,138	Older Than Solar System
ChL-1 (ID)	5,137	Older Than Solar System
Ch3 (IC)	5,220	Older Than Solar System
Ch3 (ID)	5,227	Older Than Solar System
ChD (IC)	5,103	Older Than Solar System
ChD (ID)	5,099	Older Than Solar System

More Extraordinary Examples

Below are dates calculated from Pb207/Pb206 isotopic ratios listed in standard geology journals. Most ratios give dates of 5 billion years old. The model age put forward is radically different.

Reference	Maximum	Minimum	Model Age
Earth And Planetary Science Letters, Volume 245 (2006) 743–761	4,976	4,894	0.300
Earth And Planetary Science Letters, Volume 240 (2005) 605– 620	5,049	4,743	1
Geochimica Et Cosmochimica Acta 72 (2008) 5799–5818	5,008	4,871	3
Earth And Planetary Science Letters, Volume 267 (2008) 236–246	5,002	4,852	4
Mineralogy And Petrology (1993) 47, Pages 103-126	4,990	4,985	5
Journal of Petrology, 2011, Volume 52, Number 6, Pages 1143-1183	4,977	4,966	7
Earth And Planetary Science Letters, Volume 165 (1999) 117–127	4,975	4,933	10
Earth And Planetary Science Letters, Volume 306 (2011) 86–97	6,214	4,924	10
Geochimica Et Cosmochimica Acta, 2008, Volume 72, Pages 2067 - 2089	8,126	282	10
Journal of Petrology, August 2007, Pages 1-29	5,090	5,018	10
Journal of Petrology, 2007, Volume 48, Number 12, Pages 2261-2287	4,987	4,952	16
Chemical Geology, Volume 93 (1991) Pages 231-243	4,964	4,909	20
Earth And Planetarv Science Letters, Volume 46 (1980) Pages 221-232	5,040	4,712	20
Earth And Planetary Science Letters, Volume 301 (2011) 469–478	6,194	970	20
Geochemistry And Geophysics Geosystems, 2006, Vol 7, Number 8, Pages 1-29	5,009	4,992	20
Journal Of Petrology, 2004, Volume 45, Number 3, Pages 555-607	4,994	4,919	20
Journal Of Petrology, 2011, Volume 52, Number 2, Pages 401-430	4,968	4,920	20
Gondwana Research 11 (2007) Page 382–395	5,072	5,054	30
Journal Of Petrology, 1998, Volume 39, Number 4, Pages 711-748	5,020	4,994	30
Journal Of Petrology, 2004, Volume 45, Number 5, Pages 1069–1088	4,986	4,911	30
Journal Of Petrology, April 11, 2012, Pages 1-32	4,971	4,881	30
Journal Of Petrology, September, 2006, Page 1 Of 35	5,021	4,945	30
Journal of Petrology, 2000, Volume 41, Number 7, Pages 951-966	5,316	5,010	50
Journal of Petrology, 2010, Volume 51, Number 5, Pages 993-1026	5,008	5,004	50
Precambrian Research, 1999, Volume 95, Pages 167 - 185	5,010	4,968	55
Journal Of Petrology, 1998, Volume 39 Number 11-12, Pages 1847–1864	5,019	4,807	60
Journal Of Petrology, 1999, Volume 40, Number 6, Pages 873–908	5,174	4,987	60
Journal Of Petrology, 2007, Volume 48, Number 4, Pages 661-692	5,024	5,020	60
Ore Geology Reviews 31 (2007) Pages 337–359	5,010	4,976	60

Reference	Maximum	Minimum	Model Age
Earth And Planetary Science Letters, Volume 134 (1995) 169-185	5,016	4,953	66
Earth And Planetary Science Letters, Volume 59 (1982) Pages 327-342	5,061	5,004	70
Journal Of Petrology, 1997, Volume 38, Number 1, Pages 115-132	4,949	4,888	72
Journal of Petrology, April 2008, Pages 1-28	5,053	4,980	85
Chemical Geology, Volume 211 (2004) Pages 87–109	5,053	4,980	100
Earth and Planetary Science Letters, Volume 34 (1977) 419-431	5,057	4,838	100
Earth and Planetary Science Letters, Volume 76 (1985) Pages 57-70	5,061	4,898	100
Earth and Planetary Science Letters, Volume 244 (2006) Page 251–269	5,150	5,040	100
Earth and Planetary. Science Letters, 94 (1989) Pages 78-96	5,163	5,014	100
Geological Society Of London, Special Publications, 2004, Vol 229, Pages 133-150.	6,489	4,930	100
Isotope Geoscience, 1 (1983) Pages 23-38	5,043	4,760	100
Journal Of Petrology, 1998, Volume 39, Number 7, Pages 1285-1306	5,076	4,533	100
Precambrian Research, 1988, Volume 38, Pages 147 - 164	5,074	4,940	120
Journal Of Petrology, 2005, Volume 46, Number 4, Pages 829-858	5,065	4,996	140
Geochimica et Cosmochimica Acta, 1992, Volume 56, Pages 347-368	5,236	4,822	150
Contributions Mineral Petrology (1984) 85, Pages 376-390	5,066	4,987	155
Mereoritics And Planetary Science, 2000, Volume 35, Pages 341 - 346	4,987	4,905	200
Geochimica et Cosmochimica Acta, Vol 62, Number 16, Pages 2823–2835, 1998	5,123	4,900	220
Uranium-Lead Zircon Ages, U.S. Geological Survey Report	4,982	4,024	250
Earth and Planetary Science Letters, Volume 183 (2000) 93-106	5,235	5,077	267
Chemical Geology, Volume 236 (2007) Pages 27–41	5,014	3,978	280
Earth and Planetary Science Letters, Volume 104 (1991) 1-15	5,145	4,947	300
Earth and Planetary Science Letters, Volume 94 (1989) Pages 239	5,016	668	316
Earth and Planetary Science Letters, Volume 94 (1989) Pages 236-244	5,016	668	320
http://pubs.usgs.gov/of/2008/1142/pdf/OF08-1142_508.pdf	4,997	4,799	416
Mineralium Deposita (1999) 34, Pages 273-283	5,046	5,029	456
Journal Of Petrology, 1999, Volume 40, Number 9, Pages 1399-1424	5,007	4,988	471
Geochimica et Cosmochimica Acta, Vol 62, Number 21/22, Pages 3527–3540, 1998	5,000	3,924	500
Geochimica et Cosmochimica Acta, Volume 61, Number 23, Pages 5005-5022, 1997	5,026	3,831	600
Chinese Journal Of Geochemistry, Volume 16, Number 1, 1997, Pages 80-85	5,237	5,065	700

Conclusion

Prominent evolutionist Brent Dalrymple states: "Several events in the formation of the Solar System can be dated with considerable precision." [26] Looking at some of the dating it is obvious that precision is much lacking. He then goes on: "Biblical chronologies are historically important, but their credibility began to erode in the eighteenth and nineteenth centuries when it became apparent to some that it would be more profitable to seek a realistic age for the Earth through observation of nature than through a literal interpretation of parables." [27] The Bible believer who accepts the creation account literally has no problem with such unreliable dating methods. Much of the data in Dalrymple's book is selectively taken to suit and ignores data to the contrary.

References

1 Age and mineralogy of supergene uranium, Geomorphology, 2010, Volume 117, Pages 44 - 65
2 Reference 2, Page 58
3 http://www.creationismonline.com/Isoplot/Isoplot.html
https://www.bgc.org/isoplot
4 207Pb–206Pb and 40Ar–39Ar ages from SW Montana, Precambrian Research, 2002, Volume 117, Pages 119 - 143
5 Reference 4, Page 128
6 Reference 4, Page 133
7 Uranium-Thorium-Lead Dating, Mereoritics And Planetary Science, 2000, Volume 35, Pages 341 - 346
8 Reference 7, Page 342
9 Reference 7, Page 343, 344
10 Pb–Pb dating of chondrules, Chemical Geology, 2009, Volume 259, Pages 143 - 151
11 Reference 10, Page 145
12 Reference 10, Page 147
13 Pb–Pb dating constraints, Geochimica et Cosmochimica Acta, 2007, Volume 71, Pages 1583 - 1604
14 Reference 13, Page 1596
15 U–Th–Pb dating of hydrothermal ore deposits, Chemical Geology, 2010, Volume 270, Pages 56 - 67
16 Reference 15, Page 65
17 Reference 15, Page 66
18 Reference 15, Page 62
19 U–Th–Pb dating of Yucca Mountain, Nevada, Geochimica et Cosmochimica Acta, 2008, Volume 72, Pages 2067 - 2089
20 Reference 19, Pages 2072, 2073
21 Reference 19, Pages 2080
22 40Ar/39Ar and U-Th-Pb Dating, Earth and Planetary Science Letters, 1983, Volume 62, Pages 132 - 146
23 Reference 22, Pages 139
24 Isotopic Lead Investigations, Geochimica et Cosmochimica Acta, 1976, Volume 40, Pages 635 – 643
25 Reference 24, Page 638
26 The Age Of The Earth, By G. Brent Dalrymple, 1991, Stanford University Press, Stanford, California, Page 10.
27 Reference 26, Page 23

Chapter 6

The Secrets Of Radiometric Dating

The U-Th-Pb Systematics of Allanite

The author of this article claims that allanite minerals are useful to radiometric dating [1] and has a table [2] with U/Pb, Pb/Pb and Th/Pb ratios with no dates in the table. If we use Isoplot [3] we can calculate dates from the ratios and obtain an eleven billion year spread of dates. There seems to be a massive difference between what the author says and what the dates say.

Table 1	206Pb/238U	208Pb/232Th	207Pb/206Pb
Average	1,277	2,023	1,474
Maximum	2,850	10,619	2,707
Minimum	23	6	-464
Difference	2,828	10,613	3,171

Systematics of CAMP Tholeiites

These rock samples from north America are supposed to be Jurassic [4] in age (200 million years old) The Th/Pb and U/Pb ratios via Isoplot [5] give dates as old as 19 billion years. The author conveniently left this embarrassment out of his article.

Table 2

Rock	Sample	208Pb/238U	208Pb/232Th	87Rb/86Sr
Formation	Number	Age (Ma)	Age (Ma)	Age (Ma)
Talcott flow	HB64	8,713	11,569	202
Higganum Dike	HB87	8,033	9,731	201
Holyoke flow	HB56	7,239	9,456	201
Buttress dike	HB102	8,358	12,068	201
Bridgeport Dike	HB98	10,213	19,026	201
NMB lower flow	NS1	8,338	9,273	218
	NS6	6,878	7,438	217
	NS19	6,841	7,125	306
	NS23	7,086	7,467	259
NMB Middle flow	NS7	6,301	7,029	185
	NS12	7,878	8,520	172
NMB Upper flow	NS8	8,260	8,654	202
	NS9	7,300	7,578	201
	NS13	8,073	9,569	184
	NS15	8,059	8,299	191
GAV Middle flow	GAV81	7,291	7,984	198
GAV Lower flow	GAV162	8,561	10,000	202
	GAV174	7,286	9,582	201
	GAV181	7,549	9,305	201
	GAV193	6,268	6,403	201
Shelburne dyke	NS27	7,650	8,567	201

	NS28	6,810	7,006	201
Palisades sill	NEW3	7,226	8,677	201
	NEW136	7,623	8,703	201
	NEW16	7,367	8,790	197
	NEW17	7,367	8,792	197
	NEW18	7,796	8,602	195
Orange Mt	NEW69	7,001	8,866	201
	NEW133	4,916	13,928	204
Preakness	NEW52	6,461	8,810	201
	NEW68	6,764	8,623	201
Hook Mt	NEW73	6,267	8,307	201
	NEW74	6,290	7,475	201
	NEW134	6,275	7,465	201
Mt Zion Church	CUL6	7,264	10,763	201
Hickory Grove	CUL13	7,135	11,797	201
Sander	CUL25	6,279	9,448	201
	CUL28	6,636	9,043	201
Rapidan sill	CUL8	7,312	7,600	201
	CUL9	7,830	8,783	201
Belmont Sill	CUL67	7,061	8,488	201
	Average	7,314	9,137	204
	Maximum	10,213	19,026	306
	Minimum	4,916	6,403	172
	Difference	5,297	12,623	134

Peridotites from Attawapiskat

These diamond deposits in Canada have a Rhenium depletion age of 3.6 billion years. [6] The model age has a range of 53.6 billion years. [7] The author offers no explanation.

Table 3

TRD (Ga)	TMa (Ga)
1.43	-0.31
	-5.2
3.73	-0.46
2.03	-0.32
3.59	-9.07
1.58	2.32
-	-8.51
1.55	-1.44
2.47	3.72
2.65	-48.93

Ultra-high temperature granulite's

The author of this article claims that the rock formation has a Rb/Sr age of 323 million years [8] backed up by Pb/Pb dating table [9]. If we run another table with uncalculated dates [10] through Isoplot we get dates 10 billion years older.

Table 4

Age 207Pb/206Pb	Age 206Pb/238U
5,011	
5,010	
4,994	10,306
4,746	1,363
4,780	1,703
4,748	1,626
4,971	8,163
4,911	3,640
4,824	2,906
4,683	1,738
4,947	4,877
4,411	843
4,195	702
4,002	636

Dykes in Orogenic Peridotite

These samples from China have been dated with U/Pb, Pb/Pb and Th/Pb [11] to be 400 million years old. Another table [12] in the articles gives a ten billion year spread of dates.

Table 5

87Rb/86Sr TCHUR (Ma)	147Sm/144Nd TDM (Ma)	176Lu/177Hf TDM (Ma)
-7,685	1,239	493
804	103	846
-7,693	1,280	-2,919
-8,503	1,276	579
-7,331	1,303	454
1,376	60	339
-7,338	1,383	1,763
-7,330	1,233	459
2,144	78	471
-7,335	1,311	-413

Websterite Xenoliths From Kimberley

Between the oldest [143.38 billion years] [13] and the youngest [-117.98 billion years] there is a 261.81-billion-year range. We can get the Osmium isotope [13] ratios and calculate more dates.

The author claims the true age is 3 billion years: "The Kimberley kimberlite cluster is situated in the western part of the Kaapvaal Craton in southern Africa. At 2.9 Ga two main domains of the Kaapvaal Craton, the Kimberley Block to the west, consisting of 3.2–2.8 Ga old gneisses, and the Witwatersrand Block to the east, were juxtaposed close to the Colesberg Lineament. Age constraints on the assembly of the Kimberley and Witwatersrand Blocks are provided by sediment deformation in the upper Witwatersrand basin at 2.7–2.9 Ga and the deposition of the Ventersdorp flood basalts at 2.7 Ga." [14]

Table 6

	TRD [Ga]	TMA [Ga]	Os/Os	TRD [Ga]	TMA [Ga]	Os/Os	TRD [Ga]	TMA [Ga]	Os/Os
	3.25	3.93	3.22	2.69	-11.15	2.42	3.02	3.34	2.94
	1.56	2.11	1.12	2.74	5.93	2.55	2.81	2.96	2.68
	-3.47	14.58	-4.85	2.69	2.81	2.53	3.05	4.35	2.96
	2.61	5.20	2.38	2.64	3.58	2.45	2.74	5.02	2.55
	2.60	4.87	2.38	3.19	-1.24	2.81	3.05	6.18	2.94
	-1.34	0.11	-8.21	1.54	-1.54	0.93	2.59	143.83	2.32
	3.14	0.06	-7.62	2.73	6.60	2.53	2.74	3.27	2.58
	-0.60	0.29	-1.87	0.90	1.45	0.31	0.59	2.64	-0.10
	-0.86	-15.02	-1.84	2.24	21.67	1.89	-115.96	7.05	-62.46
	2.81	3.05	2.68	2.39	2.85	2.15	-117.98	90.44	-62.12
							2.70	0.45	2.53
Average	0.97	1.918	-1.26	2.38	3.10	2.06	-19.15	24.50	-9.38
Maximum	3.25	14.58	3.22	3.19	21.67	2.81	3.05	143.83	2.96
Minimum	-3.47	-15.02	-8.21	0.90	-11.15	0.31	-117.98	0.45	-62.46
Difference	6.72	29.6	11.42	2.29	32.82	2.50	121.03	143.38	65.42

Java Subduction Zone Lavas

Dated in 1998, [15] the geological formation is allegedly Pliocene or Miocene in age: "The two linear correlations shown in Fig. 3 correspond to rocks that have distinct ages. The steeper correlation comprises Miocene rocks (9.5 Ma) whereas the shallower correlation is defined by rocks of Pliocene age (4.5 to 2 Ma). The different values obtained for the heterogeneous samples lie on both observed correlations, indicating that the heterogeneity seen in the whole data set is also expressed on the scale of a single sample." [16] If we put some isotope ratios into Excel, we get a 6,000 million year spread of dates.

Table 7

Rock Sample	Site	Rock type	K/Ar Age (Ma)	187Os/188Os Age	207Pb/206Pb Age
PK120-I	Pongkor	andesite	2	-356	
PK120-II				-193	
PK120-III				-198	
PK120-IV				-155	

PK252-I	Pongkor	andesite	2	-120	
PK252-II				-113	
CT33-I	Ciawitali	andesite	3	-37	5,486
CT33-II				-50	
CT33-III				-99	
CR3-I	Cirotan	basalt	4.5	-71	5,480
CR3-II				-82	
CR3-III				-91	
CT2	Ciawitali	andesitic basalt	3	-19	5,481
CP17	Citorek	dacitic ignimbrite	3.6	-41	5,492
CP15-I	Citorek	dacitic ignimbrite	3.6	-108	5,492
CP15-II				-112	
CT28	Ciawitali	andesite	3	-21	5,490
CP13	Cipanglesseran	dacite		-151	5,488
CR1-1	Cirotan	rhyolitic ignimbrite	9.5	-150	5,482
CR1-II				-217	
CR1-III				-113	
CR6-I	Cirotan	rhyodacitic tuff	9.5	-586	5,487
CR6-II				-559	
CR20-I	Cirotan	rhyolitic tuff	9.5	-33	5,483
				-33	
CR20-II				-34	
CR4	Cirotan	basalt	9.5	-191	

The South Qinling Orogen

Tables 1 (carbonates) and 2 (zircons) in this article [18] have very consistent dates of 230 million and 700 million years old respectively. If we calculate dates from the undated ratios in table 5 in this article [19] we get a 116 billion year spread of dates.

Table 8

Sample Number	206Pb/238U Age	207Pb/206/Pb Age	208Pb/232Th Age	207Pb/U235 Age
MY-0	1,868	4,591	39,740	3,469
Ap	1,057	4,321	5,590	2,675
MY-0s	4,007	4,126	9,790	4,080
MY-01	10,262	5,020	73,826	6,347
Bt	323	4,208	11,642	1,519
Ap	644	4,644	17,353	2,393
MY-5	2,152	4,659	71,518	3,667
Mn	694	3,707	116,860	1,897
Ct	517	3,993	25,392	1,797

MY-5s	1,553	4,595	42,667	3,258
MY-7	662	4,164	105,275	2,121
Bt	175	2,721	24,576	546
Ct	514	3,885	10,435	1,731
MY-7s	4,109	4,629	1,012	4,454
MY-8s	26,541	4,934	28,020	
MY-9s	32,766	4,941	0	
Maximum	**32,766**	**5,020**	**116,860**	**6,347**
Minimum	**175**	**2,721**	**0**	**546**
Difference	**32,591**	**2,299**	**116,860**	**5,802**

Xinjie Layered Intrusion, SW China

The mineral samples from southwest China were dated to be 262 million years old: "The samples yielded a Re–Os isochron with an age of 262 ± 27 Ma and an initial 187Os/188Os of 0.12460. The age is in good agreement with the previously reported U–Pb zircon age, indicating that the Re–Os system remained closed for most samples since the intrusion emplacement." [20] If we calculate dates from the Osmium 187/188 ratios, we get dates 3,184 million years younger. [21]

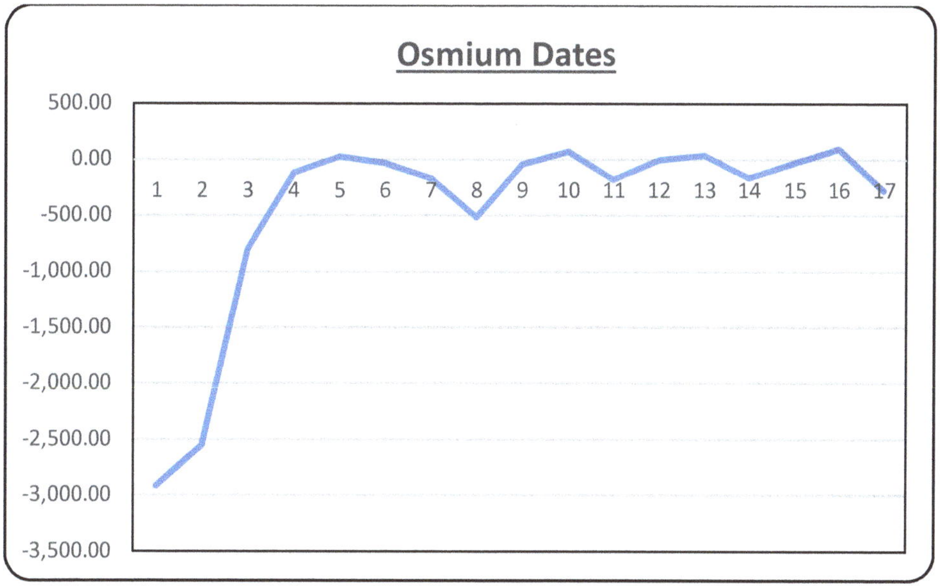

Table 9

Rock Sample	187Os/188Os Million Years
HZK411-04	-2,922.46
HZK411-06	-2,555.84
HZK411-129	-806.23
HZK411-157	-123.73
HZK411-162	21.93
HZK411-166	-35.35

HZK411-167	-172.83
HZK411-183	-524.72
HZK411-191	-50.08
HZK411-196	66.12
HZK411-205	-184.29
HZK411-212	-7.53
HZK411-217	31.75
HZK411-218	-169.56
HZK411-234	-28.81
HZK411-239	93.95
HZK411-243	-282.49
HZK411-246	36.66

References

1 The U–Th–Pb Systematics of Allanite, Geochimica et Cosmochimica Acta, Volume 135 (2014), Pages 1
2 Reference 1, pages 6, 7
3 http://www.creationismonline.com/Isoplot/Isoplot.html
https://www.bgc.org/isoplot
4 Systematics of CAMP Tholeiites, Journal of Petrology, 2014, Volume 55, Number 1, Pages 133-180
5 Reference 4, pages 140-142
6 Peridotites from Attawapiskat, Journal of Petrology, 2014, Volume 55, Number 9, Pages 1829
7 Reference 6, page 1843
8 Ultra-high temperature granulites, Journal of Petrology, 2001, Volume 42, Number 11, Pages 2015
9 Reference 8, page 2016
10 Reference 8, page 2018
11 Dykes in Orogenic Peridotite, Journal of Petrology, 2014, Volume 55, Number 12, Pages 2363
12 Reference 12, page 2367
13 Websterite Xenoliths From Kimberley, Geochimica et Cosmochimica Acta, Volume 72, (2008) 5727
14 Reference 13, page 5723
15 Java Subduction Zone Lavas, Earth and Planetary Science Letters, Volume 168, (1999), Pages 65–77
16 Reference 15, page 69
17 Reference 15, page 67
18 The South Qinling Orogen, Geochimica et Cosmochimica Acta, Volume 143, (2014), Pages 94, 195
19 Reference 18, page 198, 199
20 Xinjie Layered Intrusion, SW China, Geochimica et Cosmochimica Acta, Volume 75, (2011), Pages 1621
21 Reference 20, page 1628

Chapter 7

Meteorite Dating

History Of The Acapulco Meteorite

This meteorite was dated in 1997 by scientists [1] from France and Germany. Some of the dates [2] are older than the Solar System. We shall soon see that this is quite common for dating these rocks.

Table 1

Maximum Age	11,421	Million Years
Minimum Age	3,481	Million Years
Average Age	4,964	Million Years
Age Difference	7,940	Million Years
Difference	328%	Percent
Standard Deviation	1,723	Million Years

Potassium Argon Dating of Iron Meteorites

This article summarized meteorite dating in 1967. [3] Even 40 years later things are no better. In the opening paragraph he states that the iron meteorite from Weekeroo Station is date at ten billion years old. He then continues: "The formation or solidification ages of iron meteorites have never been well determined." [4] He then cites earlier dating which produced an age of seven billion years. [5] The author concludes with the following remark: "The ages found by us are typical of the great ages found for most iron meteorites. From these, in conjunction with the Strontium: Rubidium data of Wasserburg et al. on silicate inclusions in this meteorite, we conclude that the potassium: argon dating technique as applied to iron meteorites gives unreliable results." [6]

Table 2

Meteorite	Age
Sample	Billion Years
Neutron Activation	10.0
Stoenner and Zahringer	7.0
Muller and Ziihringer's	6.3
Wasserburg, Burnett	4.7
K-1	8.5
K-2	9.3
B-1	6.5
G-1	10.4

Pb Isotopic age of the Allende Chondrules

The meteorite was dated in 2007 using the $^{206}Pb/^{238}U$ dating method. [14] Over ten dates older than the age of the evolutionist age of the Solar System were produced and one was older [Ten Billion years] than the age of the galaxy. [15]

Table 3

Maximum Age	10,066	Million Years
Minimum Age	1,799	Million Years
Average Age	4,509	Million Years
Age Difference	8,267	Million Years
Percentage Difference	559%	Percent
Standard Deviation	1,640	Million Years

Rhenium-187-Osmium-187 in Iron Meteorites

Scientists from France used both $^{87}Sr/^{86}Sr$ and Rhenium-Osmium method were used to date this meteorite in 1998.[16] Dates in the essay [17] of the Canyon Diablo meteorite vary from one to fourteen billion years old. There is a 1,200% difference between the youngest and oldest date obtained for the one rock.

Table 4

Meteorite Name	Age Billion Years
Canyon Diablo	
Troilite 4	1.13
Leach Acetone	5.73
Leach H,O	8.31
Troilite dissolved	10.43
Metal 1	13.7

Ar-39/Ar-40 Dating of Mesosiderites

This was dated in 1990 by Scientists from the NASA Johnson Space Center, Houston, Texas.[18] All of the eleven meteorites dated gave ages older than the Solar System and three dated as being as old, or even older than the evolutionist age of the galaxy. [19] According to one table the supposed true age is just 3.5 billion years old. [20]

Table 5

Meteorite Name	Maximum Billion Years	Minimum Billion Years	Age Difference Billion Years	Percentage Difference
1. Emery	9.08	3.31	5.77	274%
2. Estherville	13.96	3.18	10.78	438%
3. Hainholz	5.48	1.55	3.93	353%
4. Lowicz	9.93	2.92	7.01	340%
5. Morristown	7.92	3.60	4.32	220%
6. Mount Padbury	5.52	3.49	2.03	158%
7. Patwar Basalt	6.14	1.80	4.34	341%
8. Patwar Gabbro	8.43	2.67	5.76	315%
9. QUE-86900	10.92	3.24	7.68	337%
10. Simondium	9.17	3.27	5.90	280%
11. Veramin	13.13	2.71	10.42	484%

40Ar-39Ar Chronology

Dated in 2009 by scientists [21] from Germany and Russia, these meteorite samples gave astounding results. Many dates were older than the evolutionist age of the Solar System, older than the galaxy and older than the Big Bang. [22] Most age results were hundreds or thousands of percent discordant.

Table 6

Sample Name	Maximum Million Years	Minimum Million Years	Age Difference Million Years	Percent Difference
Table A01. Dhofar 019 whole rock	11,679	737	10,942	1,584%
Table A02. Dhofar 019 maskelynite	10,521	818	9,703	1,286%
Table A03. Dhofar 019 pyroxene	10,730	804	9,926	1,334%
Table A04. Dhofar 019 olivine	10,487	1,778	8,709	589%
Table A05. Dhofar 019 opaque	14,917	4,420	10,497	337%
Table A06. SaU 005 whole rock	7,184	568	6,616	1,264%
Table A07. SaU 005 glass	6,235	3,247	2,988	192%
Table A08. SaU 005 maskelynite	7,432	1,344	6,088	552%
Table A10. SaU 005 olivine	13,979	3,839	10,140	364%
Table A11. Shergotty whole rock	8,542	1,112	7,430	768%
Table A15. Zagami whole rock	6,064	94	5,970	6,451%
Table A16. Zagami maskelynite	5,733	238	5,495	2,408%
Table A18. Zagami opaque	7,707	290	7,417	2,657%
Table A9. SaU 005 pyroxene	12,845	1,354	11,491	948%

Shocked Meteorites: Argon-40/Argon-39

Dated in 1997 by scientists [23] from Germany and France, these meteorite samples gave astounding results also. Many dates were older than the age of the Solar System, older than the galaxy and older than the Big Bang. [24] Most age results that were hundreds or thousands of percentage discordant.

Table 7

Sample Name	Maximum Million Years	Minimum Million Years	Difference Million Years	Percent Difference
A. Rose City (H5/S6) host rock	4,766	193	4,573	2,469
B. Rose City (H5/S6) melt	4,529	2,126	2,403	213
C. Rose City (H5/S6) host rock #1	3,876	231	3,645	1,678
D. Rose City (H5/S6) host rock #2	3,259	293	2,966	1,112
E. Travis County (H5/S4) whole rock	3,614	295	3,319	1,225
F. Yanzhuang (H6/S6) host rock	5,598	65	5,533	8,612
G. Yanzhuang (H6/S6) melt fragment	10,217	1,902	8,315	537
H. Yanzhuang (H6/S6) melt vein	7,016	1,314	5,702	534
I. Alfianello (L6/S5) whole rock	3,470	968	2,502	358
J. Bluff (L6/S6) host rock	13,348	506	12,842	2,638
K. Bluff (L6/S6) melt	3,773	554	3,219	681

L. Mbale (L5-6) whole rock	3,531	466	3,065	758
M. McKinney (L4/S4-5) whole rock	1,821	499	1,322	365
N. Ness County (L6/S6) host rock #I	5,052	987	4,065	512
O. Ness County (L6/S6) host rock #2	6,668	1,997	4,671	334
P. Paranaiba (L6/S6) host mk #I	3,332	453	2,879	736
Q. Paranaiba (L6/s6) host rock #2	5,593	3,110	2,483	180
R. Taiban (L5/S6) host rock	2,845	492	2,353	578
S. Taiban (L5/S6) melt	1,435	156	1,279	920
T. Walters (L6/S4) host rock	3,452	1,592	1,860	217
U. Walters (L6/S4) melt	4,074	2,026	2,048	201
V. Beeler (LU/S4) host rock #I	6,466	798	5,668	810
W. Beeler (LL6/S4) host rock #2	6,609	1,491	5,118	443
X. ALHA 8101 1 (eucrite) clast	3,818	375	3,443	1,018
Y. ALHA 8101 1 (eucrite) melt	2,827	244	2,583	1,159

Potassium-Argon age of Iron Meteorites

If we compare the dates below with the previous two tables [Tables 6 and 7] we see that dating done on meteorites has not improved in fifty years! The dates below [Table 8] were dating done in 1958 by scientists from Brookhaven National Laboratory, Upton, New York.[25] These dates [26] are just as stupid as the previous two tables. The choice of 4.5 billion years as an "absolute" value is purely and arbitrary choice.

Table 8

Meteorite K-Ar Dating	Age Billion Years
Mt. Ayliff	6.9
Arispe	6.8
H. H. Ninninger	6.9
Carbo	8.4
Canon Diablo I	8.5
Canon Diablo I	6.9
Canon Diablo I	6.6
Canon Diablo I	5.3
Canon Diablo II	13
Canon Diablo II	11
Canon Diablo II	10.5
Canon Diablo II	12
Toluca I	5.9
Toluca I	7.1
Toluca II	10
Toluca II	10.8
Toluca II	8.8

The Allende and Orgueil Chondrites

This rock was dated in 1976 by scientists from the United States Geological Survey, Denver, Colorado. [27] Six were dated as being over ten billion years old. [28] Two were dated as being as old as the Big Bang explosion. [28] Fifty three dates were over five billion years. [28] Below [Tables 9 and 10] we can see the strong discordance between the $^{208}Pb/^{232}Th$ and $^{206}Pb/^{238}U$ dating methods

Table 9

Pb-208/Th-232		
Maximum Age	14.40	Billion Years
Minimum Age	4.81	Billion Years
Average Age	6.40	Billion Years
Age Difference	9.59	Billion Years
Difference	299.38%	Percent
Standard Deviation	3.37	Billion Years

Table 10

Pb-206/U-238		
Maximum Age	9.86	Billion Years
Minimum Age	3.91	Billion Years
Average Age	6.02	Billion Years
Age Difference	5.95	Billion Years
Difference	252.17%	Percent
Standard Deviation	1.45	Billion Years

Precise U-Pb dating of Chondrites

This dating was done in 2005 by scientists from USA and Canada. [29] Five dates were over five billion years old. [30]

Table 11

Maximum Age	6,473	Million Years
Minimum Age	4,249	Million Years
Average Age	4,675	Million Years
Age Difference	2,224	Million Years
Difference	152%	Percent

U–Pb Ages of Angrites

This dating was done in 2007 by scientists from Australia and Canada.[31] Eight dates were older than the evolutionist age of the Solar System.[32]

Table 12

Sample Name	Pb-206/U-238 Million Years
Angra dos Reis	
4W3	5,535
5W3	5,658
Lewis Cliff 86010	
10W3a	6,072
11W3	6,625
D'Orbigny	
15R	4,842
16Ra	4,893
17R	4,695
18R	4,972
19R	5,080
20R	4,957
21W3	5,471
22W3	5,291
23W3	5,568

Argon Diffusion Properties

Dating done in 1980 of various meteorites gave many discordant values.[32] Six were dated as older than the Solar System. [33]

Table 13

Meteor's Name	Maximum Billion Years	Minimum Billion Years	Percentage Difference
Wellman	5.2	3.737	139%
Wickenburg	3.005	0.568	529%
Shaw	5.15	4.17	123%
Louisville	5.5	0.51	1,078%
Arapahoe	9.71	0.89	1,091%
Farmington	3.7	0.511	724%
Lubbock	9.4	0.12	7,833%
Orvinio	8.78	0.764	1,149%

U-Th-Pb Dating of Abee E4 Meteorite

This dating was done in 1982 by scientists from the NASA, Johnson Space Center, Houston Texas and the U.S. Geological Survey, Denver, Colorado.[35] The two table below [Table 14, 15] are a summary of Argon dating done on different meteorite samples.[36] Both sample record dates older than the evolutionist age of the solar system. The original article has undated $^{207}Pb/^{206}Pb$ ratios. If we run the through Isoplot [37] we find the ratios [38, 39] give the results in tables 16 and 17. All are much older than the evolutionist age of the solar system.

Table 14

Abee clast 2, 2, 05		
Maximum Age	7,200	Million Years
Minimum Age	3,990	Million Years
Average Age	4,640	Million Years
Age Difference	3,210	Million Years
Difference	180%	Percent
Standard Deviation	840	Million Years

Table 15

Abee clast 3, 3, 06		
Maximum Age	8,900	Million Years
Minimum Age	3,580	Million Years
Average Age	4,610	Million Years
Age Difference	5,320	Million Years
Difference	248%	Percent
Standard Deviation	1,360	Million Years

Table 16

Meteorite	Pb-206/207	Pb-206/207
Name	Ratio	Age
Abee 1	1.0992	5,370
	1.0945	5,364
	1.0947	5,364
	1.0330	5,283
Abee 2	1.1000	5,371
	1.0966	5,367
	0.8958	5,082
Abee 3	1.0976	5,368
	1.0967	5,367
	1.0708	5,333

Table 17

Meteorite	Pb-207/206	Pb-207/206
Name	Ratio	Age
Abee 1	1.0993	5,370
	1.1005	5,372
	1.0994	5,370

Abee 2	1.1005	5,372
	1.0991	5,370
Abee 3	1.0999	5,371
	1.0993	5,370
Indarch	1.1005	5,372
St. Sauveur	0.7015	4,734
Canyon Diablo	1.1060	5,379

39Ar/40Ar Ages of Eucrites

These samples were dated in 2003 by scientists from the NASA Johnson Space Center, Houston, Texas, and the Lockheed-Martin Corporation, Houston, Texas.[40] Ten of the meteorites were dated as being over five billion years old. [41]

Table 18

Meteorite Sample	Maximum Million Years	Minimum Million Years	Difference Million Years	Percent Difference
A. QUE 97053,8	9,669	3,749	5,920	257%
B. GRA 98098,26 WR	7,008	3,239	3,769	216%
C. PCA - 82502,81	5,431	3,300	2,131	164%
D. PCA - 91007,26	4,460	1,560	2,900	285%
E. Caldera	4,493	2,819	1,674	159%
F. Asuka-881388,55	4,853	3,250	1,603	149%
G. Asuka-881467,42	4,465	202	4,263	2,210%
H. GRO - 95533,7	4,096	2,823	1,273	145%
I. QUE - 97014,5	4,553	2,947	1,606	154%
J. Moama	4,484	866	3,618	517%
K. EET - 87520	5,481	2,004	3,477	273%
L. Moore County	6,742	1,827	4,915	369%
M. Serra de Mage	6,100	499	5,601	1222%
N. EET -87548	3,674	1,738	1,936	211%
O. ALH -85001,32	4,754	3,097	1,657	153%
P. Piplia Kalan	4,284	162	4,122	2644%
Q. Sioux County	4,513	2,189	2,324	206%
R. Asuka-87272,49	3,652	342	3,310	1067%
S. Macibini Glass	5,788	2,621	3,167	220%
T. QUE - 94200,13	3,724	3,169	555	117%
U. EET - 87509,24	7,496	4,026	3,470	186%
V. EET - 87509,71	4,449	3,558	891	125%

W. EET -87509,74	4,645	873	3,772	532%
X. EET - 87531,21	4,176	3,301	875	126%
Y. EET - 87503,53	5,209	3,568	1,641	145%
Z. EET - 87503,23	5,324	2,294	3,030	232%

Argon-39/Argon-40 Ages

These samples were dated in 2003 by scientists from the NASA Johnson Space Center, Houston, Texas, and the Lockheed-Martin Corporation, Houston, Texas.[42] The Monahans chondrite and halite was dated in 2001 as being over eight billion years old.[43]

Table 19

Maximum Age	8,058	Million Years
Minimum Age	3,899	Million Years
Average Age	4,474	Million Years
Age Difference	4,159	Million Years
Difference	206%	Percent

Rb-Sr Ages Of Iron Meteorites

These samples were dated in 1967 by the California Institute of Technology, Pasadena, California.[44] Even after 40 years of research and the massive improvement in laboratory equipment and computer technology, things today are just as bad as back then! Fourteen of the dates are five billion years or more.[45]

Table 20

Meteorite Rb-Sr Dating	Age Billion Years
Four Corners AM 1	8.4
	9.3
	9.1
	9.1
	8.5
	8.2
Four Corners AM 2-B1	5.0
	5.1
	4.8
Four Corners AM 2-B6	5.0
Four Corners H-1	5.0
Four Corners H-3	4.9
Four Corners N-1	5.2
Linwood H-B1	5.1
Odessa N1-8	4.9
	4.8
Toluca N-A3	5.0
	4.7
	4.9
	4.9
Colomera D6	5.1

40-Ar / 39-Ar Ages of Allende

Scientist from the Max-Planck-Institute, Heidelberg, Germany, dated these samples in 1980. [46] Seven samples were dated as being over five billion years old. [47]

Table 21

Sample Name	Maximum Million Years	Minimum Million Years	Difference Million Years	Percentage Difference
Sample 01	4,455	2,452	2,003	181%
Sample 02	5,067	3,027	2,040	167%
Sample 03	4,919	4,092	827	120%
Sample 04	4,939	4,363	576	113%
Sample 05	4,691	2,248	2,443	208%
Sample 06	4,943	4,102	841	120%
Sample 07	4,835	4,166	669	116%
Sample 08	4,776	4,207	569	113%
Sample 09	5,004	3,682	1,322	135%
Sample 10	4,505	1,871	2,634	240%
Sample 11	4,707	3,631	1,076	129%
Sample 12	5,641	4,330	1,311	130%
Sample 13	4,549	4,396	153	103%
Sample 19	5,590	4,110	1,480	136%
Sample 20	5,812	4,367	1,445	133%
Sample 21	5,784	4,256	1,528	135%
Sample 23	7,460	3,967	3,493	188%

The Fossil LL6 Chondrite

These meteorite fragments were dated in 2010 by scientists from Australia, South Africa, England and Finland. [48] Some dates are over 4,000 percent discordant. [49]The oldest dates are as old as the evolutionist age of the galaxy. [49]

Table 22

Sample Name	Maximum Age Million Years	Minimum Age Million Years	Age Difference Million Years	Percent Difference
A	2,065	164	1,902	1,263%
B	2,849	924	1,925	308%
C	2,043	177	1,867	1,157%
D	7,119	174	6,945	4,082%
E	3,889	249	3,640	1,563%
F	11,250	5,475	5,775	205%

K/Ar Age Determinations of Iron Meteorites

This was dated in 1968 and produced ages between 1.5 and 7.4 billion years. [50] Eight dates were older than the age of the Solar System. [51] Comparing dating forty years ago with the latest dating techniques shows no improvement.

Table 23

Meteorite	Maximum	Minimum	Difference	Percentage
K-Ar Dating	Billion Years	Billion Years	Billion Years	Difference
Carthage 527	6.25	3.65	2.60	171.23%
Odessa 485	7.40	4.20	3.20	176.19%
Tombigbee River 602	6.35	4.85	1.50	130.93%

The Peace River Shocked M Chondrite

The meteorite was dated by scientists from the Physics Department, Sheffield University, United Kingdom. [52] The dates listed in the original article [53] are much older than the evolutionist age of the solar system. This was done in 1988. If you compare table 23 and table 24 in my essay you will see that after 20 years of research the dating is just as bad as day one.

Table 24

Sample	Maximum	Minimum	Difference	Percent
Name	Million Years	Million Years	Million Years	Difference
TABLE 1A	3,176	190	2,986	1672%
TABLE 1B	5,006	422	4,584	1186%
TABLE 2	6,130	950	5,180	645%
TABLE 4	2,515	500	2,015	503%
TABLE 5	7,100	510	6,590	1392%

Ar-39/Ar-40 Dating of IAB Iron Meteorites

In 1979 this dating was carried out by the Department of Physics, University of California, Berkeley. [54] One of the meteorites was dated at almost ten billion years old. [55]

Table 25

Maximum Age	9,500	Million Years
Minimum Age	4,460	Million Years
Average Age	5,161	Million Years
Age Difference	5,040	Million Years
Difference	213%	Percent
Standard Deviation	1,753	Million Years

Antarctic LL-Chondrites

This sample as dated in 1990 by the Department of Earth Sciences, Faculty of Science, Kobe University, Japan. [56] Some were dated as being older than the evolutionist age of the Solar System. [57]

Table 26

Maximum Age	7,330	Million Years
Minimum Age	3,110	Million Years
Average Age	4,410	Million Years
Age Difference	4,220	Million Years
Difference	235%	Percent
Standard Deviation	950	Million Years

Single grain (U-Th)/He ages

This sample as dated in 2003 by the Department of Earth and Planetary Science, University of California, Berkeley. [58] The dating of one rock produced dates that varied by over 300 percent. [59]

Table 27

Maximum Age	4,909	Million Years
Minimum Age	1,452	Million Years
Average Age	4,091	Million Years
Age Difference	3,457	Million Years
Difference	338%	Percent

Resolution Reveals New Problems

A joint paper by scientist from Australia, USA, Denmark and France. [60] It discusses why there is discord between dating done on meteorite samples. Below is a list of the five major points discussed in the article. [61]

Table 28

Potential problem	Level of awareness and suggested actions
1	1
Presence of non-radiogenic Pb of unknown isotopic composition. The most important and common problem of all.	Recognized by most of the community. Better methods for removal of non-radiogenic Pb are required.
2	2
Deviations from closed system evolution (loss of Pb, gain or loss of U). Important and common.	Requires monitoring U–Pb concordance and studying distribution of U and radiogenic Pb.
3	3
Mis-identification of the processes that start or reset the isotopic clocks. Important and common.	Requires studying distribution of U and radiogenic Pb, improving experimental reference data set for element migration caused by diffusion, alteration and shock, and linking isotopic dating to the studies in mineralogy and petrology of meteorites.
4	4
Analytical problems (fractionation, instrument-specific etc.) and blank subtraction. Important.	Problems are widely recognized. Ongoing analytical developments help to reduce them.

5	5
Fractionation of radiogenic Pb isotopes induced by leaching of alpha recoil tracks. Potentially important.	Recognized by some "terrestrial" geochronologists, less known to meteoriticists. Detailed experimental studies are required to understand the nature and extent of fractionation.

Fission-Track Ages Of Four Meteorites

Six different meteorites were dated in 1976 by scientists from the Enrico Fermi Institute and Department of Chemistry, University of Chicago, Chicago, Illinois.[62] The dates [Table 29] varied by almost one thousand percent![63] If we look at table 30 we can see the four methods used [Fission Track, Potassium-Argon, Uranium-Helium and Rubidium-Strontium] and the discordance between them.[63]

Table 29

Sample Name	Maximum Age Billion Years	Minimum Age Billion Years	Age Difference Billion Years	Percent Difference
Bondoc	1.30	0.14	1.16	929%
Mincy	3.93	1.50	2.43	262%
Nakhla	4.40	0.77	3.63	571%
Serra	2.70	0.54	2.16	500%
Washougal	4.60	4.00	0.60	115%
Allende	4.50	3.60	0.90	125%

Table 30

Meteorite Name	Fission Track Billion Years	K-Ar Billion Years	U-He Billion Years	Rb-Sr Billion Years
Bondoc	0.14	1.30	0.60	
Mincy	1.50	3.93		
Nakhla	4.40	1.30	0.77	3.60
Serra	0.54	2.70		
Washougal	4.60	4.00		
Allende	4.50	4.40		3.60

Discordant Meteorite Ages

Many dates are highly discordant and give different ages for the one meteorite. Meteorite Dar al Gani was dated in 2004 by scientists from Italy and England.[64]

Table 31. Meteorite Dar al Gani [65]

Maximum Age	3,725	Million Years
Minimum Age	1,749	Million Years
Average Age	3,120	Million Years
Age Difference	1,976	Million Years
Difference	213%	Percent

The Kirin Chondrite was dated in 1981 by scientists from the Research School of Earth Sciences, The Australian National University. Canberra.[66]

Table 32. The Kirin Chondrite [67]

Maximum Age	4,310	Million Years
Minimum Age	520	Million Years
Average Age	3,160	Million Years
Age Difference	3,790	Million Years
Difference	828%	Percent

The Acapulco Meteorite was dated in 2003 by scientists from the Department of Earth and Planetary Science, University of California, Berkeley. [68]

Table 33. (U-Th)/He ages from Acapulco Meteorite [69]

Maximum Age	4,909	Million Years
Minimum Age	1,452	Million Years
Average Age	4,091	Million Years
Age Difference	3,457	Million Years
Difference	338%	Percent

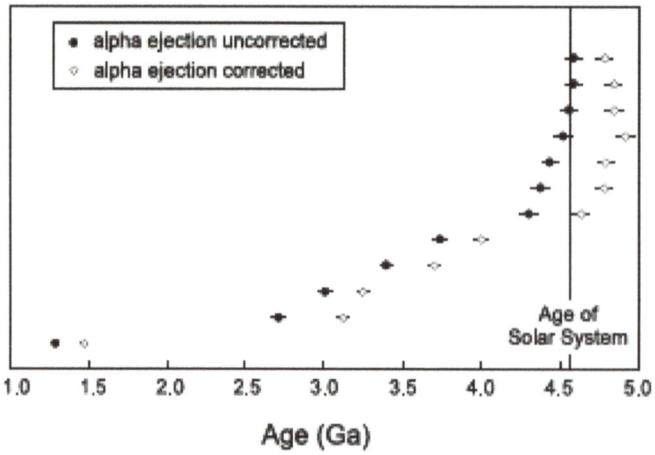

Kyoungwon Min admits that the dating of the Acapulco meteorite is extremely discordant: "Note that seven out of 12 corrected ages are older than the age of the solar system." [70] The diagram above is taken from his work. [70]

These whole rock nakhiltes were dated in 2004 by scientists from the Lunar and Planetary Laboratory, University of Arizona, Tucson, Arizona.[71]

Table 34. 40Ar-39Ar Studies of Whole Rock Nakhlites[72]

Table Number	Maximum Million Years	Minimum Million Years	Difference Million Years	Difference Percent
Table 1	1,405	262	1,143	536%
Table 2	1,409	199	1,210	708%
Table 3	1,425	761	664	187%

The Kirin Chondrite was dated in 1980 by scientists from the Research School of Earth Sciences, The Australian National University. Canberra.[73]

Table 35. History Of The Kirin Chondrite [74]

Table Number	Maximum Billion Years	Minimum Billion Years	Difference Billion Years	Difference Percent
Kirin-1	4.36	2.16	2.2	102%
Kirin-2	4.06	0.48	3.58	746%

Uranium-Thorium-Lead Dating Of Shergotty Phosphates

This dating was done in 2000 by scientists from the Department of Earth and Planetary Sciences, Hiroshima University, Japan and the Planetary Geosciences Institute, Department of Geological Sciences, University of Tennessee.[75] According to isochron diagrams in the original article, the meteorite's true age is 200 million years old. [76] If we take the list of $^{207}Pb/^{206}Pb$ ratios in this article [77] and run them through Isoplot we get the dates as shown in table 36 below.

Table 36

Sample Name	Pb-207/206 Ratio	Pb-207/206 Age
SHR04.1	0.889	5,071
SHRO5.1	0.916	5,114
SHR06.1	0.788	4,900
SHR13.1	0.876	5,051
SHRI5.1	0.833	4,979
SHR16.1	0.869	5,039
SHR19.1	0.821	4,959
SHR21.1	0.842	4,994
SHR26.1	0.922	5,123
SHR26.2	0.831	4,976
SHR27.1	0.867	5,036
SHR28.1	0.813	4,945
SHR29.1	0.827	4,969

Ion microprobe U-Th-Pb dating

This dating was done in 2000 by scientists from the Department of Earth and Planetary Sciences, Hiroshima University, Japan.[78] According to isochron diagrams in the original article, the meteorite's true age is between 1200 and 1700 million years old. [79] If we take the list of $^{207}Pb/^{206}Pb$ ratios in this article [80] and run them through Isoplot we get the dates as shown in table 37 below.

Table 37

Sample	Pb-207/206	Pb-207/206
Name	Ratio	Age
LAFA01.01	0.7907	4,905
LAFA03.01	0.3969	3,897
LAFA04.01	0.6561	4,637
LAFA04.02	0.6639	4,654
LAFA04.03	0.6898	4,710
LAFA05.01	0.7999	4,922
LAFA08.01	0.4505	4,087
LAFA09.01	0.7126	4,756
LAFA10.01	0.6506	4,625
Y-000593.1	0.9029	5,093
Y-000593.2	0.7225	4,776
Y-000593.3-1	1.0819	5,348
Y-000593.3-2	0.8453	5,000
Y-000593.4	0.7097	4,750
Y-000593.5	0.6311	4,581
Y-000749.1	0.7842	4,893
Y-000749.3	0.9092	5,103
Y-000749.4	0.7529	4,835
Y-000749.5-1	0.8569	5,019

The Chondritic Meteorite Orvinio

Scientists from Arizona, Massachusetts, New Mexico and Florida performed this dating in 2004.[81] Four of the meteorites dated to be older than the evolutionist age of the Solar System. [82] One date to be older than the Big Bang. [82] The discordance between dates varied from hundreds to thousands of percent in error. [82]

Table 38

Table	Max Age	Min Age	Difference	Percentage
Name	Million Years	Million Years	Million Years	Difference
A1	17,178	570	16,608	2,914%
A2	3,660	324	3,336	1,030%
A3	3,720	703	3,017	429%
A4	7,800	904	6,896	763%
A5	7,100	922	6,178	670%
A6	8,500	526	7,974	1,516%

Martian Meteorite Chronology

This meteorite was dated in 2011 by scientists from the Lawrence Livermore National Laboratory, Physical and Life Sciences, Institute of Geophysics and Planetary Physics, California and the Department of Earth and Planetary Sciences, University of New Mexico. [83] The article states that the meteorite's true age is 3.6 billion years. [84] If we take the list of $^{207}Pb/^{206}Pb$ ratios in this article [85] and run them through Isoplot we get the dates as shown in table 39 below.

Table 39

Sample Name	Pb-207/206 Ratio	Pb-207/206 Age
Plag(R)	0.751287431	4,832
Plag(L)	0.787456711	4,899
Px(R)	0.580150952	4,459
Px(L)	0.699212521	4,729
WR(R)	0.480536633	4,183
WR(L)	0.489632855	4,210
Ilm	0.498182294	4,236
Heated Sample		
Plag(R)	0.773980154	4,875
Plag(L)	0.640266469	4,602
Plag-rej	0.61697479	4,548
Px(R)	0.655620155	4,636
Px(L)	0.623966942	4,565
Px-rej	0.565672185	4,422
WR(R)	0.500867867	4,244
WR(L)	0.515289324	4,286
Ilm	0.498417311	4,237
NBS-981	0.913501361	5,110
Faraday–Daly	0.913967671	5,111

$^{39}Ar/^{40}Ar$ "ages" in Martian Shergottites

I downloaded this table from the official Meteoritics website. [86] Six of the meteorites were dated as being well over five billion years old. One was dated as being as old as the evolutionist age of the Milky Way Galaxy. [86]

Table 40

Sample Name	Max Age Million Years	Min Age Million Years	Difference Million Years	Percentage Difference
Los Angeles Plag	4,569	183	4,387	2,404%
Los Angeles, WR	1,270	156	1,114	714%
Los Angeles Pyx	7,432	581	6,851	1,180%
NWA-3171 Plag	2,484	203	2,281	1,121%
NWA-3171 Glass	2,056	299	1,757	588%
NWA-2975 Plag	5,709	262	5,447	2,080%

Dhofar 019 Plag	10,150	453	9,697	2,140%
Dhofar 019 WR	7,791	614	7,177	1,170%
DaG476 Plag	3,378	432	2,946	681%
DAG 476 WR	5,889	980	4,909	501%
DaG476-Px-Dark	7,975	1,746	6,229	357%
DaG476-Px-Light	4,117	391	3,726	953%
NWA-1068 WR	2,524	61	2,463	4,043%
SAU-005 WR	3,988	-0.4619	3,988	863,490%
Y-980459 WR	1,784	583	1,201	206%

Argon Dating Of Chondrites

I downloaded this table from the official Meteoritics website. [87] Four of the meteorites were dated as being well over five billion years old. One was dated as being older than the evolutionist age of the Milky Way Galaxy. [87]

Table 41

Meteorite Name	Maximum Age Billion Years	Minimum Age Billion Years	Difference Billion Years	Percentage Difference
Caddo #5	12.55	4.22	8.33	197%
EET833,5	6.82	2.21	4.60	208%
Udei Station	4.52	1.43	3.09	216%
Campo del Cielo	7.71	3.40	4.31	127%
Kendall Co.	7.59	2.06	5.53	269%

Isotopic Lead Ages Of Meteorites

This dating was done in 1973 by scientist from Switzerland and California. [88] The dates [89] below in table 42 give numerous values much older than the so-called age of the Solar System.

Table 42

Meteorite Name	206Pb/238U Million Years	207Pb/235U Million Years	207Pb/206Pb Million Years
Bruderheim-1	4126	4447	4647
Bruderheim-2	4542	4592	4628
Bruderheim-3	4959	4703	4605 4,613
Richardton-1	8615	5602	4604 4,638
Richardton-2	6834	5230	4633 4,616
Pultusk	5334	4939	4657 4,651

If we take the list of $^{207}Pb/^{206}Pb$ ratios in this article [90] and run them through Isoplot we get the dates as shown in table 43 below.

Table 43

Meteorite Name	206Pb/204Pb Amount	207Pb/204Pb Amount	207Pb/206Pb Ratio	207Pb/206Pb Age
Allende-I	1,064	1,088	1.0226	5,269
Allende-II	1,012	1,078	1.0652	5,326
Murchison	977	1,056	1.0809	5,346
	985	1,062	1.0782	5,343
Mezo-Madaras	9,449	10,384	1.0990	5,370
	9,444	10,356	1.0966	5,367
Bruderheim-I	3,562	2,683	0.7532	4,836
Bruderheim-II	3,023	2,327	0.7698	4,867
Bruderheim-III	3,275	2,469	0.7539	4,837
	3,733	2,741	0.7343	4,799
Richardton-I	2,155	1,794	0.8325	4,978
	2,187	1,796	0.8212	4,959
Richardton-II	2,228	1,827	0.8200	4,957
	2,571	2,050	0.7974	4,917
Pultusk	2,045	1,732	0.8469	5,003
	2,180	1,820	0.8349	4,982

U-Pb and ^{207}Pb-^{206}Pb ages of Eucrites

This dating was done in 2005 by scientists from the Antarctic Meteorite Research Centre, Tokyo, Japan. [91] Several dates [92] give ages much greater than the "absolute age" of 4.5 billion years for the age of the Solar System.

Table 44

Meteorite Name	Maximum Million Years	Minimum Million Years	Average Million Years
Yamato-75011	5,070	4,548	4,863
Yamato-792510	5,300	4,613	4,899
Asuka-881388	4,825	3,847	4,404
Asuka-881467	4,911	4,569	4,673
Padvalninkai	5,223	3,102	4,537

$^{40}Ar/^{39}Ar$ Dating Of Desert Meteorites

Dated in 2005 by scientists [93] from Germany and Russia, these meteorite samples gave astounding results. Many dates were older than the evolutionist age of the Solar System. [94]

Table 45

Sample Name	Million Years
Table A1. Dhofar 007 whole rock.	7,632
	6,033 5,498
Table A2. Dhofar 007 plagioclase.	7,582
	7,011 4,753 4,741
Table A3. Dhofar 300 whole rock.	9,015
	8,485 5,516 5,137
Table A5. Dhofar 300 pyroxene	8,957
	6,064 5,656 4,998 4,720
Table A5. Dhofar 300 plagioclase.	9,680
	5,793 5,721 5,395 5,237 5,035 4,788

Northwest Africa 482

These meteorites were dated in 2002 by scientists from the Lunar and Planetary Laboratory, University of Arizona, Tucson, Arizona. [95] Many dates were older than the evolutionist age of the Solar System. [96]

Table 46

Bulk Sample	Million Years
	9,670
	8,560
	8,127
	6,256
Glass Sample	**Million Years**
	9,905
	7,388
	5,708

Conclusion

Brent Dalrymple states in his anti creationist book **The Age of the Earth**: "Several events in the formation of the Solar System can be dated with considerable precision." [97]

Looking at some of the dating it is obvious that precision is much lacking. He then goes on: "Biblical chronologies are historically important, but their credibility began to erode in the eighteenth and nineteenth centuries when it became apparent to some that it would be more profitable to seek a realistic age for the Earth through observation of nature than through a literal interpretation of parables." [98]

I his book he gives a table [99] with radiometric dates of twenty meteorites. If you run the figures through Microsoft Excel, you will find that they are 98.7% in agreement. There is only a seven percent difference between the ratio of the smallest and oldest dates.

As we have seen in this essay, such a perfect fit is attained by selecting data and ignoring other data. A careful study of the latest research shows that such perfection is illusionary at best.

The Bible believer who accepts the creation account literally has no problem with such unreliable dating methods. Much of the data in Dalrymple's book is selectively taken to suit and ignores data to the contrary.

References

1 Paul Pellas, History Of The Acapulco Meteorite, Geochemica Et Cosmochemica Acta, 1997, Volume 61, Number 16, pp. 3477 - 3501

2 Reference 1, Page 3500

3 L. Rancitelli, Potassium: Argon Dating of Iron Meteorites, Science, 1967, Volume 155, Pages 999 - 1000

4 Reference 3, Page 999

5 R. W. Stoenner and J. Zahringer, Geochimica et Cosmochimica Acta, 1958, Volume 15, Page 40.

6 Reference 5, Page 1000

14 Yuri Amelin, Pb isotopic age of the Allende Chondrules, Meteoritics And Planetary Science, 2007, Volume 42, Numbers 7/8, Pages 1321 – 1335

15 Reference 14, Page 1524

16 J. L. Birck, Rhenium-187-Osmium-187 in iron meteorites, Meteoritics And Planetary Science, 1998, Volume 33, Pages 641 - 453

17 Reference 16, Page 649

18 D. D. Bogard, Ar-39, Ar-40 Dating of Mesosiderites, Geochemica Et Cosmochemica Acta, 1990, Volume 54, pages 2549 - 2564

19 Reference 18, Page 2563, 2564

20 Reference 18, Page 2551

21 Ekaterina V. Korochantseva, 40Ar-39Ar Chronology, Meteoritics And Planetary Science, 2009, Volume 44, Number 2, Pages 293 – 321

22 Reference 20, Pages 316 to 321

23 Joachim Kunz, Shocked meteorites: Argon-40-Argon-39, Meteoritics And Planetary Science, 1997, Volume 32, Pages 647 - 670

24 Reference 21, Pages 664 to 670

25 R. W. Stoenner, Potassium-argon age of iron meteorites, Geochemica Et Cosmochemica Acta, 1958, Volume 15, Pages 40 – 50

26 Reference 25, Pages 45 to 46

27 Mitsunobu Tatsumoto, The Allende and Orgueil Chondrites , Geochemica Et Cosmochemica Acta, 1976, Volume 40, pages 617 - 634

28 Reference 27, Page 627

29 Yuri Amelin, Precise U-Pb dating of Chondrites, Geochemica Et Cosmochemica Acta, 2005, Volume 69, Number 2, pages 505 – 518

30 Reference 29, Page 509

31 Yuri Amelin, U–Pb ages of angrites, Geochemica Et Cosmochemica Acta, 2008, Volume 72, Pages 221 – 232

32 Reference 31, Page 225
33 D. D. Bogard, Ar Diffusion Properties, Meteorites, Geochemica Et Cosmochemica Acta, 1980, Volume 44, Pages 1667 - 1682
34 Reference 31, Pages 1670, 1671
35 D. D. Bogard, U-Th-Pb dating of Abee E4 Meteorite, Earth and Planetary Science Letters, 1983, Volume 62, Pages 132 – 146
36 Reference 35, Page 134, 135
37 http://www.creationismonline.com/Isoplot/Isoplot.html
https://www.bgc.org/isoplot
38 Reference 35, Page 139
39 Reference 35, Page 142
40 D. D. Bogard, 39Ar/40Ar Ages of Eucrites, Meteoritics And Planetary Science, 2003, Volume 38, Number 5, Pages 669 – 710
41 Reference 39, Pages 699 to 710
42 D. D. Bogard, Argon-39/Argon-40 Ages, Meteoritics And Planetary Science, 2001, Volume 36, Pages 107 - 122
43 Reference 42, Pages 120-122
44 D. S. Burnett, Rb-Sr Ages Of Iron Meteorites, Earth and Planetary Science Letters, 1967, Volume 2, Pages 397 - 408
45 Reference 44, Pages 401, 402
46 Elmar K. Jessberger, 40-Ar/39-Ar Ages of Allende, Icarus, 1980, Volume 42, pages 380 - 405
47 Reference 46, Pages 390 – 403
48 F. Jourdan, The Fossil LL6 Chondrite, Geochemica Et Cosmochemica Acta, 2010, Volume 74, Pages 1734 – 1747
49 Reference 48, Page 1738-1739
50 W. Kaiser, K/Ar Age Determinations of Iron Meteorites, Earth and Planetary Science Letters, 1968, Volume 4, pages 84 - 88
51 Reference 50, Page 86
52 P. McConville, The Peace River shocked M chondrite, Geochemica Et Cosmochemica Acta, 1988, Volume 52, Pages 2487 - 2499
53 Reference 52, Pages 2489, 2490, 2493, 2494
54 Sidney Niemeyer, Ar-39/Ar-40 dating of IAB iron meteorites, Geochemica Et Cosmochemica Acta, 1979, Volume 43, Pages 1829 - 1840
55 Reference 54, Page 1834
56 Osamu Okano, Antarctic LL-chondrites, Geochemica Et Cosmochemica Acta, 1990, Volume 54, Pages 3509 - 3523
57 Reference 57, Page 3510
58 Kyoungwon Min, Single grain (U-Th)/He ages, Acapulco meteorite, Earth and Planetary Science Letters, 2003, Volume 209, pages 323 - 336
59 Reference 57, Page
60 Yuri Amelin, Resolution Reveals New Problems, Geochemica Et Cosmochemica Acta, 2009, Volume 73, Pages 5212 – 5223
61 Reference 60, Page 5215
62 Eugene A. Carver, Fission-track ages of four meteorites, Geochemica Et Cosmochemica Acta, 1976, Volume 40, Pages 467 - 477
63 Reference 62, Page 475
64 Luigi Folco, Meteorite Dar al Gani 896, Geochemica Et Cosmochemica Acta, 2004, Volume 68, Number 10, Page 2383
65 Reference 64, Page 2383
66 T. Mark Harrison, The Kirin Chondrite, Geochemica Et Cosmochemica Acta, 1981, Volume 45, Pages 2514
67 Reference 66, Page 2514
68 Kyoungwon Min, (U-Th)/He ages from Acapulco meteorite, Earth And Planetary Science Letters, 2003, Volume 209, Pages 328
69 Reference 66, page 328
70 Reference 66, page 332

71 Timothy D. Swindle, 40Ar-39Ar Studies of Whole Rock Nakhlites, Meteoritics And Planetary Science, 2004, Volume 39, Number 5, Pages 764 – 766
72 Reference 66, page 764-766
73 Sungshan Wang, History Of The Kirin Chondrite, Earth And Planetary Science Letters, 1980, Volume 49, Pages 117 - 131
74 Reference 73, page 120
75 Uranium-Thorium-Lead Dating Of Shergotty Phosphates, Mereoritics And Planetary Science, 2000, Volume 35, Pages 341-346
76 Reference 75, Page 343, 344
77 Reference 75, Page 342
78 Ion microprobe U-Th-Pb Dating, Meteoritics & Planetary Science, 2004, Volume 39, Number 12, Pages 2033–2041
79 Reference 78, Page 2035, 2037
80 Reference 78, Page 2036
81 The Chondritic Meteorite Orvinio, Meteoritics & Planetary Science, 2004, Volume 39, Number 9, Pages 1475–1493
82 Reference 82, Page 1488 – 1493
83 Martian Meteorite Chronology, Meteoritics & Planetary Science, 2011, Volume 46, Number 1, Pages 35–52
84 Reference 84, Page 41
85 Reference 84, Page 47
86 D. Bogard, ^{39}Ar-^{40}Ar "ages" and origin of excess ^{40}Ar in Martian shergottites http://meteoritics.org/Online%20Supplements/MAPS1080_Electronic-Annex.doc
87 Meteoritics & Planetary Science, Volume 40, Issue 2, February 2005 http://meteoritics.org/Online%20Supplements/Ar-XeData_Bogard.xls
88 G. R. Tilton, Isotopic Lead Ages Of Meteorites, Earth And Planetary Science Letters, 1973, Volume 19 Pages 321-329
89 Reference 89, Page 328
90 Reference 89, Page 323
91 U-Pb and ^{207}Pb-^{206}Pb ages of Eucrites, Geochimica et Cosmochimica Acta, 2005, Volume 69, Number 24, Pages 5847–5861.
92 Reference 92, Pages 5852 – 5853
93 Meteoritics & Planetary Science, 2005, Volume 40, Number 9/10, Pages 433–1454
94 Reference 94, Pages 1452 – 1454
95 Meteoritics & Planetary Science, 2002, Volume 37, Pages 1797-1813
96 Reference 96, Page 1806
97 The Age Of The Earth, By G. Brent Dalrymple, 1991, Stanford University Press, Stanford, California, Page 10.
98 Reference 98, Page 23
99 Reference 98, Page 287
100 Reference 98, Page 342

Chapter 8

Rocks With Negative Dates

By Paul Nethercott, August 2013

Introduction

A 40Ar/39Ar Geochronological Study

Rock samples from the Lower Onverwacht Volcanics in Barberton Mountain Land, South Africa were dated in 1992 by geologists from the Department of Physics, University of Toronto, and the Department of Geological Sciences, Queen's University, Kingston, Ontario, Canada. [1] The youngest date was -4.5 x 10^{10} million years. [2] How can a rock that exists in the present have formed 45,000 trillion years in the future? Such a proposition is illogical.

Table 1

Sample Number	Age, Million Years
B40-A, Third Run	-45,000,000,000
	-310,000
B40-E	-56,112
	386
	2,663
	2,667
	2,672
	2,943
	3,321
	3,313
	3,299
Kt-17b, First Run	6,555
	6,296
	4,969
	5,117
	6,164
	5,228
Kt-17b, Second Run	6,848
	6,479
	5,731
KT-17B, Plagioclase Concentrate	6,204
	6,904
	6,560
	6,544
	5,105
B56-A, First Run	7,810
	4,864
	4,890
B56-A, Second Run	5,597

Timing of Precambrian Melt Depletion

These rocks from Wyoming were dated [3] in 2003 using the Rubidium/Strontium and Neodymium/Samarium method. The rock samples [Tables 2 & 3] gave ages [4] between -2 and 50 billion years old! Since the Earth exists in the present how can rocks have formed in the future? How can a rock be 35 billion years older than the Big Bang explosion? The author admits some of the dates are negative: "That complete equilibrium was not achieved during this interaction is shown by the fact that the garnet–clinopyroxene tie lines for the different radiometric systems in the same sample do not provide ages that agree, and in the case of two of the Williams samples the Sm–Nd tie lines provide negative ages." [5]

Table 2

Billion Years	Billion Years
-1.24	6
-1.24	7.46
-0.22	47.37
4.54	49.63

There is a 51,970 million year spread of dates between the youngest [Negative] and the oldest [Positive] ages.

Table 3

Billion Years	Billion Years
-2.34	-4.24
-1.75	-1.47
-0.98	-1.14
-0.86	-0.84
4.47	2.51

If we run the Lead 207/206 ratios [6] through Isoplot we find that the rocks are 5 billion years old.

Table 4

Average	4,935
Maximum	5,118
Minimum	4,421

The author claims that the true age is just 2.6 billion years old: "The mean TMA of these five samples is 2.86 Ga (or 3.07 Ga without the apparently younger sample HK1-24) and given the lower bound mean TRD age of 2.61 Ga, a depletion age in the late Archean seems likely." [7]

Ion Microprobe U-Pb Dating

These rocks from Japan were dated [8] in 2001 using the Rubidium/Strontium and Potassium/Argon method. If we run the isotopic ratios through Isoplot [9] and use formulas listed in standard geology books [10] we find that the rock samples [11] gave ages between 5 billion years and negative years old! Since the Earth exists in the present how can rocks have formed in the future? How can a rock be older than the Earth? The author admits some of the dates are negative: "Though a negative age has no practical use, it does suggest that it is younger than 0.12 Ma." [12]

Table 5

Table 2 Data	Age 206Pb/238U	Age 207Pb/206Pb	Age Ratio
Average	62	4,710	76
Maximum	631	5,135	8
Minimum	0	3,771	3771

Table 6

Table 3 Data	Age 206Pb/238U	Age 207Pb/206Pb	Age Ratio
Average	0.88	4,742	5,388
Maximum	2.91	4,978	1,710
Minimum	0.25	4,479	17,916

The Long Valley Rhyolitic

These rocks from California were dated [13] in 1997 using the Rubidium/Strontium and Potassium/Argon method. The rock samples gave ages between 1 million years and negative years old! Since the Earth exists in the present how can rocks have formed in the future? The author admits some of the dates are negative:

"The negative ages are a clear indication that some phases have not reached Sr isotope equilibration with their current host glass." [14]

"In contrast, feldspars from the second group yield mineral ages that are geologically unreasonable ranging from close to the eruption age of the Bishop Tuff to negative ages." [15]

Rn-Generated 206Pb

These rocks from South Africa were dated [16] in 1998 using the Uranium/Lead method. When we run the ratios [17] through Isoplot the rock samples gave ages between 543 and 6,400 million years old! Since the Earth exists in the present how can rocks have formed in the future? How can a rock be older than the Earth? According to the article the true age is between 2 and 2.6 billion years old: "Assigning a 2.02 Ga age of mineralization and constructing secondary isochrons for paragenetically early galena and chalcopyrite, ages of the source uraninite are calculated as 2.6-2.4 Ga." [18]

Table 7

Age Pb 207/206	Age Pb 207/206
6451	5799
6330	5763
6315	5735
6217	5723
6109	5711
6009	4966

The author admits some of the dates are negative: "Analyses lying even farther to the fight, with the implication of implausibly young and even negative ages, force us to consider alternative explanations for this subsidiary array." [19]

40Argon/39 Argon Age of a Tholeiitic Basalt

These rocks from California were dated [20] in 2006 using the Argon method. The rock samples gave ages [21] between 2,357 and -579 thousand years old! Since the Earth exists in the present how can rocks have formed in the future?

Table 8

Sample	Minimum	Maximum	Difference	Ratio
Cinder Butte	-579.3	56.7	636	1,022%
Andesite of Sugarloaf Peak	14.7	589.5	636	4,010%
Little Potato Butte	-51.6	585.9	637.5	1,135%
Andesite of Potato Butte 1	-386.3	164.5	550.8	235%
Andesite of Potato Butte 2	-289.6	2357.4	2647	814%
Hat Creek Basalt 1	10	2950	2647	29,500%
Hat Creek Basalt 2	-89.3	92.4	181.7	103%

The author admits some of the dates are negative: "The Ar isotopic data, when cast on an inverse isochron diagram, indicate that the first two steps are enriched in 36Ar and thus yield negative ages. These first two steps are most likely influenced by low-temperature alteration of the sample." [22]

Isotopic Systematics of Ultramafic Xenoliths

These rocks from North China were dated [23] in 2007 using the Rubidium/Strontium and Uranium/Lead methods. The rock samples gave ages [24] between -3 and 9 billion years old! Since the Earth exists in the present how can rocks have formed in the future? How can a rock be 4.5 billion years older than the Earth? The author admits some of the dates are negative: "The Nd model ages for the individual data points are variable, from ~2.8 Ga to negative ages, consistent with our earlier observation that REE patterns for all the samples display some degree of secondary metasomatic overprinting by LREE-enriched silicate melts." [25]

If we run the isotopic ratios [24] through Isoplot we get the ages listed in the table In his article there is a **12,698** million year spread of dates between the youngest [Negative] and the oldest [Positive] ages.

Table 9

Million Years	Million Years
-3,209	965
-1,747	2,803
136	4,383
530	7,935
600	

Table 10

207Pb/206Pb	206Pb/238U
5,049	9,489
5,035	1,821
5,034	338
5,029	95
5,012	
5,009	
5,006	
5,004	

Re-Os, Sm-Nd, and Rb-Sr Isotope Evidence

These rocks from Uganda were dated [26] in 1993 using the Rubidium/Strontium and Neodymium/Samarium methods. Since the Earth exists in the present how can rocks have formed in the future? How can a rock be 6 billion years older than the Earth? The author admits some of the dates are negative:

"If Re-Os model ages are calculated using the conventional model age approach, i.e., using the measured Re/Os and osmium isotope composition in comparison to some model for bulk-Earth osmium isotope evolution, several peridotites yield negative ages, or ages that are considerably older than the Earth. This indicates that some peridotites cannot have evolved as closed systems." If we run the Osmium isotope ratios [27] through Microsoft Excel, we get the following results.

Table 11. 187Os/186Os Ages

Million Years	Million Years
-1,584	-6.46
-1,504	-1.58
-478	-0.73
-35	2.23
-19	2.78

The rock samples below gave ages [28] between -1.5 and 11 billion years old!

Table 12

Sm-Nd	Rb-Sr	% Ratio
258	5,454	2,114
959	6,245	651
434	12,716	2,930
2,038	1,351	66
1,157	4,026	348

Table 13

Re/Os	Sm/Nd	Rb/Sr
5.5	3.2	8.3
11	3	0.99
6.9	3	
6.6	2.7	
6 Negative	4 Negative	7 Negative

There is a 14,300 million year spread of dates between the youngest [Negative] and the oldest [Positive] ages.

Conclusion

Yuri Amelin states in the journal Elements that radiometric dating is extremely accurate: "However, four 238U/235U-corrected CAI dates reported recently (Amelin et al. 2010; Connelly et al. 2012) show excellent agreement, with a total range for the ages of only 0.2 million years – from 4567.18 ± 0.50 Ma to 4567.38 ± 0.31 Ma." [29-31]

To come within 0.2 million years out of 4567.18 million years means an accuracy of 99.99562%. Looking at some of the dating it is obvious that precision is much lacking. The Bible believer who accepts the creation account literally has no problem with such unreliable dating methods. Much of the data in radiometric dating is selectively taken to suit and ignores data to the contrary.

References

1 A 40Ar/39Ar Geochronological Study, Precambrian Research, 1992, Volume 57, Pages 91-119
2 Reference 2, Page 109
3 Timing of Precambrian Melt Depletion, Lithos, Volume 77, 2004, Pages 453-472
4 Reference 3, page 458, 460
5 Reference 3, page 466
6 Reference 3, page 459
7 Reference 3, page 463
8 Ion Microprobe U-Pb Dating, Journal of Volcanology and Geothermal Research, Volume 117, 2002, Pages 285-296
9 http://www.creationismonline.com/Isoplot/Isoplot.html
https://www.bgc.org/isoplot
10 Principles of Isotope Geology, Second Edition, By Gunter Faure, Published By John Wiley And Sons, New York, 1986. Pages 120 [Rb/Sr], 205 [Nd/Sm], 252 [Lu/Hf], 266 [Re/OS], 269 [Os/OS].
11 Reference 8, page 288, 290
12 Reference 8, page 291
13 The Long Valley Rhyolitic, Geochimica et Cosmochimica Acta, 1998, Volume 62, Number 21/22, Pages 3561-3574
14 Reference 13, page 3567
15 Reference 13, page 3569
16 Rn-Generated 206Pb, Mineralogy and Petrology, 1999, Volume 66, Pages 171-191
17 Reference 16, page 182, 183
18 Reference 16, page 171
19 Reference 16, page 176
20 40Ar/39Ar Age of a Tholeiitic Basalt, Quaternary Research, Volume 68, 2007, Pages 96-110
21 Reference 20, pages 101, 102
22 Reference 20, pages 103
23 Isotopic Systematics of Ultramafic Xenoliths, Chemical Geology, Volume 248, 2008, Pages 40-61
24 Reference 23, page 46
25 Reference 23, page 54
26 Re-Os, Sm-Nd, and Rb-Sr Isotope Evidence, Geochemica et Cosmochimica Acta, 1995, Volume 59, Number 5, Pages 959-977
27 Reference 26, pages 970, 971
28 Reference 31, pages 963
29 Dating the Oldest Rocks in the Solar System, Elements, 2013, Volume 9, Pages 39-44
30 Amelin, Earth and Planetary Science Letters, 2010, Volume 300, Pages 343-350
31 Connelly, Science, 2012, Volume 338, Pages 651-655

Chapter 9

The Neodymium-Samarium Dating Method

Pb, Sr, and Nd Isotopic Features

According to the article [1] this rock formation in China was dated in 2001 by scientists from China. According to the essay the true age is: "They define a Rb-Sr isochron age of 286 Ma. Pb isotopic compositions for bitumen and crude oil from Karamay, Liaohe, and Tarim all show features of crust–mantle mixing." [1] The Neodymium/Samarium dating method gives the following dates: "Thus, the Nd isotopic compositions strongly show an influence from depleted mantle (286 Ma)." [2]A Neodymium/Samarium Isochron gives more dating information "143Nd/144Nd and 147Sm/144Nd ratios vary within 0.51157 to 0.51197 and 0.0778 to 0.153, respectively, and yield old, depleted mantle Nd model ages of 1.5 to 3.2 Ga." [3] Several tables [4] in the essay [tables one to six] have isotopic ratios which can be calculated. As we can see below the Lead 207/206 dating method gives wildly discordant dates. How can both methods be so at variance with each other?

1. Lead 207/206 Age Dating Summary

Table 1	207Pb/206Pb	87Rb/86Sr
Dating Summary	Age	Age
Average	5,009	3,758
Maximum	5,029	24,661
Minimum	4,982	182
Difference	47	24,479

2. Lead 207/206 Age Dating Summary

Table 2	207Pb/206Pb	87Rb/86Sr
Dating Summary	Age	Age
Average	4,995	646
Maximum	5,097	702
Minimum	4,845	565
Difference	252	138

3. Lead 207/206 Age Dating Summary

207Pb/206Pb	Table 3	Table 4	Table 5	Table 6
Dating Summary	Age	Age	Age	Age
Average	4,151	5,060	5,027	5,079
Maximum	5,018	5,063	5,066	6,471
Minimum	1,776	5,053	4,987	4,978
Difference	3,242	9	79	1,493

Sources of Labrador Sea Sediments

According to the article [5] this rock formation in Labrador was dated in 2002 by scientists from Canada. According to the essay the true age is 8,600 years old: "The newly acquired Pb isotopic data allow us to better constrain the different source areas that supplied clay-size material during the last deglaciation, until 8.6 kyr (calendar ages)." [5] A table [6] in the essay has Carbon-14 dates alongside isotopic ratios which can be calculated. As we can see below the Lead 207/206 dating method gives wildly discordant dates. How can both methods be so at variance with each other?

4. Lead 207/206 Versus Carbon-14 Age Dating Summary

Dating Summary	Carbon 14 Age Years	Calibrated Age Years	207Pb/206Pb Million Years	Carbon 14 Age Dating Ratio	Calibrated Age Dating Ratio
Average	11,656	13,114	4,967	456,448	408,945
Maximum	22,190	26,064	4,982	636,961	584,938
Minimum	7,792	8,485	4,944	223,722	190,469
Difference	14,398	17,579	38	413,239	394,469

If we run the isotopic ratios give in standard geology magazines through the computer program Isoplot [7] we find that the Uranium/Thorium/Lead isotopic ratios in the rocks disagree radically with the Rubidium/Strontium ages. The U/Th/Pb ratios give ages older than the evolutionist age of the Earth, Solar System, Galaxy and Universe. How can Earth rocks be dated as being older than the Big Bang?

If we use isotopic formulas [8-11] given in standard geology text we can arrive at ages from the Rubidium/Strontium and Neodymium/Samarium ratios. The formula for Rubidium/Strontium age is given as:

$$t = \frac{2.303}{\lambda}\log\left(\frac{(87Sr/86Sr)-(87Sr/86Sr)_0}{(87Rb/86Sr)}+1\right) \qquad [1]$$

Where t equals the age in years. λ equals the decay constant. (87Sr/86Sr) = the current isotopic ratio. $(87Sr/86Sr)_0$ = the initial isotopic ratio. (87Rb/86Sr) = the current isotopic ratio. The same is true for the formula below.

$$t = \frac{2.303}{\lambda}\log\left(\frac{(143Nd/144Nd)-(143Nd/144Nd)_0}{(147Sm/144Nd)}+1\right) \qquad [2]$$

Here are examples of isotopic ratios taken from several articles in major geology magazines which give absolutely absurd dates.

Rocks of the Central Wyoming Province

These rock samples were dated in 2005 by scientists from the University of Wyoming. [12] If we run the Rubidium/Strontium and Neodymium/Samarium isotope ratios [13] from the article through Microsoft Excel we get the following values:

5. Ages Dating Summary

Dating Summary	Age 87Rb/86Sr	Age 147Sm/144Nd	Age 207Pb/206Pb	Age 208Pb/232Th	Age 206Pb/238U
Average	2,863	2,869	5,123	17,899	11,906
Maximum	2,952	2,954	5,294	38,746	18,985
Minimum	2,630	2,631	4,662	6,650	7,294
Std Deviation	38	39	152	9,754	3,298

The Uranium/Lead dates [14] are up to sixteen billion years older than the Rubidium/Strontium and Neodymium/Samarium dates. The Thorium/Lead dates are up to thirty-six billion years older. The so-called true age is just a guess.

Correlated Nd, Sr And Pb Isotope Variation

According to the article [15] this specimen [Walvis Ridge, Walvis Bay] was dated in 1982 by scientists from the Massachusetts Institute of Technology, and the Department of Geochemistry, University of Cape Town, South Africa. According to the article [16] the age of the sample is 70 million years. If we run the various isotope ratios [16] from the article through Microsoft Excel, we get the following values respectively:

6. Age Dating Summary

Summary	Pb207/Pb206	147Sm/144Nd	87Rb/86Sr
Average	5,033	70	64
Maximum	5,061	70	93
Minimum	5,004	69	0
Difference	57	140	93

A Depleted Mantle Source For Kimberlites

According to the article [17] this specimen [kimberlites from Zaire] was dated in 1984 by scientists from Belgium. According to the article [18] the age of the samples is 70 million years. If we run the various isotope ratios [19] from the article through Microsoft Excel we get the following values respectively:

7. Age Dating Summary

Summary	207Pb/206Pb	206Pb/238U	87Rb/86Sr	147Sm/144Nd
Average	4,977	4,810	86	72
Maximum	5,017	10,870	146	80
Minimum	4,909	1,391	50	63
Difference	108	9,478	196	17

The 207Pb/206Pb maximum age is 34 times older than the 87Rb/86Sr maximum age. The 206Pb/238U maximum age is 74 times older than the 147Sm/144Nd maximum age. There is a 10.8-billion-year difference between the oldest and youngest age attained.

Sm-Nd Isotopic Systematics

According to the article [20] this specimen [Enderby Land, East Antarctic] was dated in 1984 by scientists from the Australian National University, Canberra, and the Bureau of Mineral Resources, Canberra. According to the article [20] the age of the sample is 3,000 million years. If we run the Rubidium/Strontium isotope ratios [21] from the article through Microsoft Excel, we get the following values respectively:

8. Rubidium/Strontium Age Dating Summary

Average	-873
Maximum	3,484
Minimum	-25,121
Difference	28,605

There is almost a 30-billion-year difference between the oldest and youngest dates.

Strontium, Neodymium And Lead Compositions

According to the article [22] this specimen [Snake River Plain, Idaho] was dated in 1985 by scientists from the Geology Department, Rice University, Houston, Texas, the Earth Sciences Department, Open University, England and the Geology Department, Ricks College, Idaho. According to the article [22] the age of the sample is 3.4 billion years. If we run the various isotope ratios [23] from the article through Microsoft Excel, we get the following values respectively:

9. Age Dating Summary

Summary	Pb207/Pb206	Pb207/Pb206	87Rb/86Sr
Average	5,143	5,138	40,052
Maximum	5,362	5,314	205,093
Minimum	4,698	4,940	1,443
Difference	664	374	203,650

The Lead isotope ratios from two different tables give dates 200 billion years younger than the Rubidium/Strontium isotope ratios. The Average age of the Rubidium/Strontium isotope ratios is 40 billion years. Below we can see some of the maximum ages and how stupid they are.

10. 87Rb/86Sr, Maximum Ages

Age	Age
Million Years	Million Years
205,093	11,974
189,521	11,908
188,777	9,960
95,450	9,101
52,643	7,124
13,119	6,022
12,220	5,089

Sr, Nd, and Os Isotope Geochemistry

According to the article [24] this specimen [Camp Creek area, Arizona] was dated in 1987 by scientists from The University of Tennessee, the University of Michigan, the University of California, Leeds University, and the University of Chicago. According to the article [25] the age of the samples is 120 million years. If we run the various isotope ratios [26] from two different tables in the article through Microsoft Excel we get the following values respectively:

11. Rubidium/Strontium and Sm/Nd Age Dating Summary

Summary	87Rb/86Sr	87Rb/86Sr	147Sm/144Nd	147Sm/144Nd
Average	310	103	120	159
Maximum	1,092	207	123	400
Minimum	0	0	120	119
Difference	1,092	207	3	281

The author's choice of 120 million years is just a guess.

Pb, Nd and Sr Isotopic Geochemistry

According to the article [27] this specimen [Bellsbank kimberlite, South Africa] was dated in 1991 by scientists from the University Of Rochester, New York, Guiyang University in China, and the United States Geological Survey, Colorado. According to the article [67] the age of the samples is just 1 million years. If we run the various isotope ratios [68] from two different tables in the article through Microsoft Excel we get the following values respectively:

12. Age Dating Summary

Table	207Pb/206Pb	206Pb/238U	208Pb/232Th	87Rb/86Sr
Summaries	Age	Age	Age	Age
Average	5,057	5,092	10,182	-1,502
Maximum	5,120	8,584	17,171	0
Minimum	5,002	0	0	-3,593
Difference	118	8,584	17,171	3,593

In tables 9 to 12 we can see some of the astounding spread of dates [million of years]. The oldest date is over 17 billion years old. The youngest is less than negative 3.5 billion years. The difference between the two is over 20 billion years. According to the article the true age of the rock is just one million years old!

13. 208Pb/232Th, Maximum Ages

Age	Age	Age	Age
17,171	13,322	9,737	7,968
15,343	13,202	9,707	7,830
15,299	13,001	9,049	7,250
15,136	11,119	8,420	6,972
15,054	10,873	8,419	6,628
13,476	10,758	8,368	6,577

14. 206Pb/238U, Maximum Ages

Age	Age	Age
8,584	6,656	5,576
7,975	6,654	5,520
7,314	6,518	5,285
7,184	6,448	5,159
6,861	5,758	5,099

15. Pb 207/206, Maximum Ages

Age	Age	Age	Age
5,120	5,067	5,060	5,049
5,109	5,066	5,059	5,045
5,097	5,066	5,051	5,044
5,077	5,065	5,050	5,044
5,067	5,062	5,050	5,033
5,067	5,060	5,050	5,022

16. 87Rb/86Sr, Minimum Ages

Age	Age	Age	Age
-3,593	-2,981	-1,917	-1,323
-3,231	-2,725	-1,611	-1,245
-3,089	-2,050	-1,499	-1,229
-3,067	-1,926	-1,370	-1,194

Sr, Nd, and Pb isotopes

According to the article [30] this specimen [eastern China] was dated in 1992 by scientists from the University Of Rochester, New York, Guiyang University in China, and the United States Geological Survey, Colorado. According to the article: "Observed high Th/U, Rb/Sr, 87Sr/86 Sr and Delta 208, low Sm/Nd ratios, and a large negative Nd in phlogopite pyroxenite with a depleted mantle model age of 2.9 Ga, support our contention that metasomatized continental lower mantle lithosphere is the source for the EMI component." [30] If we run the various isotope ratios [31] from two different tables in the article through Isoplot we get the following values respectively:

17. Age Dating Summary

Dating	232Th/208Pb	206Pb/238U	207Pb/206Pb
Summaries	Age	Age	Age
Average	14,198	7,366	5,014
Maximum	94,396	22,201	5,077
Minimum	79	1,117	4,945
Difference	94,317	21,083	131

If the true age is 2.9 billion years why so much discordance? In tables 14 and 15 we can see some of the astounding spread of dates [million of years]. The oldest date is over 94 billion years old. The youngest is 79 million years. The difference between the two is over 94 billion years. The oldest date is 1,194 times older than the youngest. According to the article the true age of the rock is 2.9 billion years old!

18. 208Pb/232Th, Maximum Ages

Age	Age	Age	Age
94,396	39,267	10,595	8,171
90,683	26,266	10,284	7,789
74,639	18,334	9,328	7,638
58,153	16,357	8,821	7,375
55,324	14,250	8,771	7,317
45,242	11,215	8,403	5,759

19. 206Pb/238U, Maximum Ages

Age	Age	Age	Age
22,201	9,878	7,348	5,746
21,813	9,656	7,335	5,700
19,320	9,054	7,249	5,218
16,656	8,242	7,202	5,201
16,200	8,044	7,019	5,163
14,748	7,996	6,923	5,159
13,607	7,590	6,848	5,099
11,256	7,422	6,292	4,812

An Extremely Low U/Pb Source

According to the article [32] this specimen [lunar meteorite] was dated in 1993 by scientists from the United States Geological Survey, Colorado, the United States Geological Survey, California and The National Institute of Polar Research, Tokyo. According to the article: "The Pb-Pb internal isochron obtained for acid leached residues of separated mineral fractions yields an age of 3940 ± 28 Ma, which is similar to the U-Pb (3850 ± 150 Ma) and Th-Pb (3820 ± 290 Ma) internal isochron ages. The Sm-Nd data for the mineral separates yield an internal isochron age of 3871 ± 57 Ma and an initial 143Nd/I44Nd value of 0.50797 ± 10. The Rb-Sr data yield an internal isochron age of 3840 ± 32 Ma." [32] If we run the various isotope ratios [33] from two different tables in the article through Isoplot we get the following values respectively:

20. Rubidium/Strontium Age Dating Summary

Average	3,619
Maximum	5,385
Minimum	721
Difference	4,664

21. Uranium Age Dating Summary

Table	207Pb/206Pb	206Pb/238U	208Pb/232Th	207Pb/235U
Summaries	Age	Age	Age	Age
Average	4,673	8,035	10,148	4,546
Maximum	5,018	56,923	65,286	8,128
Minimum	3,961	1,477	2,542	2,784
Difference	1,057	55,445	62,744	5,344

The article claims that the Rubidium/Strontium age is 3.8 billion years for this meteorite. If that is the true age why are all the Uranium/Thorium/Lead dates [76] so stupid? Or are they right and the Rubidium/Strontium is wrong?

22. 208Pb/232Th, Maximum Ages

Age	Age	Age	Age
65,286	14,430	9,094	5,401
33,898	14,410	6,520	5,396
25,013	13,107	6,166	5,365
22,178	12,738	6,121	5,098
21,204	11,641	5,671	5,035
17,611	11,174	5,408	4,678

23. 206Pb/238U, Maximum Ages

Age	Age	Age	Age
56,923	10,895	6,764	5,777
27,313	10,278	6,670	5,625
17,873	9,653	6,449	5,602
13,680	8,009	6,436	5,278
13,623	7,395	6,070	5,147

The 72 Ma Geochemical Evolution

According to the article [34] this specimen [Madeira Archipelago] was dated in 2000 by scientists from Germany. The average Lead date is 705 times older than the average Rubidium date. The true age is claimed to be 430 million years old. [34] If we run the various isotope ratios [35] from two different tables in the article through Isoplot we get the following values respectively:

24. Age Dating Summary

Table	207Pb/206Pb	87Rb/86Sr	147Sm/144Nd
Summaries	Age	Age	Age
Average	4,938	7	10
Maximum	5,199	55	164
Minimum	4,898	-4	0
Difference	302	59	164

If the true age is 430 million years than none of the dating methods are even vaguely close. The oldest date is 731 times older than the youngest.

Temporal Evolution of the Lithospheric Mantle

According to the article [36] this specimen from the Eastern North China Craton was dated in 2009 by scientists from China, USA and Australia. Various tables [37] in the essay have either calculated dates or ratios which can be calculated. As we can see below they are all at strong disagreement with each other. There is a spread of dates over a 32 billion year range.

25. Age Dating Summary

Table	147Sm/144Nd	176Lu/176Hf	187Re/188Os	87Rb/86Sr
Summaries	Age	Age	Age	Age
Average	291	-220	1,048	9
Maximum	3,079	4,192	20,710	22
Minimum	-3,742	-9,369	-11,060	0
Difference	6,821	13,561	31,770	22

Geochemistry Of The Jurassic Oceanic Crust

According to the article [38] this specimen from the Canary Islands was dated in 1998 by scientists from Germany. According to the essay: "An Sm–Nd isochron gives an age of 178 ± 17 Ma, which agrees with the age predicted from paleomagnetic data." [38] The article places the age in the late Cretaceous period. Various tables [39] in the essay have isotopic ratios which can be calculated. As we can see below they are all at strong disagreement with each other. There is a spread of dates over a 350 billion year range! None of the Lead or Rubidium based dating methods even come vaguely close to a Jurassic age.

26. Age Dating Summary

Dating	87Rb/86Sr	207Pb/206Pb
Summary	Age	Age
Average	-149,488	4,974
Maximum	51,967	5,024
Minimum	-299,346	4,845
Difference	351,313	179

Origin Of The Indian Ocean-Type Isotopic Signature

According to the article [40] this rock formation in the Philippine Sea plate was dated in 1998 by scientists from Department of Geology, Florida International University in Miami. According to the essay the true age is: "Spreading centers in three basins, the West Philippine Basin (37-60 Ma), the Parece Vela Basin (18-31 Ma), and the Shikoku Basin (17-25 Ma) are extinct, and one, the Mariana Trough (0-6 Ma), is active (Figure 1)." [40] Numerous table and charts affirm this as the true age. [41] Two tables [42] in the essay have isotopic ratios which can be calculated. As we can see below they are all at radical disagreement with each other. There is a spread of dates of over 13 billion years! None of the Uranium/Lead based dating methods even come vaguely close to the so called true age. The oldest date is 706 times older than the youngest date.

27. Age Dating Summary

Dating Summary	Age 87Rb/86Sr	Age 147Sm/144Nd	Age 207Pb/206Pb	Age 206Pb/238U	Age 208Pb/232Th
Average	42	41	4,960	4,260	8,373
Maximum	55	54	4,989	7,093	13,430
Minimum	19	20	4,921	1,904	3,065
Difference	37	33	68	5,188	10,365

Sr, Nd, and Pb isotopes in Proterozoic Intrusives

According to the article [43] this specimen from the Grenville Front in Canadian Labrador was dated in 1986 by scientists from Lunar and Planetary Institute, Texas, the United States Geological Survey, and the Geological Survey of Canada. According to the essay: "We report Sr, Nd, and Pb isotopic compositions of mid-Proterozoic anorthosites and related rocks (1.45-1.65 Ga) and of younger olivine diabase dikes (1.4 Ga) from two complexes on either side of the Grenville Front in Labrador." [43] The article places the age in the pre Cambrian period. Various tables [44] in the essay have isotopic ratios which can be calculated. As we can see below they are all at strong disagreement with each other. If the Uranium/Lead dating method is used to test or calibrate the other methods then they are totally wrong.

28. Age Dating Summary

Dating Summary	Age 87Rb/86Sr	Age 207Pb/206Pb
Average	1,437	5,135
Maximum	1,503	5,218
Minimum	1,395	4,931
Difference	108	287

Age and Isotopic Relationships

According to the article [45] this rock formation in Antarctica was dated in 1992 by scientists from California and Germany. According to the essay the true age is: "Nevertheless, concordant Ph-Pb model ages of pyroxene separates were obtained *(20')*: 4.55784 ± 52 Ga for LEW and 4.55780 ± 42 Ga for ADOR." [45] Several tables [46] in the essay have isotopic ratios which can be calculated. As we can see below they are all at disagreement with each other. The two on the far right show how discordant the best dating evolutionist can offer.

29. Age Dating Summary

Dating Summary	Age 87Rb/86Sr	Age 207Pb/206Pb	Age 207Pb/206Pb	Age 147Sm/144Nd	Age 147Sm/144Nd
Average	4,556	4,707	5,007	4,452	902
Maximum	4,610	5,002	5,110	4,497	1,428
Minimum	4,518	4,558	4,960	4,397	536
Difference	92	444	150	101	891

The Beni Bousera Ultramafic Complex of Northern Morocco

According to the article [47] this rock formation in Morocco was dated in 1995 by scientists from New York. According to the essay the true age is: "The data are presented in Table 5. Garnet-clinopyroxene two-point Sm-Nd isochrons from samples Ga and Ii yield ages of 23.0 ± 7.3 m.y. and 20.1 ± 6.9 m.y." [48] Several tables [49] in the essay have isotopic ratios which can be calculated. As we can see below the Rhenium/Osmium gives wildly discordant dates.

30. Rhenium/Osmium Age Dating Summary

Average	-272,455
Maximum	-124,882
Minimum	-361,842
Difference	236,960

Implications for Banda Arc Magma Genesis

According to the article [50] this rock formation in the Banda Arc, East Indonesia was dated in 1995 by scientists from University of Utrecht, the Royal Holloway University of London, the Free University of Amsterdam and Comell University. According to the essay the true age is: "In summary, the western part of New Guinea is characterised by Phanerozoic rocks (600 Ma) in contrast to the northern part of Australia, which is dominated by Proterozoic rocks (2200-1400 Ma)." [51] Several tables [52] in the essay have isotopic ratios which can be calculated. As we can see below the Lead 207/206 dating method gives wildly discordant dates. How can both methods be so at variance with each other?

31. Lead 207/206 Age Dating Summary

Average	4,971
Maximum	4,991
Minimum	4,933
Difference	57

The Petrogenesis of Martian Meteorites

According to the article [53] these two meteorite samples was dated in 2002 by scientists from the University of New Mexico, the Johnson Space Center, Texas and the Lockheed Engineering and Science Company, Texas. According to the essay the true age based on Neodymium/Samarium dating is 173 and 166 million years old. [53] A table [54] in the essay has Rubidium/Strontium isotopic ratios which can be calculated. As we can see below Rubidium/Strontium dating method gives wildly discordant dates. The Table 1 summary is the rock that is supposed to be 173 million year old. The Table 2 summary is the rock that is supposed to be 166 million year old. How can both methods be so at variance with each other?

32. Rubidium/Strontium Age Dating Summary

Dating Summary	87Rb/86Sr Table 1	87Rb/86Sr Table 2
Average	579	240
Maximum	3,233	697
Minimum	170	74
Difference	3,063	624

Conclusion

Brent Dalrymple states in his anti creationist book The Age of the Earth: "Several events in the formation of the Solar System can be dated with considerable precision." [55]

Looking at some of the dating it is obvious that precision is much lacking. He then goes on: "Biblical chronologies are historically important, but their credibility began to erode in the eighteenth and nineteenth centuries when it became apparent to some that it would be more profitable to seek a realistic age for the Earth through observation of nature than through a literal interpretation of parables." [56]

I his book he gives a table [57] with radiometric dates of twenty meteorites. If you run the figures through Microsoft Excel, you will find that they are 98.7% in agreement. There is only a seven percent difference between the ratio of the smallest and oldest dates. As we have seen in this essay, such a perfect fit is attained by selecting data and ignoring other data. A careful study of the latest research shows that such perfection is illusionary at best. The Bible believer who accepts the creation account literally has no problem with such unreliable dating methods. Much of the data in Dalrymple's book is selectively taken to suit and ignores data to the contrary.

References

1 Pb, Sr, and Nd Isotopic Features, Geochimica et Cosmochimica Acta, 2001, Volume 65, Number 15, Pages 2555–2570
2 Reference 53, Pages 2559
3 Reference 53, Pages 2560
4 Reference 53, Pages 2558, 2561-2566
5 Sources of Labrador Sea Sediments, Geochimica et Cosmochimica Acta, 2002, Volume 66, Number 14, Pages 2569
6 Reference 57, Pages 2572-2573
7 http://www.creationismonline.com/Isoplot/Isoplot.html
https://www.bgc.org/isoplot
8 Radioactive and Stable Isotope Geology, By H.G. Attendon, Chapman And Hall Publishers, 1997. Page 73 [Rb/Sr], 195 [K/Ar], 295 [Re/OS], 305 [Nd/Nd].
9 Principles of Isotope Geology, Second Edition, By Gunter Faure, Published By John Wiley And Sons, New York, 1986. Pages 120 [Rb/Sr], 205 [Nd/Sm], 252 [Lu/Hf], 266 [Re/OS], 269 [Os/OS].
10 Absolute Age Determination, Mebus A. Geyh, Springer-Verlag Publishers, Berlin, 1990. Pages 80 [Rb/Sr], 98 [Nd/Sm], 108 [Lu/Hf], 112 [Re/OS].
11 Radiogenic Isotope Geology, Second Edition, By Alan P. Dickin, Cambridge University Press, 2005. Pages 43 [Rb/Sr], 70 [Nd/Sm], 205 [Re/OS], 208 [Pt/OS], 232 [Lu/Hf].
12 Rocks of the Central Wyoming Province, Canadian Journal Of Earth Science, 2006, Volume 43, Pages 1419
13 Reference 27, Page 1436-1437
14 Reference 27, Page 1439
15 Correlated N D, Sr And Pb Isotope Variation, Earth and Planetary Science Letters, Volume 59, 1982, Pages 327
16 Reference 45, Pages 330, 331
17 A Depleted Mantle Source For Kimberlites, Earth and Planetary Science Letters, Volume 73, 1985, Pages 269
18 Reference 47, Pages 270
19 Reference 47, Pages 271, 273

20 Sm-Nd Isotopic Systematics, Earth and Planetary Science Letters, Volume 71, 1984, Pages 46
21 Reference 50, Pages 49
22 Strontium, Neodymium And Lead Compositions, Earth and Planetary Science Letters, Volume 75, 1985, Pages 354-368
23 Reference 52, Pages 356, 363
24 Sr, Nd, and Os isotope geochemistry, Earth and Planetary Science Letters, Volume 99, 1990, Pages 362
25 Reference 63, Pages 364
26 Reference 63, Pages 365, 368
27 Pb, Nd and Sr isotopic geochemistry, Earth and Planetary Science Letters, Volume 105, 1991, Pages 149
28 Reference 66, Pages 154, 160
29 Reference 66, Pages 156, 157
30 Sr, Nd, and Pb isotopes, Earth and Planetary Science Letters, Volume 113, 1992, Pages 107
31 Reference 68, Pages 110
32 An extremely low U/Pb source, Geochimica et Cosmochimica Acta, 1993, Volume 57, Pages 4687-4702
33 Reference 75, Pages 4690, 4691
34 The 72 Ma Geochemical Evolution, Earth and Planetary Science Letters, Volume 183, 2000, Pages 73
35 Reference 77, Pages 76-79
36 Temporal Evolution of the Lithospheric Mantle, Journal Of Petrology, 2009, Volume 50, Number 10, Pages 1857
37 Reference 108, Pages 1873, 1874, 1877, 1879, 1880
38 Geochemistry of Jurassic Oceanic Crust, Journal Of Petrology, 1998, Volume 39, Number 5, Pages 859–880
39 Reference 115, Pages 867, 868
40 Origin of the Indian Ocean-type isotopic signature, Journal Of Geophysical Research, 1998, Volume 103, Number B9, Pages 20,963
41 Reference 134, Pages 20965, 20969
42 Reference 134, Pages 20968, 20969
43 Sr, Nd, and Pb isotopes in Proterozoic Intrusives, Geochimica et Cosmochimica Acta, 1986, Volume 50, Pages 2571-2585
44 Reference 43, Pages 2575, 2577
45 Age and Isotopic Relationships, Geochimica et Cosmochimica Acta, 1992, Volume 56, Pages 1673-1694
46 Reference 43, Pages 1676, 1678, 1684, 1686, 1687
47 The Beni Bousera Ultramafic Complex of Northern Morocco, Geochimica et Cosmochimica Acta, 1996, Volume 60, Number 8, Pages 1429
48 Reference 47, Pages 1434
49 Reference 47, Pages 1442
50 Implications for Banda Arc Magma Genesis, Geochimica et Cosmochimica Acta, 1995, Volume 59, Number 12, Pages 2573-2598
51 Reference 50, Pages 2588
52 Reference 50, Pages 2580-2581
53 The Petrogenesis of Martian Meteorites, Geochimica et Cosmochimica Acta, 2002, Volume 66, Number 11, Pages 2037–2053
54 Reference 53, Pages 2040-2041
55 The Age Of The Earth, By G. Brent Dalrymple, 1991, Stanford University Press, Stanford, California, Page 10.
56 Reference 55, Page 23
57 Reference 55, Page 287

Chapter 10

Rocks Older Than The Galaxy

Age Of Uranium Mineralization

These rocks were dated [1] in from the Gas Hills in Wyoming were dated in 1979 using the Uranium-Lead method. The rock sample GH-B1 was dated giving ages [2] between -1,240 and 12,000 million years old!

Table 1

Table 3	Table 4	Table 5
Million Years	Million Years	Million Years
11,780	7,232	5,060
-190	4,654	4,830
-200	4,355	-34
-220	3,540	-160
-310	-290	-240
-340	-340	-260
-420	-550	-500
-530		-610
-530		-650
-1,240		

"These systematics are similar to those observed by Ludwig for the Shirley Basin uranium ores, for which preferential loss of radioactive daughters in the U decay chain was shown to be the dominant cause of apparent-age discordance." [3]

"The trends of apparent age and discordance of the total ore, uraninite-coffinite, and pyrite analyses for the Gas Hills and Crooks Gap ores are very similar to those reported for the Shirley Basin uranium ores." [4]

Another group of rock samples were dated [5] giving absurd values. Many had negative ages! Some were older than the Solar System. How can Earth rocks be older than the Solar System?

Table 2

Million Years	Million Years
7,323	-340
4,830	-500
5,060	-550
-240	-610
-290	-650

Table 3

Sample Name	Maximum Age Million Years	Minimum Age Million Years	Difference Million Years	Difference Percentage
CG-A4	7,323	-340	7,663	-2,253%
CG-A5	4,654	-550	5,204	-946%
CG-A1	4,355	-290	4,645	-1,601%

A rock sample number GH-A6 was dated [6] as being between 5,870 million and negative 650 million years old. Looking at positive dates above zero and ignoring negative ages what do we find? The oldest is 5,870 million years old and the youngest [7] is 8 million years old. One is 733 times older than the other. Using a table [14] in the essay which has the $^{206}Pb/^{204}Pb$ and $^{207}Pb/^{204}Pb$ we can easily work out the $^{207}Pb/^{206}Pb$ ratios in the sample.

Table 4

Sample Number	207Pb/206Pb Ratio	207Pb/206Pb Million Years
GH-B3	0.462	4,123
GH-B3	0.480	4,181
GH-B6	0.316	3,549
GH-D2407	0.332	3,628
GH-D2407	0.413	3,958
GH-D2407	0.407	3,936
CG-A6	0.351	3,712
CG-A6	0.363	3,763

If we run the $^{207}Pb/^{206}Pb$ ratios through Isoplot [8] sample is over 3,500 million years old. The dates are not put beside the ratios in the original essay. The author states in the opening paragraph of his essay that the rock formation is only "inclusion of all samples increases the observed range to 12 to 41 million years." [16] In the first paragraph he admits that the isotopic composition has been contaminated over time producing anomalous dates. His choice of this narrow range is purely guesswork. Looking at all the dates it is just random whichever you pick.

Shocked Meteorites: Argon-40/Argon-39

Joachim Kunz [44] from the Max Plank Institute in Heidelberg, Germany did this dating in 2009 using the Argon-40/Argon-39 dating method. If we look at the appendix [45] at the end of his article we find many dates older than the Solar Stem and Galaxy.

Table 5

Sample Name	Million Years
F. Yanzhuang. Host rock	5,598
G. Yanzhuang. Melt fragment	10,217
	5,423
	5,503
H. Yanzhuang. Melt vein	7,016
J. Bluff. Host rock	13,348
	10,938

	6,272
N. Ness County. Host rock #1	5,052
O. Ness County. Host rock #2	6,668
	5,576
Q. Paranaiba. Host rock #2	5,593
V. Beeler. Host rock #1	6,466
W. Beeler. Host rock #2	6,609

Potassium-Argon Age Of Iron Meteorites

This dating [46] was done in 1958. Even dating done fifty years later is giving dates just as absurd. The opening paragraph of the article states:

"Under the usual assumptions accepted for this method, ages have been calculated and found to be close to 10 billion years, which is about twice the reported age of stone meteorites, and also higher than the supposed age of the universe." [47] The data in Table 6 below was taken from the data in [48] the original essay.

Table 6

Meteorite	Age
K-Ar Dating	Billion Years
Mt. Ayliff	6.9
Arispe	6.8
H. H. Ninninger	6.9
Carbo	8.4
Canon Diablo I	8.5
Canon Diablo I	6.9
Canon Diablo I	6.6
Canon Diablo I	5.3
Canon Diablo II	13
Canon Diablo II	11
Canon Diablo II	10.5
Canon Diablo II	12
Toluca I	5.9
Toluca I	7.1
Toluca II	10
Toluca II	10.8
Toluca II	8.8

The Allende and Orgueil Chondrites

This dating was done in 1976 by scientists [49] from the United States Geological Survey, Denver, Colorado. The data in Table 7 below was taken from Pb-206/U-238 and Pb-208/Th-232 dating [50] summary in the original essay. Thirty one of the dates below are older than the age of the Solar System. Four are over ten billion years. One date is older than the Big Bang explosion date.

Table 7

Pb-206/U-238	Pb-208/Th-232
Billion Years	**Billion Years**
9.86	16.49
8.95	14.4
8.82	11.7
7.82	10.40
7.80	10.40
7.75	10.1
6.66	9.86
6.50	9.55
6.50	9.15
6.44	7.52
6.42	6.99
6.35	6.40
6.33	5.44
6.05	5.35
5.73	5.15
5.73	4.81

African Peridotite Xenoliths

These kimberlites of southern Africa were dated in 1989 using Rhenium-Osmium dating method. [17] Some of the ages [18] are older than the Solar System and galaxy.

Table 8

5.6	Billion Years
12.6	Billion Years

If we insert the Osmium ratios listed in article [19] into Microsoft Excel use the dating formula listed in Gunter Faure's book [20] we get the dates listed in table 9.

$$t = \frac{1.04 - (^{187}Os \div ^{186}Os)}{0.050768} \times 10^{9}$$

Table 9

Average	889
Maximum	2,659
Minimum	-3,309

Osmium/Osmium dating

"TMA varies from 0.11 to 5.7 Ga with three samples having Re/Os that is too high to explain their measured 187Os/186Os." [21]

The Siberian Craton

Xenoliths from kimberlites intruding [22] the Siberian craton were dated in 1995 using the Re-Os, Sm-Nd, and Rb-Sr dating methods. The results in Table 10 were acquired using Rubidium-Strontium [23] isotope dating as being between 5 and 13 billion years old. The dates in Table 11 were obtained using Rhenium-Osmium [24] dating method.

"If Re/Os model ages are calculated using the conventional model age approach, i.e., using the measured Re/OS and osmium isotope composition in comparison to some model for bulk-Earth osmium isotope evolution, several peridotites yield negative ages, or ages that are considerably older than the Earth" [25]

Table 10

5.45	Billion Years
6.24	Billion Years
12.71	Billion Years

Table 11

5.5	Billion Years
11.0	Billion Years
6.9	Billion Years
6.6	Billion Years

Table 12

Statistics	Billion Years
Average	-144,339
Maximum	2,777
Minimum	-1,584,857

Osmium/Osmium Ratio Dating

History Of The Acapulco Meteorite

This well known meteorite was dated in 1997 by scientists [26] from France and Germany. According to the dates in Table 13 given [27] below, the meteorite is older than the galaxy. Even if we take into account the given uncertainty levels listed is the essay, [26] the rocks could still be 8.6 billion years old.

Table 13

Maximum Age	11,421	Million Years
Minimum Age	3,481	Million Years
Average Age	4,964	Million Years
Age Difference	7,940	Million Years
Difference	328%	Percent
Standard Deviation	1,723	Million Years

Potassium/Argon Dating of Iron Meteorites

The Weekeroo Station iron meteorite was dated [28] in 1967 using the Potassium-Argon dating method. The author of the article begins with the following remarks:

"The formation or solidification ages of iron meteorites have never been well determined. The most direct method seems to be that of Stoenner and Zahringer, who measured the potassium and argon contents by neutron-activation analysis. Their data, however, indicated ages of from about 7 to 10 billion years, whereas the age of the solar system is generally well accepted at about 4.7 billion years. Fisher later confirmed these data, but concluded that they were evidence of an unexplained potassium: argon anomaly rather than that they indicated true ages. From Muller and Zahringer's more recent data they conclude that a Potassium/Argon age of about 6.3 billion years can be assigned to many iron meteorites." [29]

The author of the article then concludes with the following remarks:

"The ages found by us are typical of the great ages found for most iron meteorites. From these, in conjunction with the Strontium/Rubidium data of Wasserburg on silicate inclusions in this meteorite, we conclude that the Potassium: Argon dating technique as applied to iron meteorites gives unreliable results. One may derive ad hoc possible explanations of the discord between the silicate and iron-phase ages, such as shock emplacement of these inclusions within the metal matrix without disturbing the potassium: argon ratios in the metal, but we feel that such mechanisms are unlikely." [30]

The essay lists a number of dates in the opening paragraph. The last four in table below are taken from Table 1 in the original essay.

Table 14

Meteorite Sample	Billion Years
Stoenner and Zahringer	10.0
Stoenner and Zahringer	7.0
Muller and Zahringer's	6.3
Wasserburg, Burnett	4.7
K-1	8.5
K-2	9.3
B-1	6.5
G-1	10.4

Stabilisation of Archaean Lithosphere

The Rhenium-Osmium isotope method was used [31] to date these rocks in 1995. The data [32] in the table below give absurd ages:

Table 15

Sample Name	Billion Years
PHN-2600	8.5
F-865	10.2
PHN-2825	15.6
PHN-5239	11.1

The author tries to explain such dating errors: "For example, several of the peridotite Re/Os model ages calculated using measured 187Re/188Os (TM) either give geologically unreasonable ages or do not intersect the Bulk Earth evolution line at all. Walker reasoned that the highly refractory compositions of Kaapvaal peridotites could have led to complete removal of Re during formation." [33]

Pb Isotopic age of the Allende Chondrules

Professor Yuri Amelin from The Australian National University did the research in 2007. [34] More than ten dates are older than the age of the Solar System. One is as old as the Galaxy. [35]

Table 16

Million Years	Million Years
10,066	5,396
6,945	5,345
5,956	5,336
5,604	5,180
5,526	5,147
5,462	4,950

If we run some of the isotopic ratios listed in the online supplement [36] through Isoplot we get the following dates:

Table 17

238U/ 206Pb	207Pb/ 235U	208Pb/232Th
10,066	5,731	5,947
6,945	5,202	5,920
5,956	4,956	5,860
5,604	4,864	5,735
5,526	4,832	5,636
5,462	4,826	5,335
5,396	4,807	5,265

Rhenium-187/Osmium-187 In Iron Meteorites

The ^{187}Rhenium/^{187}Osmium method and Potassium-Argon method were used to date these meteorite [37] fragments in 1997. Four of the dates were older than the Solar System and two were older than the Galaxy. [38]

Table 18

Canyon Diablo Meteorite	Billion Years
Leach Acetone	5.73
Leach H,O	8.31
Troilite dissolved	10.43
Metal 1	13.7

Ar-39/Ar-40 Dating of Mesosiderites

Donald Bogard from the Johnson Space Center in Houston, Texas performed this dating [36] in 1990 using the Argon dating method. The table below is a summary from the appendix [37] in the original essay. Three dates are as old or older than the Galaxy. Eleven are older than the Solar System.

Table 19

Meteorite Name	Maximum Age Billion Years	Minimum Age Billion Years	Age Difference Billion Years
1. Bondoc	4.02	3.20	0.82
2. Emery	9.08	3.31	5.77
3. Estherville	13.96	3.18	10.78
4. Hainholz	5.48	1.55	3.93
5. Lowicz	9.93	2.92	7.01
6. Morristown	7.92	3.60	4.32
7. Mount Padbury	5.52	3.49	2.03
8. Patwar Basalt	6.14	1.80	4.34
9. Patwar Gabbro	8.43	2.67	5.76
10. QUE-86900	10.92	3.24	7.68
11. Simondium	9.17	3.27	5.90
12. Veramin	13.13	2.71	10.42

40Ar-39Ar Chronology

Ekaterina V. Korochantseva from Heidelberg, Germany did this dating in 2009. [41] Below is a mathematical summary of the appendix [42] given in the original magazine article.

Table 20

Sample Name	Maximum Age	Minimum Age	Average Age	Age Difference
Table A01. Dhofar 019 whole rock	11,679	737	2,883	10,942
Table A02. Dhofar 019 maskelynite	10,521	818	2,674	9,703
Table A03. Dhofar 019 pyroxene	10,730	804	3,694	9,926
Table A04. Dhofar 019 olivine	10,487	1,778	4,549	8,709
Table A05. Dhofar 019 opaque	14,917	4,420	8,453	10,497
Table A06. SaU 005 whole rock	7,184	568	1,653	6,616
Table A07. SaU 005 glass	6,235	3,247	4,242	2,988
Table A08. SaU 005 maskelynite	7,432	1,344	3,899	6,088
Table A10. SaU 005 olivine	13,979	3,839	6,559	10,140
Table A11. Shergotty whole rock	8,542	1,112	2,995	7,430
Table A15. Zagami whole rock	6,064	94	2,276	5,970
Table A16. Zagami maskelynite	5,733	238	1,202	5,495
Table A18. Zagami opaque	7,707	290	1,525	7,417
Table A9. SaU 005 pyroxene	12,845	1,354	4,763	11,491

(Ages in million so years)

In Table 21 we can see below that 44 dates are older than the age of the Solar System and nine are over ten billion years.

Table 21

Sample Name	Million Years	Sample Name	Million Years
Table A05. Dhofar 019	14,917	Table A02. Dhofar 019	7,233
Table A09. SaU 005	13,979	Table A06. SaU 005	7,184
Table A18. Zagami	12,845	Table A02. Dhofar 019	7,168
Table A01. Dhofar 019	11,679	Table A03. Dhofar 019	6,857
Table A03. Dhofar 019	10,730	Table A09. SaU 005	6,680
Table A02. Dhofar 019	10,521	Table A05. Dhofar 019	6,482
Table A04. Dhofar 019	10,487	Table A04. Dhofar 019	6,451
Table A02. Dhofar 019	10,322	Table A07. SaU 005	6,235
Table A03. Dhofar 019	10,142	Table A07. SaU 005	6,192
Table A05. Dhofar 019	9,669	Table A14. Shergotty	6,064
Table A05. Dhofar 019	9,613	Table A09. SaU 005	5,874
Table A01. Dhofar 019	9,260	Table A04. Dhofar 019	5,771
Table A05. Dhofar 019	9,148	Table A07. SaU 005	5,745
Table A04. Dhofar 019	9,111	Table A15. Zagami	5,733
Table A10. SaU 005	8,542	Table A03. Dhofar 019	5,693
Table A01. Dhofar 019	8,507	Table A08. SaU 005	5,608
Table A09. SaU 005	8,323	Table A07. SaU 005	5,598
Table A03. Dhofar 019	8,197	Table A08. SaU 005	5,575
Table A05. Dhofar 019	7,987	Table A07. SaU 005	5,414
Table A17. Zagami	7,707	Table A18. Zagami	5,403
Table A04. Dhofar 019	7,610	Table A05. Dhofar 019	5,391
Table A08. SaU 005	7,432	Table A07. SaU 005	5,389

The author explains the radically absurd ages as contamination: "The temperature extractions above 1380 C display apparent ages exceeding the age of the solar system that is indicative of the presence of excess argon." [43]

Ultra-high Excess Argon in Kyanites

These rocks from Japan were dated in 2005 using [44] the Argon 40 isotope method. The opening paragraph of this article states:

"A laser fusion Ar-Ar technique applied on single crystals of kyanite from river sands of the Kitakami Mountain region of northeast Japan yielded ages of up to 16 Ga, more than three times the age of the earth. Although the age values are geologically meaningless, the ultra-high excess argon in kyanites is unique and hitherto unreported. We interpret this to be an artifact of ultra-high argon pressure derived from radiogenic argon in potassium-rich phases such as phengites during the Barrovian type retrogression of the ultra-high pressure rocks in this region." [45]

"In this study, we report the results from fusion Ar-Ar technique on single crystals of kyanite recovered from river sands in the Kitakami region. However, the kyanites yielded ages that are two to three times older than the age of the earth." [46]

Table 22

Sample	Billion Years
Ky6	7.7
Ky7	11.1
Ky8	15.1
Ky9	9.9
Ky11	16.3
Ky13	11.1

Conclusion

Prominent evolutionist Brent Dalrymple states: "Several events in the formation of the Solar System can be dated with considerable precision." [47]

Looking at some of the dating it is obvious that precision is much lacking. He then goes on: "Biblical chronologies are historically important, but their credibility began to erode in the eighteenth and nineteenth centuries when it became apparent to some that it would be more profitable to seek a realistic age for the Earth through observation of nature than through a literal interpretation of parables." [48]

The Bible believer who accepts the creation account literally has no problem with such unreliable dating methods. Much of the data in Dalrymple's book is selectively taken to suit and ignores data to the contrary.

References

1 **Kenneth R. Ludwig, Age Of Uranium Mineralization, Economic Geology, 1979, Volume 74, Pages 1654 – 1668**
2 **Reference 8, Page 1661**
3 **Reference 8, Page 1658**
4 **Reference 8, Page 1664**
5 **Reference 8, Page 1662**
6 **Reference 8, Page 1663**
7 **Reference 8, Page 1658**
8 http://www.creationismonline.com/Isoplot/Isoplot.html
https://www.bgc.org/isoplot
9 **Joachim Kunz, Shocked meteorites: Argon-40/Argon-39, Meteoritics And Planetary Science, 1997, Volume 32, Pages 647 – 670**
10 **Reference 9, Pages 664-670**
11 **R. W. Stoenner, Potassium/Argon age of iron meteorites, Geochemica Et Cosmochemica Acta, 1958, Volume 15, Pages 40-50**
12 **Reference 11, Page 40**
13 **Reference 11, Pages 45, 46**
14 **Mitsunobu Tatsumoto, The Allende and Orgueil Chondrites , Geochemica Et Cosmochemica Acta, 1976, Volume 40, pages 617 – 634**
15 **Reference 14, Page 627**
16 **Reference 8, Page 1654**
17 **R. J. Walker, African Peridotite Xenoliths, Geochimica et Cosmochimica Acta, 1989, Volume 53, Page 1583-1595**
18 **Reference 17, Page 1591**
19 **Reference 17, Page 1588**
20 **Principles Of Isotopic Geology, Gunter Faure, John Wiley Publishers, New York, 1986, Page 269**
21 **Reference 16, Page 1590**
22 **D. G. Pearson, The Siberian Craton, Geochimica et Cosmochimica Acta, 1995, Volume 59, Number 5, Page 959-977**
23 **Reference 22, Page 970**
24 **Reference 22, Page 971**
25 **Reference 22, Page 968**
26 **Paul Pellas, History Of The Acapulco Meteorite, Geochemica Et Cosmochemica Acta, 1997, Volume 61, Number 16, pp. 3477 – 3501**
27 **Reference 26, Page 3500**
28 **L. Rancitelli, Potassium: Argon Dating of Iron Meteorites, Science, 1967, Volume 155, Pages 999 - 1000**
29 **Reference 28, Page 999**
30 **Reference 28, Page 1000**
31 **D. G. Pearson, Stabilisation of Archaean lithosphere, Earth and Planetary Science Letters, 1995, Volume 134, Pages 341-357**
32 **Reference 31, Page 344**
33 **Reference 31, Page 348**
34 **Yuri Amelin, Pb isotopic age of the Allende chondrules, Meteoritics And Planetary Science, 2007,**

Volume 42, Numbers 7/8, Pages 1321 – 1335
35 Reference 34, Page 1324
36 http://onlinelibrary.wiley.com/doi/10.1111/j.1945-5100.2007.tb00577.x/suppinfo
37 J. L. Birck, Rhenium-187/Osmium-187 in iron meteorites, Meteoritics And Planetary Science, 1998, Volume 33, Pages 641-453
38 Reference 37, Page 649
39 D. D. Bogard, Ar-39/Ar-40 Dating of Mesosiderites, Geochemica Et Cosmochemica Acta, 1990, Volume 54, Pages 2549 – 2564
40 Reference 39, Page 2563, 2564
41 Ekaterina V. Korochantseva, 40Ar-39Ar Chronology, Meteoritics And Planetary Science, 2009, Volume 44, Number 2, Pages 293-321
42 Reference 41, Pages 316-321
43 Reference 41, Page 298
44 T. Itaya, Ultra-high Excess Argon in Kyanites, Gondwana Research, 2005, Volume 8, Number 4, Pages 617-621
45 Reference 44, Page 617
46 The Age Of The Earth, By G. Brent Dalrymple, 1991, Stanford University Press, Stanford, California, Page 10.
47 Reference 46, Page 23

Chapter 11

Rocks Older Than The Solar System

Examining The Thorium Lead Dating Method

SHRIMP Uranium/Lead Geochronology

These rocks from Western Australia were dated [1] in 2001 using the Uranium/Lead and Thorium/Lead dating methods. The article claims that the true age is 3 billion years old. [1] If we put the ratios from a table [2] in the article into Microsoft Excel and run the values through Isoplot we get ages between 2 million and 24 billion years old! How can a rock be 10 billion years older than the Big Bang explosion? Of all the samples, 18 are older than the Earth, 3 are older than the Galaxy and 2 are older than the Universe. There is a 24 billion year spread of dates between the youngest and the oldest ages.

Table 1

Statistics	208Pb/232Th	207Pb/206Pb	206Pb/238U	207Pb/235U
Average	5,075	3,027	1,303	1,294
Maximum	24,344	6,495	2,941	2,940
Minimum	8	869	5	2
Difference	24,336	5,627	2,935	2,938

Table 2

Statistics	208Pb/232Th	207Pb/206Pb	206Pb/238U	207Pb/235U
Average	1,989	2,688	2,793	2,729
Maximum	23,355	2,688	2,793	2,729
Minimum	56	2,651	2,558	2,618
Difference	23,300	37	236	111

Table 3

Statistics	208Pb/232Th	207Pb/206Pb	207Pb/235U
Average	1,834	2,716	2,098
Maximum	11,964	3,347	3,351
Minimum	0.1	2,490	59
Difference	11,964	857	3,291

The Beverley Uranium Deposit

These rocks from the North Flinders Ranges, South Australia., were dated [3] in 2010 using the Uranium/Lead and Thorium/Lead dating methods. The article claims that the true age is 400 million years old. [4] If we put the ratios from a table [5] in the article into Microsoft Excel and run the values through Isoplot we get ages between 1 million and 20 billion years old! How can a rock be 5 billion years older than the Big Bang explosion? Of all the samples, 6 are older than the Earth, 3 are older than the Galaxy and 2 are older than the Universe. There is a 20 billion year spread of dates between the youngest and the oldest ages. In table 5 we can see the percentage difference between the Thorium dates and the other three dating ratios used. The difference is almost 600,000 percent!

Table 4

Statistical Summary	Age 207/206	Age 206Pb/238U	Age 207Pb/235U	Age 208Pb/232Th
Average	737	3	3	3,758
Maximum	2,429	7	7	20,583
Minimum	9	0.1934	1	52
Difference	2,420	7	6	20,531

Table 5

Statistical Summary	Ratio 207Pb/206Pb	Ratio 206Pb/238U	Ratio 207Pb/235U
Average	25,841%	95,107%	91,073%
Maximum	137,220%	580,693%	571,750%
Minimum	654%	1,260%	800%
Difference	136,565%	579,433%	570,950%

Isotopic Systematics of the Goalpara Ureilite

This meteorite was dated [6] in 1994 using the Uranium/Lead and Thorium/Lead dating methods. The article claims that the true age is 4.55 billion years old. [6] If we put the ratios from a table [7] in the article into Microsoft Excel and run the values through Isoplot we get ages between 5 and 173 billion years old! How can a rock be 160 billion years older than the Big Bang explosion? Of all the samples, 123 are older than the Earth, 77 are older than the Galaxy and 71 are older than the Universe. There is a 168 billion year spread of dates between the youngest and the oldest ages.

Table 6

Statistics	207Pb/206Pb	206Pb/238U	208Pb/232Th
Average	5,056	27,406	87,825
Maximum	5,279	51,612	173,633
Minimum	4,979	4,929	17,658
Difference	300	46,683	155,976

Uranium-Thorium-Lead Isotope Data

These rocks from the Marble Bar area of the Pilbara Craton, Western Australia, were dated [8] in 2011 using the Uranium/Lead and Thorium/Lead dating methods. The article claims that the true age is 3.4 billion years old. [8] If we put the ratios from a table [9] in the article into Microsoft Excel and run the values through Isoplot [10] we get ages between 5 and 100 billion years old! How can a rock be 85 billion years older than the Big Bang explosion? Of all the samples, 45 are older than the Earth, 23 are older than the Galaxy and 17 are older than the Universe. There is a 75 billion year spread of dates between the youngest and the oldest ages.

Table 7

Statistics	207 Pb /206Pb	208Pb/232Th	207Pb/235U	206Pb/238U
Average	5,325	56,976	7,319	15,192
Maximum	5,403	100,601	10,054	31,005
Minimum	5,222	24,980	5,795	7,138
Difference	181	75,622	4,259	23,868

Table 8

208Pb/232Th	207Pb/235U	206Pb/238U
100,601	10,054	31,005
84,457	8,230	20,343
73,968	8,143	19,584
67,423	7,763	17,306
58,353	7,658	17,088
57,116	7,027	13,410
55,311	6,977	13,022
51,607	6,682	11,479
44,439	6,661	11,353
39,090	6,521	10,652
26,361	6,313	9,926
24,980	5,795	7,138

Uranium, Thorium and Lead Geochronology

These rocks from the Kola Peninsula in Russia were dated [11] in 2011 using the Uranium/Lead and Thorium/Lead dating methods. The article claims that the true age is 350 million years old. [11] If we put the ratios from a table [12] in the article into Microsoft Excel and run the values through Isoplot we get ages between 269 and 5,140 million years old! There is an 1,100 percent difference between some dates. That percentage difference equals almost 5,000 million years!

Table 9

Statistics	207Pb Age/232Th Age	238U Age/232Th Age	238U/206Pb Age	207Pb/206Pb Age
Average	859%	255%	1,054	3,381
Maximum	1275%	1165%	5,140	4,741
Minimum	361%	74%	269	1,318
Difference	914%	1092%	4,871	3,423

The Uranium, Thorium and Lead Compositions

These rocks from the Morocco and France were dated [13] in 2007 using the Uranium/Lead and Thorium/Lead dating methods. If we put the ratios from a table [14] in the article into Microsoft Excel and run the values through Isoplot we get ages between 2 and 92 billion years old! How can a rock be 75 billion years older than the Big Bang explosion? Of all the samples, 53 are older than the Earth, 13 are older than the Galaxy and 6 are older than the Universe. There is a 90 billion year spread of dates between the youngest and the oldest ages.

Table 10

Statistics	207Pb/206Pb	208Pb/232Th	206Pb/238U
Average	4,955	15,609	4,873
Maximum	5,090	92,494	18,639
Minimum	4,871	1,939	1,437
Difference	219	90,556	17,202

Rubidium/Strontium and Uranium/Lead Systematics

These rocks from the Kola Peninsula in Russia were dated [15] in 2011 using the Uranium/Lead and Thorium/Lead dating methods. The article claims that the true age is 2075–2100 million years old. [15] If we put the ratios from a table [16] in the article into Microsoft Excel and run the values through Isoplot we get ages between 2 and 10 billion years old! Of all the samples, 45 are older than the Earth, 23 are older than the Galaxy and 17 are older than the Universe. There is a 75 billion year spread of dates between the youngest and the oldest ages.

Table 11

Statistics	207Pb/206Pb	206Pb/238U	206Pb/238U	87Sr/86Sr
Average	5,020	7,253	8,177	2,185
Maximum	5,102	10,539	10,283	3,436
Minimum	4,834	2,814	5,303	1,739
Difference	267	7,725	4,980	1,697

Cu-Pb-Zn-Ag Mineralisation

These rocks from the Democratic Republic of Congo were dated [17] in 2009 using the Uranium/Lead and Thorium/Lead dating methods. The article claims that the true age is 520 million years old. [18] If we put the ratios from a table [19] in the article into Microsoft Excel and run the values through Isoplot we get ages between 0.1 and 200 billion years old! How can a rock be 185 billion years older than the Big Bang explosion? Of all the samples, 96 are older than the Earth, 42 are older than the Galaxy and 35 are older than the Universe. There is a 198 billion year spread of dates between the youngest and the oldest ages.

Table 12

Statistics	208Pb/232Th	207Pb/206Pb	206Pb/238U	207Pb/235U
Average	52,321	4,856	11,884	5,775
Maximum	199,319	6,275	48,496	12,150
Minimum	882	3,056	174	848
Difference	198,437	3,219	48,322	11,302

Uranium-Lead Age Of Baddeleyite

This meteorite was dated [20] in 2011 using the Uranium/Lead and Thorium/Lead dating methods. The article claims that the true age is 4.1 billion years old. [21] If we put the ratios from a table [22] in the article into Microsoft Excel and run the values through Isoplot we get ages between 0.1 and 165 billion years old! How can a rock be 150 billion years older than the Big Bang explosion?_Of all the samples 11 are older than the Universe. There is a 125 billion year spread of dates between the youngest and the oldest ages.

Table 13

Statistics	Pb 207/206	207Pb/235U	206Pb/238U	207Pb/235U	Pb206/U238	Pb208/232Th
Average	4,042	2,209	1,047	833	222	101,231
Maximum	5,112	4,517	3,306	2,515	297	165,469
Minimum	2,689	681	238	161	183	40,297
Difference	2,423	3,836	3,068	2,353	114	125,172

Table 14

Pb208/232Th	Pb208/232Th
165,469	102,437
150,399	82,898
143,322	74,124
137,057	47,131
127,166	43,247

Mesozoic Lithosphere Destruction

These rocks from the North China Craton were dated [23] in 2001 using the Uranium/Lead and Thorium/Lead dating methods. The article claims [24] that the true age is 125 million years old. If we put the ratios from a table [25] in the article into Microsoft Excel and run the values through Isoplot we get ages between 5 and 44 billion years old! How can a rock be 30 billion years older than the Big Bang explosion? Of all the samples, 40 are older than the Earth, 15 are older than the Galaxy and 12 are older than the Universe. There is a 40 billion year spread of dates between the youngest and the oldest ages.

Table 15

Statistics	Pb 207/206	206Pb/238U	207Pb/235U	Pb208/232Th
Average	5,056	7,431	35,683	11,303
Maximum	5,098	14,282	44,683	27,208
Minimum	5,047	5,871	33,524	8,258
Difference	51	8,411	11,159	18,950

If we use isotopic formulas [26-29] given in standard geology text we can arrive at ages from the Rb/Sr and Nd/Sm ratios listed in the article. The formula for Rb/Sr age is given as:

$$t = \frac{2.303}{\lambda}\log\left(\frac{(87Sr/86Sr)-(87Sr/86Sr)_0}{(87Rb/86Sr)}+1\right) \quad [1]$$

Where t equals the age in years. λ equals the decay constant. (87Sr/86Sr) = the current isotopic ratio. $(87Sr/86Sr)_0$ = the initial isotopic ratio. (87Rb/86Sr) = the current isotopic ratio. The same is true for the formula below.

$$t = \frac{2.303}{\lambda}\log\left(\frac{(143Nd/144Nd)-(143Nd/144Nd)_0}{(147Sm/144Nd)}+1\right) \quad [2]$$

If we put the ratios from this table [30] in the article into Microsoft Excel and use these formulas we get ages between 116 and 125 million years old! The Uranium/Lead ratios give ages between 5 billion and 44 billion years old!

Table 16

Method/Sample	FC1-1	FC1-2	FC5-1	FC6-1	FC6-2	FC7	FC4
Pb207/206	5,047	5,047	5,051	5,051	5,049	5,051	5,098
206Pb/238U	6,050	6,658	5,871	6,407	6,539	6,212	14,282
207Pb/235U	33,767	34,765	33,524	34,380	34,588	34,071	44,683
Pb208/232Th	8,402	8,396	8,725	8,774	9,358	8,258	27,208
Rb/Sr	124	126	124	126	126	124	116
Nd/Sm	125	126	126	125	125	125	116

Middle Atlas Peridotite Xenoliths

These rocks from Morooco were dated [31] in 2009 using the Uranium/Lead and Thorium/Lead dating methods. If we put the ratios from a table [32] in the article into Microsoft Excel and run the values through Isoplot we get ages between 3 and 14 billion years old! How can a rock be as old as the Big Bang explosion? Of all the samples, 3 are older than the Earth, 1 are older than the Galaxy and 1 are older than the Universe. There is a 6 billion year spread of dates between the youngest and the oldest ages.

Table 17

Statistics	208Pb/232Th	207Pb/206Pb	206Pb/238U
Average	9,493	4,939	5,056
Maximum	14,557	4,996	6,419
Minimum	4,429	4,882	3,693
Difference	10,127	114	2,727

A Precise 232Th/208Pb Chronology

These rocks from Inner Mongolia were dated [33] in 1993 using the Uranium/Lead and Thorium/Lead dating methods. The article claims that the true age is 555 million years old. [33] If we put the ratios from a table [34] in the article into Microsoft Excel and run the values through Isoplot we get ages between 400 million and 55 billion years old! How can a rock be 40 billion years older than the Big Bang explosion? Of all the samples, 170 are older than the Earth, [34] are older than the Galaxy and 19 are older than the Universe. There is a 75 billion year spread of dates between the youngest and the oldest ages.

Table 18

Statistics	207Pb/206Pb	208Pb/232Th	206Pb/238U
Average	5,068	764	9,321
Maximum	8,077	5,699	54,790
Minimum	3,586	402	4
Difference	4,491	5,297	54,787

Age of the MET 78008 Ureilite

This meteorite was dated [35] in 1994 using the Uranium/Lead and Thorium/Lead dating methods. The article claims that the true age is 4.56 billion years old. [36] If we put the ratios from a table [36] in the article into Microsoft Excel and run the values through Isoplot we get ages between 5 and 90 billion years old! How can a rock be 65 billion years older than the Big Bang explosion? Of all the samples, 63 are older than the Earth, 32 are older than the Galaxy and 29 are older than the Universe. There is a 75 billion year spread of dates between the youngest and the oldest ages.

Table 19

Statistics	207Pb/206Pb	206Pb/238U	208Pb/232Th
Average	5,077	15,565	47,442
Maximum	5,327	30,179	90,595
Minimum	4,963	7,496	14,271
Difference	364	22,683	76,324

Table 20

Statistics	206Pb/238U	207Pb/206Pb
Average	11,520	4,495
Maximum	25,513	4,576
Minimum	4,283	4,411
Difference	21,229	166

Conclusion

Yuri Amelin states in the journal Elements that radiometric dating is extremely accurate: "However, four 238U/235U-corrected CAI dates reported recently (Amelin et al. 2010; Connelly et al. 2012) show excellent agreement, with a total range for the ages of only 0.2 million years – from 4567.18 ± 0.50 Ma to 4567.38 ± 0.31 Ma." [37-39]

To come within 0.2 million years out of 4567.18 million years means an accuracy of 99.99562%. Looking at some of the dating it is obvious that precision is much lacking. The Bible believer who accepts the creation account literally has no problem with such unreliable dating methods. Much of the data in radiometric dating is selectively taken to suit and ignores data to the contrary.

References

1 **SHRIMP U–Pb Geochronology, International Earth Science, 2002, Volume 91, Pages 406-432**
2 **Reference 1, pages 414, 416, 423**
3 **The Beverley Uranium Deposit, Economic Geology, 2011, Volume 106, Pages 835-867**
4 **Reference 3, pages 846**
5 **Reference 3, pages 866**
6 **Isotopic Systematics of the Goalpara Ureilite, Gcochimica et Cosmochimtca Acta, 1995, Volume 59, Number 2, Pages 381-390**
7 **Reference 6, page 384**
8 **U–Th–Pb Isotope Data, Earth and Planetary Science Letters, 2012, Volume 319-320, Pages 197-206**
9 **Reference 8, page 199**
10 http://www.creationismonline.com/Isoplot/Isoplot.html
https://www.bgc.org/isoplot
11 **U–Th–Pb Geochronology, Gondwana Research, 2012, Volume 21, Pages 728–744**
12 **Reference 11, page 735**
13 **The U, Th and Pb Compositions, Geochimica et Cosmochimica Acta, 2009, Volume 73, Pages 469–488**
14 **Reference 13, page 475, 476**
15 **Rb–Sr and U–Pb Systematics, Lithology and Mineral Resources, 2011, Volume 46, Number 2, Pages 151-164**
16 **Reference 15, page 156, 158**
17 **Cu–Pb–Zn–Ag Mineralisation, Mineral Deposita, 2010, Volume 45, Pages 393-410**
18 **Reference 17, page 393, 394**
19 **Reference 17, page 397, 398**
20 **Uranium-Lead Age Of Baddeleyite, Journal Of Geophysical Research, 2011, Volume 116, Page 1-12**
21 **Reference 20, page 7**
22 **Reference 20, page 6**
23 **Mesozoic Lithosphere Destruction, Contributions Mineral Petrology, 2002, Volume 144, Pages 241-253**
24 **Reference 23, page 243**
25 **Reference 23, page 246**
26 **Radioactive and Stable Isotope Geology, By H.G. Attendon, Chapman And Hall Publishers, 1997. Page 73 [Rb/Sr], 195 [K/Ar], 295 [Re/OS], 305 [Nd/Nd].**
27 **Principles of Isotope Geology, Second Edition, By Gunter Faure, Published By John Wiley And Sons, New York, 1986. Pages 120 [Rb/Sr], 205 [Nd/Sm], 252 [Lu/Hf], 266 [Re/OS], 269 [Os/OS].**
28 **Absolute Age Determination, Mebus A. Geyh, Springer-Verlag Publishers, Berlin, 1990. Pages 80 [Rb/Sr], 98 [Nd/Sm], 108 [Lu/Hf], 112 [Re/OS].**
29 **Radiogenic Isotope Geology, Second Edition, By Alan P. Dickin, Cambridge University Press, 2005. Pages 43 [Rb/Sr], 70 [Nd/Sm], 205 [Re/OS], 208 [Pt/OS], 232 [Lu/Hf].**
30 **Reference 23, page 245**
31 **Middle Atlas Peridotite Xenoliths, Geochimica et Cosmochimica Acta, 2010, Volume 74, Pages 1417-1435**
32 **Reference 31, page 1425**
33 **A Precise 232Th-208Pb Chronology, Geochimica et Cosmochimica Acta, 1994, Volume 58, Number 15, Pages 3155-3169**
34 **Reference 33, page 3160-3163**

35 Age of the MET 78008 Ureilite, Geochimica et Cosmochimica Acta, 1995, Volume 59, Number 11, Pages 2319-2329
36 Reference 35, page 2324
37 Dating the Oldest Rocks in the Solar System, Elements, 2013, Volume 9, Pages 39-44
38 Amelin, Earth and Planetary Science Letters, 2010, Volume 300, Pages 343-350
39 Connelly, Science, 2012, Volume 338, Pages 651-655

Chapter 12

Rocks Older Than The Earth

Pre Cambrian Earth Rocks

This dating [1] was done in 2005 at the Heidelberg University in Germany. The author comments on the cause for such absurd dates:

"The bulk 40Ar/36Ar ratio is more radiogenic than atmospheric composition, indicating—in addition to an atmospheric component—the presence of a slight but detectable contribution of an excess 40Ar component, i.e., 40Ar trapped from an external source, because it cannot be due to in situ decay of 40K. This circumstance is indicated by the very high apparent ages (up to 5 Ga) of the irradiated type I shungite (Appendix A1)." [2]

Below we can see some of the dates [3] given in the article. Several dates are older than the theory of evolution allows:

Table 1

Sample Temperature Centigrade	Age Million Years	Error Million Years
820	4,964	239
850	4,916	114
880	5,269	120
910	5,804	123
940	5,425	109
970	4,843	114
1070	5,054	205

Ages from 4,843 to 5,804 million years old.

Mount Isa, Queensland

These rocks were dated in 2006 by Mark Kendrick [4] from the University of Melbourne. The data in tables 2 to 7 shows ages [5] of Earth rocks from 4,700 to 10,000 million years old. Dalrymple leaves these dates out of his books [6,7].

Table 2

Sample Eloise Mine	Million Years	Age Category
Cr-2	5,620	Older Than Solar System
Cr-3	5,511	Older Than Solar System
300	6,127	Older Than Solar System
1400	5,370	Older Than Solar System
Total	4,804	Older Than Earth

Ages from 4,804 to 5,620 million years old.

Table 3

Sample	Million	Age
Eloise Mine	Years	Category
250	6,442	Older Than Solar System
350	6,393	Older Than Solar System
450	4,931	Older Than Earth
1200	4,760	Older Than Earth
Total	4,777	Older Than Earth

Ages from 4,760 to 6,442 million years old.

Table 4

Sample	Million	Age
Eloise Mine	Years	Category
200	7,412	Older Than Solar System
250	9,969	Older Than Galaxy
300	8,655	Older Than Solar System
350	5,871	Older Than Solar System
400	6,568	Older Than Solar System
450	6,060	Older Than Solar System
1200	5,201	Older Than Solar System
1300	4,805	Older Than Earth
1400	5,049	Older Than Solar System
Total	5,601	Older Than Solar System

Ages from 4,805 to 9,969 million years old.

Table 5

Sample	Million	Age
Osborne Mine	Years	Category
300	7,715	Older Than Solar System

Table 6

Sample	Million	Age
Railway Fault	Years	Category
200	5,176	Older Than Solar System
350	4,759	Older Than Earth

Table 7

Sample	Million	Age
Railway Fault	Years	Category
Cr	4,844	Older Than Earth
Cr	4,883	Older Than Earth
Cr	5,418	Older Than Solar System
Cr	5,238	Older Than Solar System

Ages from 4,844 to 5,418 million years old.

Broken Hill, New South Wales

These rocks were dated [8] in 1981 using the ^{40}Ar / ^{39}Ar dating method. According to the dates obtained, many of the rocks are older than the Earth and Solar System. Some of the rocks are as old as the galaxy itself. The author of the article comments:

"It has been argued already that the high initial ages in the release patterns of both hornblende and plagioclase can be translated into a concentration of excess 40Ar. Concentrations for those samples analysed by the 40Ar / 39Ar spectrum method are given in Table 5, and can be used to estimate the partition coefficient of Ar between hornblende and plagioclase." [9]

"Excess 40Ar was incorporated into minerals during the 520-Ma event at a temperature of about 350°C." [10]

There is no way of proving this assumption. It is just an excuse for such ridiculous ages of geological system that supposedly formed between 1,600 and 500 million years ago. [11] The data in tables 8 to 14 shows ages [12] greater than the age of the Solar System.

Table 8

Temperature	Age	Age
40Ar/39Ar	Million Years	Category
Plagioclase		
700	7,473	Older Than Solar System
650	5,753	Older Than Solar System
B80	6,185	Older Than Solar System
1230	5,244	Older Than Solar System
1250	5,191	Older Than Solar System
FUSE	5,721	Older Than Solar System
Hornblende		
470	5,050	Older Than Solar System
530	4,802	Older Than Earth

Ages from 4,802 to 7,473 million years old.

Table 9

Temperature	Age	Age
40Ar/39Ar	Million Years	Category
TF	5,170	Older Than Solar System
350	6,931	Older Than Solar System
430	7,015	Older Than Solar System
490	6,611	Older Than Solar System
540	6,167	Older Than Solar System
590	5,050	Older Than Solar System
1060	4,637	Older Than Earth
1080	4,929	Older Than Earth
1100	5,171	Older Than Solar System
1200	6,037	Older Than Solar System
FUSE	7,010	Older Than Solar System

Ages from 4,637 to 7,015 million years old.

Table 10

Temperature	Age	Age
40Ar/39Ar	Million Years	Category
Clinopyroxene		
1040	4,704	Older Than Earth
1090	4,970	Older Than Earth
1070	4,989	Older Than Earth
1120	4,767	Older Than Earth
FUSE	5,373	Older Than Solar System

Ages from 4,704 to 5,373 million years old.

Table 11

Temperature	Age	Age
40Ar/39Ar	Million Years	Category
TF	6,730	Older Than Solar System
350	7,317	Older Than Solar System
440	5,055	Older Than Solar System
520	4,861	Older Than Earth
580	5,075	Older Than Solar System
650	4,973	Older Than Earth
930	5,409	Older Than Solar System
970	6,795	Older Than Solar System
1000	7,587	Older Than Solar System
1030	6,960	Older Than Solar System
1060	6,799	Older Than Solar System
1070	6,511	Older Than Solar System
1090	7,257	Older Than Solar System
1140	7,823	Older Than Solar System
1170	7,666	Older Than Solar System
1300	9,588	Older Than Solar System
1380	8,432	Older Than Solar System
FUSE	7,234	Older Than Solar System

Ages from 4,861 to 9,588 million years old.

Table 12

Temperature	Age	Age
40Ar/39Ar	Million Years	Category
Plagioclase		
710	7,653	Older Than Solar System
770	6,484	Older Than Solar System
800	7,367	Older Than Solar System
820	6,709	Older Than Solar System
Hornblende		
550	5,068	Older Than Solar System
620	4,777	Older Than Earth

Ages from 4,777 to 7,653 million years old.

Table 13

Temperature	Age	Age
40Ar/39Ar	Million Years	Category
Plagioclase		
360	5,748	Older Than Solar System
550	5,459	Older Than Solar System
840	5,998	Older Than Solar System
Hornblende		
960	9,681	Older Than Solar System
960	9,582	Older Than Solar System
990	9,852	Older Than Solar System
Muscovite		
560	9,521	Older Than Solar System

Ages from 5,459 to 9,852 million years old.

The data in table 14 shows [13] ages older than the Earth and Solar System.

Table 14

Sample	Mineral	Age
Number	Type	Million Years
79-173	Plagioclase	5,800
79-173	Hornblende	5,300
79-459	Hornblende	5,500
79-459	Plagioclase	7,000
79-461	Hornblende	5,500
79-461	Plagioclase	7,300

Ages from 5,300 to 7,300 million years old.

Ages In The Allende Meteorite

This dating was done in 1983 [14] and gave ages between 2,990 and 8,880 million years old. [15] The author discusses the problem and proposed solutions:

"The existence in the Allende meteorite of coarse-grained Ca-Al-rich inclusions (CAI) with 40Ar/39Ar apparent ages exceeding the age of the solar system was reported by Jessberger and Dominik [1] and Jessberger et al. [2] and confirmed by Herzog et al. [3]." [16]

Table 15

Sample	Age A	Age B
Name	Million Years	Million Years
EGG 1		
700	5,070	
1000	5,190	
1200	4,730	
1650	4,570	
Total	4,860	4,800
EGG 2		
700	7,370	
1000	4,670	

1200	3,430	
1650	4,510	
Total	4,470	4,470
EGG 3		
700	8,880	
1000	6,450	
1200	2,990	
1650	5,660	
Total	5,930	5,020

Ages from 2,990 to 8,880 million years old.

Below [Table 16] we can see some more dating [17] that was done on the same meteorite by Herzog in 1980. He give three possible reasons [18] why the dates are in such conflict with the standard evolutionary model:

1 "The coarse-grained Ca-Al-rich inclusions are really older than 4.6 G.y., associated with in situ decay of K in pre-solar dust."

2 "The excess Argon 40 and Argon 36 could be due to atmospheric contamination."

3 "The excess 40 and the trapped 36 may have come from the degassing of matrix and/or rim material sometime in the interval 3.6 - 4.1 G.y. ago."

Table 16

Mineral System	Age Million Years	Error Million Years
Vein	8,500	700
Spinel	6,900	800
Vein	5,250	140
Spinel	6,400	500
Bulk	5,120	20
Bulk	5,100	100
01. Skel.	6,290	10

Ages from 5,100 to 8,500 million years old.

U-Th-Pb, Sm-Nd And Rb-Sr Model Ages

Below we can see some more dating [19] that was done on some Moon rocks by Oberli in 1978. Oberli states [20] that the U-Th-Pb data is concordant but the Neodymium dates are uncertain. Again it is just an arbitrary choice he makes as to which date is certain and which date is not.

Table 17

Sample Number	Pb-206/Pb-207 Million Years	Pb-206/U-238 Million Years	Pb-208/Th-232 Million Years	Nd-143/Nd-144 Million Years	Rb-87/Sr-86 Million Years
66075, 11D	5,371	7,794	8,280		
66075, 11	5,358	7,740	8,375	4,530	4,240

Ages from 4,240 to 8,375 million years old.

Gerontology Of The Allende Meteorite

This article appeared [21] in Nature magazine in 1979. Jessberger admits that the wildly discordant ages cannot be due to normal processes:

“In the Allende meteorite several elements are found to have an isotopic composition that cannot be due to radioactive or spallation or fractionation processes.” [22]

“In the most widely accepted theory a supernova triggered the collapse of the solar nebula, and the anomalously high ages would be due to an enhanced 40K/39K isotopic ratio produced in the explosive carbon burning shell of the supernova? In another, controversial interpretation these ages could have chronological significance, as here the presolar grains are relicts from various old stellar nucleosynthetic and condensation processes unrelated to the formation of the Solar System.” [22]

He then quotes several [23, 24, 25] science journals for an explanation. He thinks the ages could be residue from an ancient supernova or contamination for pre galactic dust not related to the formation of the Solar System. Again, like Oberli his solution is totally unprovable. How would you test such a hypothesis? Some of the dates are older than the galaxy. How do we know that Earth rocks have not been contaminated in such a way? During the formation of the Solar System, the Earth might have absorbed such materials. His choice of “true” ages is just guess and not provable science.

Table 18

Meteorite Sample 17	Age Million Years	Error Million Years	Age Million Years	Error Million Years
500	7,680	80	4,960	420
580	5,830	80	4,600	160
660	5,350	40	4,970	60
740	5,090	20	4,970	40
820	5,080	40	4,990	60
890	5,210	40	5,210	40
950	4,970	60	4,970	60
1,010	4,970	30	4,970	30
1,070	5,340	40	5,340	40
1,130	5,540	20	5,430	40
1,200	6,210	100	5,250	240
1,280	5,190	190	1,460	1,480
1,380	7,200	590	2,670	5,650
Total	5,500	20	5,120	60

Ages from 1,460 to 7,680 million years old.

Table 19

Meteorite Sample 18	Age Million Years	Error Million Years	Age Million Years	Error Million Years
450	11,010	60	4,520	2,240
580	8,060	140	4,470	500
670	7,500	40	4,970	160
750	6,310	30	4,900	90
830	5,370	20	5,130	60
900	4,960	40	4,960	40
970	4,900	40	4,900	40
1,040	4,890	40	4,890	40
1,110	4,900	30	4,900	30

1,190	4,820	20	4,820	20
1,300	5,370	100	5,370	100
Total	6,050	40	5,080	50

Ages from 4,470 to 11,010 million years old.

Conclusion

Brent Dalrymple states:

"Several events in the formation of the Solar System can be dated with considerable precision." [26]

Looking at some of the dating it is obvious that precision is much lacking. He then goes on:

"Biblical chronologies are historically important, but their credibility began to erode in the eighteenth and nineteenth centuries when it became apparent to some that it would be more profitable to seek a realistic age for the Earth through observation of nature than through a literal interpretation of parables." [27]

The Bible believer who accepts the creation account literally has no problem with such unreliable dating methods. Much of the data in Dalrymple's book is selectively taken to suit and ignores data to the contrary.

References

1 Argon isotope fractionation, By Mario Trieloff, Geochimica et Cosmochimica Acta, 2005, Volume 69, Number 5, Pages 1253–1264

2 Reference 1, Page 1254

3 Reference 1, Page 1263

4 Evaluation of 40Ar–39Ar quartz ages, By M.A. Kendrick, Geochimica et Cosmochimica Acta, 2006, Volume 70, Pages 2562–2576

5 Reference 4, Pages 2573-2575

6 Ancient Earth, Ancient Skies, The Age Of Earth And Its Cosmic Surroundings, G. Brent Dalrymple, Stanford University Press, Stanford, California, 2004 https://archive.org/download/B-001-001-783/B-001-001-783.pdf

7 How Old is the Earth? A Response to "Scientific" Creationism, by G. Brent Dalrymple, 2006 http://www.talkorigins.org/faqs/dalrymple/contents.html

8 Excess 40Ar in metamorphic rocks from Broken Hill, By T. Mark Harrison, Earth and Planetary Science Letters, 1981, Volume 55, Pages 123 - 149

9 Reference 8, Page 141

10 Reference 8, Page 147

11 Reference 8, Page 124

12 Reference 8, Page 128 – 133

13 Reference 8, Page 137

14 Ages in Allende Inclusions, By I. M. Villa, Earth and Planetary Science Letters, 1983, Volume 63, Pages 1 – 12

15 Reference 14, Page 5

16 Reference 14, Page 1

17 39Ar -40Ar Systematics Of Allende Inclusions, Page 3, By G. F. Herzog http://www.lpi.usra.edu/meetings/lpsc1980/pdf/1155.pdf

18 Reference 17, Page 2.

19 U-Th-Pb, Sm-Nd And Rb-Sr Model Ages, Page 833, By F. Oberli http://www.lpi.usra.edu/meetings/lpsc1978/pdf/1289.pdf

20 Reference 19, Pages 832, 834

21 Gerontology of the Allende meteorite, By Elmar K. Jessberger, Nature, 1979, Volume 277, Pages 554 - 556

22 Reference 21, Page 554

23 Cameron, A, G. W. & Truran. J. W. Icarus, 1977, Volume 30, Page 447.

24 Clayton D, D, Nature, 1975, Volume 257, Page 36.
25 Clayton D. D., Earth Planetary Science Letters, 1977, Volume 36, Page 381.
26 The Age Of The Earth, By G. Brent Dalrymple, 1991, Stanford University Press, Stanford, California, Page 10.
27 Reference 26, Page 23

Chapter 13

Rocks Older Than The Universe

Trillion-Year-Old Rocks!

These rocks from Black Hills, South Dakota were dated in 1970 giving ridiculous dates. The oldest [Trillion Years!] is 60 times older than the Big Bang explosion. The article simply says: "**Anomalous age data for pegmatite minerals.**"[1]

Table 1

		Rb-Sr Date	Rb-Sr Date
Sample/Mines Mineral	Mineral Type	Million Years	Billion Years
Hugo Mine	Albite	7,100	7
Hugo Mine	Apatite	900,000	900
Hugo Mine	Lithiophyllite	53,000	53
Tin Mountain	Montebraeite	36,000	36
Tin Mountain	Apatite	75,000	75
Bob Ingersoll Mine	Montebrasite	81,000	81
Bob Ingersoll Mine	Apatite	460,000	460

Rocks 18 billion Years Old

This rock was from the Great Northern Peninsula, Newfoundland. It was dated in 1974. As the article says: "The most striking of these is the consistent pattern of anomalously high apparent ages obtained for high temperature fractions (i.e. fraction s corresponding to temperatures > 925-950°C). These anomalously high apparent ages almost certainly reflect the presence of excess radiogenic argon." [2] The table in the article [3] lists 11 rock samples with radical discordant dates. The first two rocks have internal ages varying between the "youngest" and "oldest" by a factor of 2000% and 1000% respectively.

Table 2

Maximum Age	Minimum Age	Difference	Difference
Million Years	Million Years	Million Years	Percentage
18,620	651	17,969	2,760%

Rocks 80 Billion Years Old!

Some of these rocks have been dated to be five times older than the Big Bang explosion! [4, 5] These rocks from Yucca Mountain, Nevada were dated in 2008 by U–Th–Pb dating method.

Table 3

Sample	Pb-206/U-238	Pb-208/Th-232	Error	Difference
Number	Million Years	Million Years	Million Years	Percentage
HD2059Pb4-Cc	1,738	12,900	4,040	7,963
HD2089APb1-Cc1	7,940			
HD2089APb1-Cc2a	6,372			
HD2089APb1-Cc2b	7,504			
HD2089APb1-Cc2c	6,292			
HD2089APb1-Cc3	4,423	28,600	7,700	647
HD2177Pb1-Cc	20,209	1,555	140	7,296
HD2233Pb1-Ch2	8	82,030	180,500	1,986,199
HD2233Pb2-Ch2	7	57,900	40,800	1,153,386

As we can see form the table below that some of the dates are almost 2 million percent discordant. That means that the dating methods can give ages for the same rock that vary by a factor of 20,000. One part of the rock is dated as being 20,000 times older than another.

Table 4

Sample Number	Difference Percentage	Sample Number	Difference Percentage
HD2098Pb3-Cc	1,094	HD2059Pb4-Cc	7,963
HD2074Pb2-Cc1	1,224	HD2062Pb1-Cc	12,772
HD2055Pb11-Cc	1,246	HD2074Pb1-Cc3	44,828
HD2062Pb2-Cc	1,311	HD2089APb1-Cc1	49,625
HD2055Pb12-Op	1,467	HD2089APb1-Cc2b	50,027
HD2055Pb12-Cc	1,584	HD2089APb1-Cc2c	69,911
HD2089APb2-Cc	1,970	HD2155Pb1-Cc	121,400
HD2109Pb1-Cc	2,083	HD2055Pb11-Op	195,100
HD2065Pb4-Cc	2,691	HD2233Pb2-Ch2	1,153,386
HD2177Pb1-Cc	7,296	HD2233Pb1-Ch2	1,986,199

Rocks 22 Billion Years Old

This dating was done in 1990 on rocks from the Ouzzal granite unit in Algeria. Maluski used Argon dating and it gave dates over 22 billion years old.[6]

Table 5

Sample Name	Maximum Age Million Years	Minimum Age Million Years	Average Age Million Years	Age Difference Million Years	Percent Difference
A. TEK 58 plagioclase 1	13,435	1,800	7,043	11,635	746%
B. TEK 58 plagioclase 2	8,071	2,446	6,024	5,625	329%
C. TEK 58 plagioclase 3	15,407	1,214	3,857	14,193	1,269%
D. TEK 58 plagioclase 4	10,776	1,800	4,650	8,976	598%
E. TEK 58 pyroxene	11,621	5,744	9,909	5,877	202%
F. TEK 58 biotite	4,522	1,700	2,147	2,822	266%
G. TEK 58 garnet	22,090	3,716	11,685	18,374	594%

We can see in table 6 on the next page some of the extremely discordant dates.

Table 6

A. Plagioclase 1 Million Years	B. Plagioclase 2 Million Years	C. Plagioclase 3 Million Years	D. Plagioclase 4 Million Years	E. Pyroxene Million Years	G. Garnet Million Years
5,062	5,008	6,045	5,360	9,150	7,361
6,027	5,410	7,995	5,564	9,276	8,311
6,303	5,712	11,804	6,424	9,564	8,906
6,489	5,739	15,407	6,452	9,684	10,232
7,492	5,892		7,318	9,874	10,310
9,228	5,983		7,689	9,899	10,790
11,783	6,453		10,776	9,943	11,448
13,263	6,785			10,097	11,568
13,287	6,939			10,102	11,961
13,435	7,372			10,314	12,780
	7,779			10,521	13,750
	8,071			10,578	14,689
				10,610	16,224
				10,617	19,945
				10,685	20187
				10,729	20,272
				10,736	20,742
				10,873	22,090
				10,889	
				11,041	
				11,288	
				11,382	
				11,389	
				11,396	
				11,621	

Maluski comments: "Apparent ages as old as l0 - 11 Ga are obtained between 450 and 1100 C, which implies that the excess component is widely distributed over all the sites without a preferential location. The internal age discordance is mainly due to the low amount and variability of 39Ar released at each temperature increment. This is probably because K occurs as microscopic impurities within pyroxene, the degassing of which is very irregular." [6]

Volcanic Rocks 15 Billion Years Old

The article describes Rubidium-Strontium dating of volcanic rocks in the Highwood Mountains and Eagle Buttes, Montana, U.S.A. This was performed in 1994. Ages [7] greater than the Big Bang date were obtained.

Table 7

6.46	Billion Years Old
6.83	Billion Years Old
10.8	Billion Years Old
15.5	Billion Years Old

"These extreme isotopic characteristics are accompanied by parent daughter ratios that give all the Highwood peridotites old model ages (Rb-Sr, 2.14-15.5 Ga; Sm-Nd, 2.78-6.83 Ga) compared to the other ultramafic samples."[8]

15 Billion Years Old

This article[9] refers to dating of xenoliths from the Kaapvaal craton in South Africa. These rocks were dated in 1995. Pearson's explanation is: "For example, several of the peridotite Re/Os model ages calculated using measured 187-Re/ 188-Os (TM) either give geologically unreasonable ages or do not intersect the Bulk Earth evolution line at all. Walker et al. [14] reasoned that the highly refractory compositions of Kaapvaal peridotites could have led to complete removal of Re during formation."[10]

Table 8

8.5	Billion Years Old
10.2	Billion Years Old
11.1	Billion Years Old
15.6	Billion Years Old

Moon Rocks 28 Billion Years Old

The following dating was done in 1972.[11] Table Nine[11] gives ages twice as old as the Big Bang explosion date. Table Ten[12] gives ages twice as old as the Moon and Solar System.

Table 9

Pb-207	Pb-206	Pb-207	Pb-208
Pb-206	U-238	U-235	Th-232
Billion Years	Billion Years	Billion Years	Billion Years
5.58	9.21	6.43	24.92
5.65	8.73	6.39	23.50
5.43	10.28	6.54	28.14

Table 10

Pb-207	Pb-206	Pb-207	Pb-208
Pb-206	U-238	U-235	Th-232
Billion Years	Billion Years	Billion Years	Billion Years
5.31	6.98	5.74	10.79
5.33	6.81	5.71	10.34
5.28	7.15	5.76	11.23

Rocks 23 Billion Years Old

This article describes Rubidium-Strontium dating of Precious Metal Veins of the Coeur D'Alene Mining District, Idaho. This dating[13] was done in 2002 and gave ages over 20 billion years old.

Table 11

Sample Number	Age Million Years	Difference Percentage
858-07G	4,475	
858-07H	1,727	159%
858-07L	7,816	
858-07M	1,195	554%
858-07U	971	
858-07V	2,630	171%
858-08C	1,855	
858-08D	6,105	229%
858-08AA	3,028	
858-08AB	588	415%
858-09D	1,490	
858-09E	754	98%
858-09F	2,453	
858-09G	682	259%
858-09J	719	
858-09K	2,696	274%
858-09L	395	
858-09M	1,465	270%
918-13A	278	
918-13B	2,209	694%
918-13C	23,312	
918-13D	968	2308%
918-15L	873	
918-15M	4,291	391%

The samples are in pairs. Each pair is taken from the exact same location. Some dates are between two and twenty-three times discordant for the one rock. The one dating method will give two different dates for the same rock! One date is twenty-three times older than the younger one.

Conclusion

Even though it is commonly claimed to be absolute proof of millions of years, there are many problems with radiometric dating. The recently published "**Radioisotopes & the Age of the Earth**" "**Earth's Catastrophic Past** " and other publications by young earth creationists shows that accepting a literal view of the Genesis creation account and a young age of the earth can be defended scientifically and old age successfully rebutted.

For in six days the LORD made heaven and earth, the sea, and all that in them is, and rested the seventh day: wherefore the LORD blessed the Sabbath day, and hallowed it. (**Exodus 20:8-11**

References

1 **G. H. Riley, Isotopic discrepancies in Black Hills, South Dakota, Geochemica Et Cosmochemica Acta, 1970, Volume 34, pages 721.**

2 **Vidas Stukas, 40 Ar/39ar Dating Of The Long Range Dikes, Earth and Planetary Science Letters, 1974, Volume 22, Pages 261**

3 **Reference 2, Page 260**

4 **Yuri Amelin, Natural Radionuclide Mobility, Geochemica Et Cosmochemica Acta, 2008, Volume 72, Pages 2067 – 2089**

5 Reference 4, Pages 2080, 2081
6 H. Maluski, Location of extraneous argon, Chemical Geology, 1990, Volume 80, pages 201 - 204
7 R. W. Carlson, North Western Wyoming Craton, Earth And Planetary Science Letters, 1994, Volume 126, Page 460
8 Reference 7, Page 465
9 D. G. Pearson, Xenoliths from the Kaapvaal Craton, Earth And Planetary Science Letters, 1995, Volume 134, Page 344
10 Reference 9, Page 348
11 G. J. Wasserburg, Three Apollo 14 Basalts, Earth And Planetary Science Letters, 1972, Volume 14, Pages 289.
12 Reference 11, Page 291
13 Robert J. Fleck, Age and Origin of Base and Precious Metal, Economic Geology, 2002, Volume 97, Pages 35 – 37

Chapter 14

The Osmium 187/186 Dating Method

By Paul Nethercott

April 2014

Evidence from Gorgona Island and Curacao

These rocks from Gorgona Island, Colombia and Curacao Island (Dutch Caribbean), were analysed in 1998 by scientists from the Department Of Geology, University Of Maryland. [1] The model age for Gorgona Island is 90 million years old: "Previous studies have reported K-Ar and 40Ar/39Ar ages for basalts from Gorgona. These ages range from approximately 86 to 92 Ma, averaging about 88 Ma." [2] The model age for Curacao Island is 90 million years old also: "Basaltic lavas from Curacao Lava Formation have been dated by 40Ar/39Ar step heating techniques at 88–90 Ma." [3] The article contains a table [4] with Osmium ratios that have no dates beside them. If we put the Osmium tables into Microsoft Excel and use the formulas in standard geology text books we get the values in table 1 The fifty one dates obtained from the sample ratios have a 64 billion year range from 183 million years old to 64.4 billion years old. Twelve dates are over 10 billion years old. The choice of 90 million years old as the true age is just a random guess.

Table 1	Million Years	% Discordance	Difference
Average	-9,427	10,575	9,517
Maximum	-183	71,720	64,548
Minimum	-64,458	303	273

Rocks from Southern West Greenland

These rocks from Southern West Greenland, were analysed in 1999 by scientists from The Australian National University, Canberra. [5] The model age for two sets of samples is 3,460 and 3,810 million years old. [6] The article contains a table [6] with Osmium ratios that have no dates beside them. If we put the Osmium tables into Microsoft Excel and use the formulas in standard geology text books we get twenty three dates. Only two are as old as the so called [7] model age. The choice 3,500 million years as the true age is just a random guess.

Table 2

Age Million Years	Age Million Years	Age Million Years	Age Million Years
3,348	3,231	2,541	1,457
3,344	3,112	2,272	1,372
3,318	3,056	2,012	966
3,312	2,799	1,776	783
3,272	2,722	1,487	321

187Os-186Os Systematics of Os-Ir-Ru

These rocks from southwestern Oregon were dated in 2004 by scientist from the Department of Geology, University of Maryland using the Argon 40/39 and Uranium/Lead dating methods. [8] According to the article the true age is 162 million years old: "An age of 162 Ma for the Josephine ophiolite has been established via ^{40}Ar–^{39}Ar and U–Pb geochronology of mafic portions of the ophiolite." [9] Another magazine gives the same chronology: "A rapid sequence of events, from ophiolite generation to thrust emplacement, has been determined using $^{40}Ar/^{39}Ar$ and Pb/U geochronology. Ophiolite generation occurred at 162–164 Ma, a thin hemipelagic sequence was deposited from 162 to 157 Ma, and flysch deposition took place between 157 and 150 Ma." [10] The article contains tables with Osmium 187/186 ratios that have no dates beside them. If we put the tables into Microsoft Excel and use the formula below used in standard geology text books [11-13] we can calculate dates from the undated isotopic ratios.

(1)

$$t = \frac{1.04 - ({}^{187}Os / {}^{186}Os)}{0.050768}$$

In the above formula, t = billions of years. The same date can be calculated from the Osmium 187/188 ratios. If we use another formula [14] we can convert the Osmium 187/188 ratio to the Osmium 187/186 ratio.

(2)

$$\frac{{}^{187}Os}{{}^{186}Os} \times 0.12035 = \frac{{}^{187}Os}{{}^{188}Os}$$

(3)

$$\frac{{}^{187}Os}{{}^{186}Os} = \frac{({}^{187}Os \div {}^{188}Os)}{0.12035}$$

(4)

$$t = \frac{\left(\frac{({}^{187}Os \div {}^{188}Os)}{0.12035}\right) - 1.04}{0.050768}$$

Table 3	Million Years	% Discordance	Difference
Average	-439	588	811
Maximum	637	2,351	3,808
Minimum	-3,646	115	104

We can see from table 3 the date range and percentage of discordance. There is a 4,434 million year range between the youngest and oldest dates. None of the fifteen dates even come close. Eight are impossible negative or future ages.

Determination of 187Os in Molybdenite

These rocks were analysed in 2001 by scientist from the National Research Centre of Geo Analysis, Beijing China using plasma-mass spectrometry methods. [15] The article contains a table [16] with Osmium 187/186 ratios that have no dates beside them. If we put the tables into Microsoft Excel and use the formulas in standard geology text books we get twenty seven absolutely impossible negative ages.

Table 4	Billion Years	Million Years
Average	6.78	6,783
Maximum	13.56	13,559
Minimum	3.17	3,165

186Os–187Os Systematics of Hawaiian Picrites

These Hawaiian rocks were analysed in 2009 by scientist from the Department of Geology, University of Maryland. [17] According to the article the true age is 2 billion years old: "Ratios were calculated for a model age of 2 Ga, assuming that the material had chondritic." [18] The article contains a three tables [19] with Osmium 187/186 ratios that have no dates beside them. If we put the tables into Microsoft Excel and use the formulas in standard geology text books we get the values in table 3. The fifty nine dates range between -726 and -143,516 million years old. The choice of this as the true age is just a random guess. Table three below is a summary of table two's isotopic ratios in the original article.

Table 3	Million Years	% Discordance	Difference
Average	-55,151	2,758	53,367
Maximum	-817	7,176	141,516
Minimum	-143,516	41	164

Evidence from Icelandic Picrites

These rocks from Iceland were analysed in 2007 by scientist from the NASA Johnson Space Centre, Houston, Texas. [20] According to the article the true age is 60 million years old: "The Os and He isotopes of Iceland picrites provide important insights into the Iceland plume system from 60 Ma to present." [21] The article contains two tables [22] with Osmium 187/186 ratios that have no dates beside them. If we put the tables into Microsoft Excel and use the formulas in standard geology text books we get the values in table 4. The thirty four dates are between 1,783 and -2,218 million years old. There is a four billion years range between the youngest and oldest dates. The author's choice of 60 million years as the true age is just a random guess. The difference between the so called true age [Model Age] and the calculated ratio age varies between 894 and 2,279 million years in error.

Table 4	Million Years	% Discordance	Difference
Average	-503	2,691	1,597
Maximum	1,783	3,798	2,279
Minimum	-2,219	1,490	894

186Os/187Os Systematics of Hawaiian Picrites

These rocks from Hawaii were analysed in 1999 by scientist from the Department of Geology, University of Maryland. [23] According to the article the true age is 2 billion years old: "various possible ancient recycled oceanic crustal components (upper crust, basalt, reducing sediments and metalliferous sediments) formed at 2 Ga." [24] The article contains a table [25] with Osmium 187/186 ratios that have no dates beside them. If we put the tables into Microsoft Excel and use the formulas in standard geology text books we get the values in table 5. Out of the seventy three dates, there is a 9.85 billion year range between the youngest [-8,068] and oldest [**1,785**] dates. The choice of 2 billion years as the true age is just a random guess.

Table 5	Million Years
Average	-1,187
Maximum	1,785
Minimum	-8,068

Chart 1

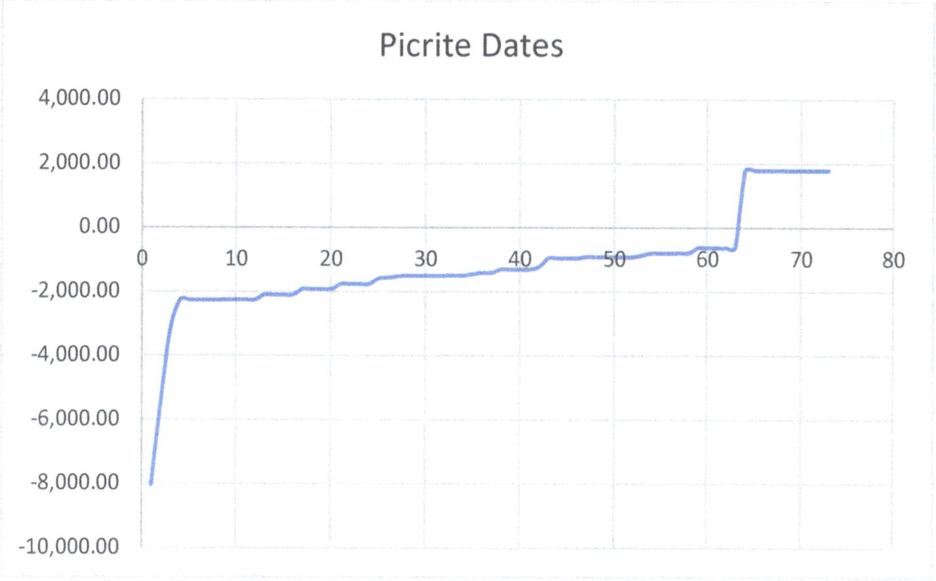

186Os/188Os and 187Os/188Os Measurements (Part 2)

These rocks were analysed in 2007 by scientist from the University of Durham. [26] According to the article the true age is 600 million years old. [27] The article contains two tables [28] with Osmium 187/186 ratios that have no dates beside them. If we put the tables into Microsoft Excel and use the formulas in standard geology text books we get the values in table 6, 7 and chart 2. Out of the sixty two dates, there is a 9.1 billion year range between the youngest [-8,075] and oldest [1,058] dates. The choice of 600 million years as the true age is just a random guess.

Table 6	Million Years	% Discordance	Difference
Average	-7,674	1,379	8,274
Maximum	-5,945	1,446	8,675
Minimum	-8,075	1,091	6,545

Table 7	Million Years	% Discordance	Difference
Average	219	2,340	499
Maximum	1,058	8,380	612
Minimum	-12	118	276

Chart 2

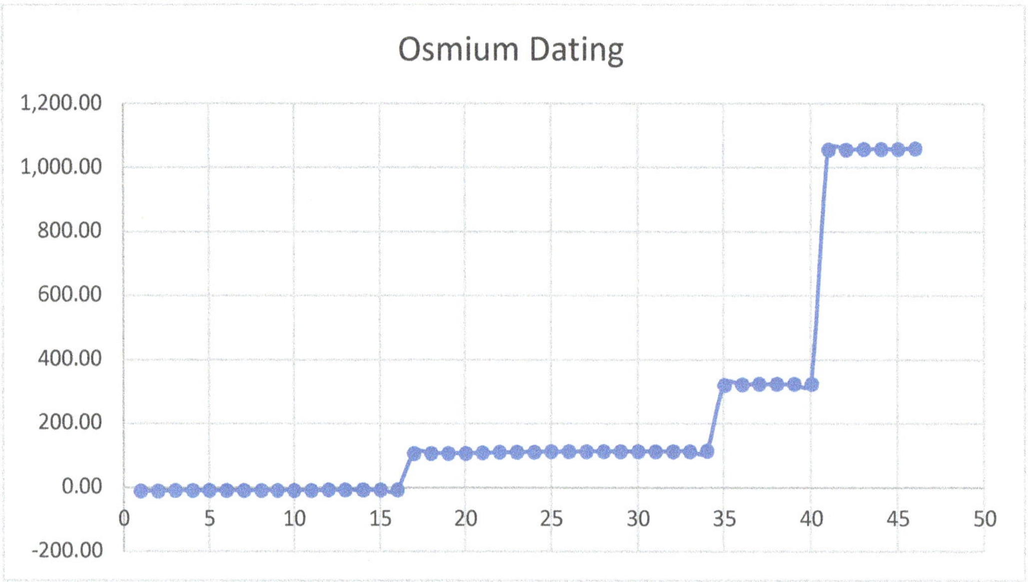

The Sudbury Igneous Complex, Ontario

These rocks from Canada were analysed in 2000 by scientist from the Department of Geology, University of Maryland. [29] According to the article the true age is 1800 million years old. "The ages agree with the canonical value of 1850 Ma for the Sudbury Igneous Complex (SIC). For Hanging Wall and Deep Zone ores at Strathcona, the age of 1780 Ma may reflect resetting by dyke activity." [29] The article contains two tables [30] with Osmium ratios that have no dates beside them. If we put the tables into Microsoft Excel and use the formulas in standard geology text books we get the values in tables 8 and 9. The forty one dates range from -128 billion years old to -2.3 trillion years old. The choice of 1.8 billion years as the true age is just a random guess.

Table 8	Million Years	% Discordance	Difference
Average	-220,265	12,337	222,065
Maximum	-152,828	17,660	317,886
Minimum	-316,086	8,590	154,628

Table 9	Million Years	% Discordance	Difference
Average	-632,140	35,219	633,940
Maximum	-128,289	132,624	2,387,235
Minimum	-2,385,435	7,227	130,089

187Os/186Os in Oceanic Island Basalts

These rocks from various islands were analysed in 1994 by scientists from Paris. [31] According to the article the true age for the samples varies from historic volcanic eruptions to eruptions 16 million years old. [32] The article contains two tables [32] with Osmium ratios that have no dates beside them. If we put the tables into Microsoft Excel and use the formulas in standard geology text books we get the values in tables 10 and 11. The so called true age for the samples ranges from 100 years old to 16 million years old. The dates obtained from the sample ratios ranges from 827 million years old to 10.7 billion years old. The choice of zero to 16 million years old as the true age is just a random guess. The ratios from the second table give ages between -800 billion and -3.5 trillion years old!

Table 10

Model Age Million Years	Model Age % Difference	Model Age Difference
0.0001	1,634,888,118	1,635
0.0001	1,497,005,988	1,497
0.0001	1,319,728,963	1,320
0.0001	1,378,821,305	1,379
0.15	853,556	1,280
1	94,548	944
1	86,669	866
0.55	175,486	965
9	38,738	3,477
16	67,094	10,719
0.0001	827,292,783	827
0.0001	846,990,230	847
4.5	39,395	1,768
4.5	37,644	1,689
0.35	489,622	1,713
0.0001	3,762,212,417	3,762
Average	704,301,410	2,168
Maximum	3,762,212,417	10,719
Minimum	37,644	827

Table 11	Billion Years	Million Years
Average	-797.944	-797,944
Maximum	-0.197	-197
Minimum	-3,564.450	-3,564,450

186Os/188Os and 187Os/188Os Measurements (Part 1)

These rocks were analysed in 2007 by scientist from the University of Durham. [33] According to the article the true age is 600 million years old. [27] The article contains a table [34] with Osmium 187/186 ratios that have no dates beside them. If we put the tables into Microsoft Excel and use the formulas in standard geology text books we get the values in table 12. Out of the twenty one dates, there is a 9.8 billion year range between the youngest [**-8,074**] and oldest [**1,783**] dates. The choice of 600 million years as the true age is just a random guess.

Table 12	Million Years	% Discordance	Difference
Average	-3,380.88	899	5,677
Maximum	1,783.58	1,446	8,675
Minimum	-8,074.99	297	2,379

The Marine 187Os/186Os Record

These rocks were analysed in 1994 by scientist from the Max-Planck-Institute, Mainz, Germany. [35] According to the article the true age is 80 million years old. [36] The article contains a table [37] with Osmium 187/186 ratios that have no dates beside them. If we put the tables into Microsoft Excel and use the formulas in standard geology text books we get the values in table 13. Out of the twenty one dates, there is a 135 billion year range between the youngest [-3,821] and oldest [-139,459] dates. The choice of 80 million years as the true age is just a random guess.

Table 13	Million Years	% Discordance	Difference
Average	-59,648	74,659	59,568
Maximum	-3,821	174,422	139,378
Minimum	-139,458	4,877	3,901

Re-Os Isotope Systematics in Black Shales

These rocks from the Himalayas were analysed in 1999 by scientists from Physical Research Laboratory, in India. [38] According to the article the true age for the samples is 550 million years old. [38] The article contains two tables [39] with Osmium ratios that have no dates beside them. If we put the tables into Microsoft Excel and use the formulas in standard geology text books we get the values in tables 14 and 15. The so called true age for the samples is 600 million years old. The thirty dates obtained from the sample ratios ranges from -264 billion years old to -1.87 trillion years old. The choice of 600 million years old as the true age is just a random guess.

Table 14	Million Years	% Discordance	Difference
Average	-497,072	82,945	497,672
Maximum	-146,254	312,170	1,873,020
Minimum	-1,872,420	24,476	146,854

Table 15	Million Years	% Discordance	Difference
Average	-435,973	72,762	436,573
Maximum	-264,143	185,611	1,113,663
Minimum	-1,113,063	44,124	264,743

A Metamorphosed Early Cambrian Crust

These rocks from the Eastern Austrian Alps, were analysed in 2002 by scientists from Germany and Austria. [40] According to the article the true age for the samples is 600 million years old. [41] The article contains a table [42] with Osmium ratios that have no dates beside them. If we put the tables into Microsoft Excel and use the formulas in standard geology text books we get the values in table 16. The so called true age for the samples is 600 million years old. The thirty three dates obtained from the sample ratios ranges from 1 billion years old to -710 billion years old. The choice of 600 million years old as the true age is just a random guess.

Table 16	Million Years	% Discordance	Difference
Average	-144,811	24,264	145,441
Maximum	1,091	118,428	710,565
Minimum	-709,965	137	131

Cameroon Volcanic Line Lavas

These rocks from the Cameroon in Africa, were analysed in 2002 by scientists from Germany. [43] According to the article the lava deposits formed in the Cenozoic Era making the so called true age for the samples 60 million years old. [43] The article contains two tables [44] with Osmium and Lead 207/206 ratios that have no dates beside them. If we put the Osmium tables into Microsoft Excel and use the formulas in standard geology text books we get the values in table 17. The so called true age for the samples is 60 million years old. The nineteen dates obtained from the sample ratios ranges from -289 million years old to 19.6 billion years old. The forty nine Lead 207/206 ratios give dates between 4800 and 5000 billion years old. The choice of 60 million years old as the true age is just a random guess.

Table 17	187Os/188Os	207Pb/206Pb	Difference
Average	-2,852	4,899	7,751
Maximum	289	4,959	4,670
Minimum	-19,613	4,837	24,450

Lens with Sub-Baltic Shield

These rocks from Sweden, were analysed in 2002 by scientists from Queens College, New York. [45] According to the article the so called true age for the samples 450 million years old. [45] The article contains a table [46] with Osmium ratios that have no dates beside them. If we put the Osmium tables into Microsoft Excel and use the formulas in standard geology text books we get the values in table 18 and chart 4. The so called true age for the samples is 450 million years old. The forty dates obtained from the sample ratios a 42 billion year range from 1,205 million years old to -40,956 million years old.

"Minimum model ages (TRD) assuming that Re addition occurred either at 450Ma or more recently (i.e. today) yield meaningless future ages in almost all cases. Model ages (TMA) that assume Re was present at the time of sulphide formation are also scattered and meaningless for most samples." [47] Seventy nine calculated dates [Chart 3] actually listed in the article [46] are between 34 billion and -58 billion years old. There is an 82 billion year age difference between the youngest and oldest dates. Forty nine dates [62%] are impossible negative or future ages. Twenty three dates [29%] are over 4.6 billion years old. Twenty dates [25%] are over 5 billion years old. Nine dates [11%] are over 11 billion years old. You can see the random spread in chart 3. The choice of 450 million years old as the true age is just a random guess.

Table 18	Million Years	% Discordance	Difference
Average	-10,204	2,381	10,692
Maximum	1,205	9,201	41,406
Minimum	-40,956	170	755

Chart 3

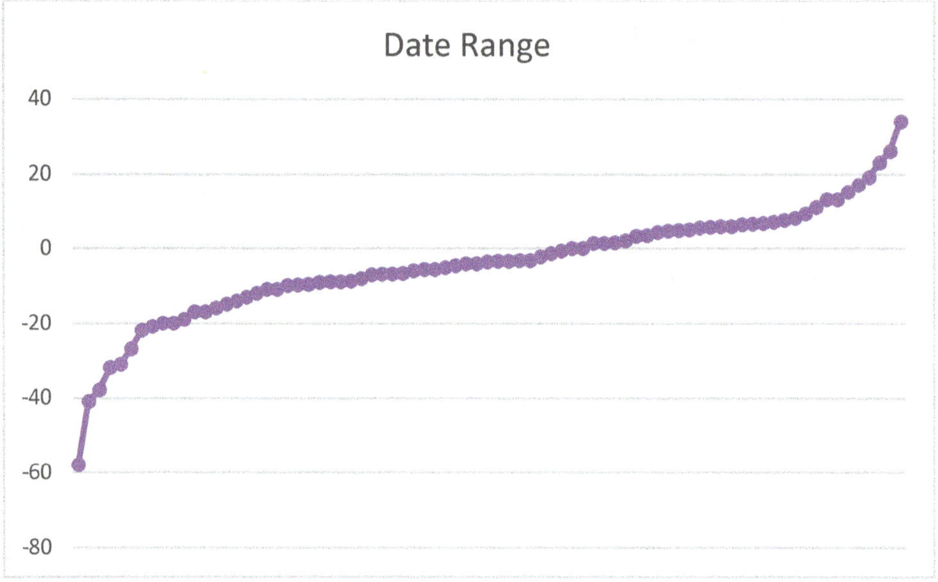

The Beni Bousera Peridotite Massif

These rocks from Morocco, were analysed in 2003 by scientists from Durham University. [48] According to the article the so called true age for the samples are between 540 and 4,000 million years old. [49] The article contains a table [49] with Osmium ratios that have no dates beside them. If we put the Osmium tables into Microsoft Excel and use the formulas in standard geology text books we get the values in table 19. The nine dates obtained from the sample ratios have a 12.867 trillion year range from -9.3 billion years old to -12.876 trillion years old.

Table 19	Million Years	% Discordance	Difference
Average	-2,027,093	313,374	2,028,605
Maximum	-9,302	2,384,647	12,877,095
Minimum	-12,876,555	1,488	9,972

Chromite Deposits Of the Ipueira

These rocks from Brazil, were analysed in 2002 by scientists from Brazil. [50] According to the article the so called true age for the samples are 2,000 million years old. [50] The article contains a table [51] with Osmium ratios that have no dates beside them. If we put the Osmium tables into Microsoft Excel and use the formulas in standard geology text books we get the values in table 20. The eleven dates obtained from the sample ratios have an 48,294 million year range from 2,662 million years old to -50,956 million years old. The choice of 2,000 million years old as the true age is just a random guess.

Table 20	Million Years	% Discordance	Difference
Average	-5,936	531	13,635
Maximum	2,662	2,648	52,956
Minimum	-50,956	151	3,010

Origin of Paleoproterozoic Komatiites

These rocks from Finnish Lapland, were analysed in 2003 by scientists from the Department Of Geology, University Of Maryland. [52] According to the article the so called true age for the samples are 2,000 million years old. [53] The article contains a table [54] with Osmium ratios that have no dates beside them. If we put the Osmium tables into Microsoft Excel and use the formulas in standard geology text books we get the values in table 21. The thirty five dates obtained from the sample ratios have an 11.07 trillion year range from 1,922 million years old to -11,068,187 million years old. The choice of 2,000 million years old as the true age is just a random guess.

Table 21	Million Years	% Discordance	Difference
Average	-487,016	24,478	489,016
Maximum	1,922	553,509	11,070,187
Minimum	-11,068,187	71	78

Evidence from 2.8 Ga Komatiites

These rocks from Kostomuksha on the Russian Finland border were analysed in 1999 by scientists from The University of Chicago. [55] The model age for the samples is 2,800 million years old. [55] The article contains a table [56] with Osmium ratios that have no dates beside them. If we put the Osmium tables into Microsoft Excel and use the formulas in standard geology text books we get fourteen dates. None are as old as the so called model age. There is a 21,701 million year range between the youngest and oldest dates. The choice 2,800 million years as the true age is just a random guess.

Table 22	Million Years	% Discordance	Difference
Average	-8,333	316	-8,978
Maximum	2,215	696	4,582
Minimum	-19,486	1	-22,286

187Os Isotopic Constraints

These rocks from Zimbabwe were analysed in 2001 by scientists from the Department of Geology, University of Maryland. [57] The model age for the samples is between 790 and 3,260 million years old. [58] The article contains a table [58] with Osmium ratios that have no dates beside them. If we put the Osmium tables into Microsoft Excel and use the formulas in standard geology text books we get the values in table 23. Out of the twenty three dates ten [43%] are impossible future or negative ages. Thirteen [56%] are over 6 billion years old. Nine [34%] are over 12 billion years old. The choice of the true age is just a random guess.

Table 23	Million Years	% Discordance	Difference
Average	-54,043	2,900	56,272
Maximum	2,372	25,627	458,724
Minimum	-456,934	171	48

References

1 Geochimica et Cosmochimica Acta, 1999, Volume 63, Number 5, Pages 713-728, Evidence from Gorgona Island and Curacao
2 Reference 1, page 714
3 Reference 2, page 716
4 Reference 3, page 717
5 Geochimica et Cosmochimica Acta, 2002, Volume 66, Number 14, Pages 2615–2630, Rocks from Southern West Greenland
6 Reference 5, page 2620
7 Reference 5, page 2621-2622
8 Earth and Planetary Science Letters, Volume 230 (2005), Pages 211– 226, 187Os–186Os systematics of Os–Ir–Ru
9 Reference 8, page 221
10 Journal Geophysical Research, Volume 99, 1994, Pages 4293–4321. http://onlinelibrary.wiley.com/doi/10.1029/93JB02061/abstract

11 Principles of Isotope Geology, Second Edition, By Gunter Faure, Published By John Wiley And Sons, New York, 1986, Page 269

12 Introduction to Geochemistry: Principles and Applications, Page 241 By Kula C. Misra, Wiley-Blackwell Publishers, 2012 http://books.google.com.au/books?id=ukOpssF7zrIC&printsec=frontcover

13 Radioactive and Stable Isotope Geology, Issue 3 By H. G. Attendorn, Robert Bowen, Page 298 Chapman and Hall Publishers, London, 1997 http://books.google.com.au/books?id=-bzb_XU7OdAC&printsec=frontcover

14 http://www.geo.cornell.edu/geology/classes/Geo656/656notes03/656%2003Lecture11.pdf

15 Talanta, 2001, Volume 55, Pages 815–820, Determination of 187Os in molybdenite

16 Reference 15, page 819

17 Geochimica et Cosmochimica Acta, Volume 75 (2011) Pages 4456–4475, 186Os–187Os systematics of Hawaiian Picrites

18 Reference 17, page 4467

19 Reference 17, pages 4459, 4460, 4467, 4471

20 Geochimica et Cosmochimica Acta, Volume 71 (2007) Pages 4570–4591, Evidence from Icelandic picrites

21 Reference 20, pages 4587

22 Reference 20, pages 4574, 4581

23 Earth and Planetary Science Letters, Volume 174 (1999) Pages 25-42, 186Os/187Os systematics of Hawaiian picrites

24 Reference 23, pages 35

25 Reference 23, pages 28, 29

26 Chemical Geology, 2008, Volume 248, Pages 394–426, Accurate 186Os/188Os and 187Os/188Os measurements

27 Reference 26, pages 404

28 Reference 26, pages 398, 410, 411

29 Geochimica et Cosmochimica Acta, Volume 66 (2002) Number 2, Pages 273-290, The Sudbury Igneous Complex, Ontario

30 Reference 29, pages 278, 279

31 Earth and Planetary Science Letters, Volume 129 (1995) Pages 145-161, 187Os/186Os in oceanic island basalts

32 Reference 31, pages 146, 154

33 Chemical Geology, Volume 248 (2008) Pages 363–393, 186Os/188Os and 187Os/188Os Measurements (Part 1)

34 Reference 33, page 380

35 Earth and Planetary Science Letters, Volume 130 (1995) Pages 155-167, The Marine 187Os/186Os Record

36 Reference 35, page 155

37 Reference 35, page 156

38 Geochimica et Cosmochimica Acta, Volume 63 (1999) No. 16, Pages 2381-2392, Re-Os isotope systematics in black shales

39 Reference 38, page 2384, 2385

40 Journal Of Petrology, 2004, Volume 45, Number 8, Pages 1689-1723, A Metamorphosed Early Cambrian Crust

41 Reference 40, page 1717

42 Reference 40, page 1708

43 Journal Of Petrology, 2005, Volume 46, Number 1, Pages 169-190, Cameroon Volcanic Line Lavas

44 Reference 43, page 177, 178

45 Journal Of Petrology, 2004, Volume 45, Number 2, Pages 415-437, Lens with Sub-Baltic Shield

46 Reference 45, page 432

47 Reference 45, page 431

48 Journal Of Petrology, 2004, Volume 45, Number 2, Pages 439–455, The Beni Bousera Peridotite Massif

49 Reference 48, page 449

50 Journal Of Petrology, 2003, Volume 44, Number 4, Pages 659–678, Chromite Deposits Of the Ipueira

51 Reference 50, page 667

52 Journal Of Petrology, 2006, Volume 47, Number 4, Pages 773-789, Origin of Paleoproterozoic Komatiites

53 Reference 52, page 773

54 Reference 52, page 778

55 Earth and Planetary Science Letters, 2001, Volume 186, Pages 513-526, Evidence from 2.8 Ga Komatiites

56 Reference 55, page 516

57 Geochimica et Cosmochimica Acta, 2002, Volume 66, Number 18, Pages 3317-3325, 187Os isotopic constraints

58 Reference 57, page 3318

Chapter 15

The Potassium Argon Dating Method

By Paul Nethercott

April 2014

The Long Valley Rhyolitic

These rocks from California were dated [1] in 1997 using the Rubidium/Strontium and Potassium/Argon method. The rock samples gave ages between 1 million years and negative years old! Since the Earth exists in the present how can rocks have formed in the future? The author admits some of the dates are negative:

"The negative ages are a clear indication that some phases have not reached Sr isotope equilibration with their current host glass." [2]

"In contrast, feldspars from the second group yield mineral ages that are geologically unreasonable ranging from close to the eruption age of the Bishop Tuff to negative ages." [3]

Rhenium-187/Osmium-187 In Iron Meteorites

The ^{187}Rhenium/^{187}Osmium method and Potassium-Argon method were used to date these meteorite [4] fragments in 1997. Four of the dates were older than the Solar System and two were older than the Galaxy. [5]

Table 1

Canyon Diablo Meteorite	Billion Years
Leach Acetone	5.73
Leach H_2O	8.31
Troilite dissolved	10.43
Metal 1	13.7

Geochemistry of Hornblende Gabbros

These rock samples from Sonidzuoqi (Inner Mongolia, North China) were dated in 2008 by scientist from the Chinese Academy of Sciences, Beijing using the Potassium/Argon and Uranium/Lead age dating. [6] The true age of the rock formation is supposed to be 500 million years old. "Limited hornblende K–Ar and SHRIMP U–Pb zircon ages document the Late Silurian to Early Devonian gabbroic emplacement." [6] "The Siluro-Devonian hornblende gabbros, together with a pre-490 Ma ophiolitic melange of MORB-OIB affinity, 483–471 Ma arc intrusions, 498–461 Ma trondhjemite-tonalite-granodiorite plutons, and 427–423 Ma calc-alkaline granites from the same area." [6] The article contains a table [6] that has twenty eight ratios that have no dates beside them. Out of the twenty eight dates we calculated from these ratios there is a total disagreement with the so called 'true age.' Whichever date you choose for each meteorite as the true one is just a random guess.

Table 2	207Pb/206Pb	206Pb/238U	207Pb/235U	208Pb/232Th
Average	5,011	6,612	5,422	22,967
Maximum	5,014	7,297	5,648	24,397
Minimum	5,007	5,922	5,237	20,621

Potassium Argon Dating of Iron Meteorites

This article summarized meteorite dating in 1967. [7] Even 40 years later things are no better. In the opening paragraph he states that the iron meteorite from Weekeroo Station is date at ten billion years old. He then continues: "The formation or solidification ages of iron meteorites have never been well determined." [8] He then cites earlier dating which produced an age of seven billion years. [9] The author concludes with the following remark: "The ages found by us are typical of the great ages found for most iron meteorites. From these, in conjunction with the Strontium: Rubidium data of Wasserburg et al. on silicate inclusions in this meteorite, we conclude that the potassium: argon dating technique as applied to iron meteorites gives unreliable results." [10]

Table 3

Meteorite	Age
Sample	Billion Years
Neutron Activation	10.0
Stoenner and Zahringer	7.0
Muller and Ziihringer's	6.3
Wasserburg, Burnett	4.7
K-1	8.5
K-2	9.3
B-1	6.5
G-1	10.4

Potassium-Argon age of Iron Meteorites

If we compare the dates below with the previous two tables [Tables 1 and 2] we see that dating done on meteorites has not improved in fifty years! The dates below [Table 4] were dating done in 1958 by scientists from Brookhaven National Laboratory, Upton, New York. [11] These dates [12] are just as stupid as the previous two tables. The choice of 4.5 billion years as an "absolute" value is purely and arbitrary choice.

Table 4

Meteorite	Age
K-Ar Dating	Billion Years
Mt. Ayliff	6.9
Arispe	6.8
H. H. Ninninger	6.9
Carbo	8.4
Canon Diablo I	8.5
Canon Diablo I	6.9
Canon Diablo I	6.6
Canon Diablo I	5.3
Canon Diablo II	13
Canon Diablo II	11
Canon Diablo II	10.5
Canon Diablo II	12
Toluca I	5.9
Toluca I	7.1
Toluca II	10
Toluca II	10.8
Toluca II	8.8

Fission-Track Ages Of Four Meteorites

Six different meteorites were dated in 1976 by scientists from the Enrico Fermi Institute and Department of Chemistry, University of Chicago, Chicago, Illinois. [13] The dates [Table 5] varied by almost one thousand percent! [63] If we look at table 5 we can see the four methods used [Fission Track, Potassium-Argon, Uranium-Helium and Rubidium-Strontium] and the discordance between them. [14]

Table 5

Sample Name	Maximum Age Billion Years	Minimum Age Billion Years	Age Difference Billion Years	Percent Difference
Bondoc	1.30	0.14	1.16	929%
Mincy	3.93	1.50	2.43	262%
Nakhla	4.40	0.77	3.63	571%
Serra	2.70	0.54	2.16	500%
Washougal	4.60	4.00	0.60	115%
Allende	4.50	3.60	0.90	125%

Table 6

Meteorite Name	Fission Track Billion Years	K-Ar Billion Years	U-He Billion Years	Rb-Sr Billion Years
Bondoc	0.14	1.30	0.60	
Mincy	1.50	3.93		
Nakhla	4.40	1.30	0.77	3.60
Serra	0.54	2.70		
Washougal	4.60	4.00		
Allende	4.50	4.40		3.60

Ion Microprobe U-Pb Dating

These rocks from Japan were dated [15] in 2001 using the Rubidium/Strontium and Potassium/Argon method. If we run the isotopic ratios through Isoplot [16] and use formulas listed in standard geology books [17] we find that the rock samples [18] gave ages between 5 billion years and negative years old! Since the Earth exists in the present how can rocks have formed in the future? How can a rock be older than the Earth? The author admits some of the dates are negative: "Though a negative age has no practical use, it does suggest that it is younger than 0.12 Ma." [19]

Table 7 Data	Age 206Pb/238U	Age 207Pb/206Pb	Age Ratio
Average	62	4,710	76
Maximum	631	5,135	8
Minimum	0	3,771	3771

Table 8 Data	Age 206Pb/238U	Age 207Pb/206Pb	Age Ratio
Average	0.88	4,742	5,388
Maximum	2.91	4,978	1,710
Minimum	0.25	4,479	17,916

References

1 The Long Valley Rhyolitic, Geochimica et Cosmochimica Acta, 1998, Volume 62, Number 21/22, Pages 3561-3574

2 Reference 2, page 3567

3 Reference 2, page 3569

4 J. L. Birck, Rhenium-187/Osmium-187 in iron meteorites, Meteoritics And Planetary Science, 1998, Volume 33, Pages 641-453

5 Reference 4, Page 649

6 Geochemistry of hornblende gabbros, International Geology Review, 2009, Volume 51, Number 4, Pages 345, 361

7 L. Rancitelli, Potassium: Argon Dating of Iron Meteorites, Science, 1967, Volume 155, Pages 999 - 1000

8 Reference 7, Page 999

9 R. W. Stoenner and J. Zahringer, Geochimica et Cosmochimica Acta, 1958, Volume 15, Page 40.

10 Reference 7, Page 1000

11 R. W. Stoenner, Potassium-argon age of iron meteorites, Geochemica Et Cosmochemica Acta, 1958, Volume 15, Pages 40 – 50

12 Reference 11, Pages 45 to 46

13 Eugene A. Carver, Fission-track ages of four meteorites, Geochemica Et Cosmochemica Acta, 1976, Volume 40, Pages 467 - 477

14 Reference 13, Page 475

15 Ion Microprobe U-Pb Dating, Journal of Volcanology and Geothermal Research, Volume 117, 2002, Pages 285-296

16 http://www.creationismonline.com/Isoplot/Isoplot.html
https://www.bgc.org/isoplot

17 Principles of Isotope Geology, Second Edition, By Gunter Faure, Published By John Wiley And Sons, New York, 1986. Pages 120 [Rb/Sr], 205 [Nd/Sm], 252 [Lu/Hf], 266 [Re/OS], 269 [Os/OS].

18 Reference 15, page 288, 290

19 Reference 15, page 291

Chapter 16

The Rhenium-Osmium Dating Method

Versus The Osmium 188/187 Method

By Paul Nethercott

May 2014

Pt-Re-Os Systematics

These Iron meteorites were dated in 2003 by scientist from the University of California using the Rhenium/Osmium dating methods. [1] According to the article the true age is based on Re/Os method is 0000 million years old. "The Re-Os isochron ages for the complete suites of IIAB and IIIAB irons are 4,530 +/- 50 Ma and 4,517 +/- 32 Ma, respectively, and are similar to previously reported Re-Os ages for the lower-Ni end members of these two groups. Both isochrons are consistent with, but do not require crystallization of the entire groups within 10-30 Ma of the initiation of crystallization." [1] The article contains a table with Osmium 188/187 ratios that have no dates beside them. [2] If we put the tables into Microsoft Excel and use the formula in standard geology text books we can calculate dates from the undated isotopic ratios. There is a 14,763 million year discrepancy between the supposed true age [4,530 million years ago] and the Osmium isotope ratio age [-10,233 million years future]. The article claims that the Rhenium/Osmium dating method is accurate within 50 million years [98.9 %]. [1]

Table 1	Age (Ma)	% Discordance	Difference (Ma)
Average	-3,219	3,485	7,777
Maximum	296	132,615	14,791
Minimum	-10,233	145	4,262
Difference	10,529	132,470	10,529

Re-Os, and Mo Isotope Systematics

These black shales from the Barberton Greenstone Belt, South Africa were dated in 2004 by scientist from the University Of Berne, Switzerland using the Rhenium/Osmium dating methods. [3] According to the article the true age is based on Rhenium/Osmium method is 3,250 million years old. "Re-Os data and PGE concentrations as well as Mo concentrations and isotope data are reported for suites of fine clastic sediments and black shales from the Barberton Greenstone Belt, South Africa (Fig Tree and Moodies Groups, 3.25–3.15 Ga), the Belingwe Greenstone Belt, Zimbabwe (Manjeri Formation, ca. 2.7 Ga) and shales from the Witwatersrand, Ventersdorp and Transvaal Supergroups, South Africa ranging from 2.95 to 2.2 Ga." [3] The article contains a table with Osmium 188/187 ratios that have no dates beside them. [4] If we put the tables into Microsoft Excel and use the formula in standard geology text books we can calculate dates from the undated isotopic ratios. There is a 2,413,235 million year [2.4 trillion year] discrepancy between the supposed true age [3,250 million years ago] and the Osmium isotope ratio age [2,409,985 million years future].

Table 2	Age (Ma)	% Discordance	Difference (Ma)
Average	-236,564	8,674	239,572
Maximum	-23,132	89,359	2,412,685
Minimum	-2,409,985	812	26,382
Difference	2,386,853	88,547	2,386,303

Evolution of the South China block

These mineral samples from Taiwan were dated in 2008 by scientist from the Macquarie University, Sydney using the Rhenium/Osmium dating methods. [5] According to the article the true age is based on several dating methods is 1,000 million years old. "Such sulphides yield *T*RD age peaks of 1.9, 1.7–1.6, 1.4–1.3 and 0.9–0.8 Ga, which may record the timing of melt extraction and/or metasomatic events in the mantle. These periods are contemporaneous with the major crustal events recorded by U–Pb dates and Nd and Hf model ages in the overlying crust." [5] The article contains two tables with Osmium 188/187 ratios that have no 188/187 dates beside them. [6] If we put the tables into Microsoft Excel and use the formula in standard geology text books we can calculate dates [Table 3] from the undated isotopic ratios. There is a 54,000 million year discrepancy between the supposed true age [1,000 million years ago] and the Osmium isotope ratio age [53,129 million years future]. The second table contains Rhenium depletion ages. These dates are summarized in table 4. There is a 116-billion-year difference between the youngest [-90 billion] and the oldest [25.9 billion] dates. The author's choice of true age is just a random guess.

Table 3	Series A	Series B
Average	5,317	731
Maximum	20,476	3,120
Minimum	-53,129	-3,754
Difference	73,605	6,874

Table 4	Million Years	Million Years	Million Years	Million Years
Average	1,023	395	957	-249
Maximum	3,100	25,900	3,100	10,700
Minimum	-2,900	-59,500	-3,200	-90,000
Difference	6,000	85,400	6,300	100,700

If we run the isotopic ratios give in standard geology magazines through the computer program Isoplot [7] we find that the Uranium/Thorium/Lead isotopic ratios in the rocks disagree radically with the Rhenium-Osmium ages. The U/Th/Pb ratios give ages older than the evolutionist age of the Earth, Solar System, Galaxy and Universe. How can Earth rocks be dated as being older than the Big Bang?

If we use isotopic formulas given in standard geology text we can arrive at ages from the Osmium 188/187 and 187/186 ratios. Here are examples of isotopic ratios taken from several articles in major geology magazines which give absolutely absurd dates. The article contains tables with Osmium 187/186 ratios that have no dates beside them. If we put the tables into Microsoft Excel and use the formula below used in standard geology text books [8-11] we can calculate dates from the undated isotopic ratios.

(1)

$$t = \frac{1.04 - (^{187}Os/^{186}Os)}{0.050768}$$

In the above formula, t = billions of years. The same date can be calculated from the Osmium 187/188 ratios. If we use another formula [12] we can convert the Osmium 187/188 ratio to the Osmium 187/186 ratio.

(2)

$$\frac{^{187}Os}{^{186}Os} \times 0.12035 = \frac{^{187}Os}{^{188}Os}$$

(3)

$$\frac{{}^{187}Os}{{}^{186}Os} = \frac{({}^{187}Os \div {}^{188}Os)}{0.12035}$$

(4)

$$t = \frac{1.04 - \left(\frac{({}^{187}Os \div {}^{188}Os)}{0.12035}\right)}{0.050768}$$

Isotopic Compositions Of Mantle Xenoliths

These rocks from North and Central America, Europe, southern Africa, Asia, and the Pacific region were dated in 1999 by scientist from the Department of Geology, University of Maryland using the Rhenium/Osmium dating methods. [13] According to the article the true age is based on Rhenium depletion model is between 1,550 and 1,750 million years old. [14] The article contains a table with Osmium 187/188 ratios that have no dates beside them. [15] If we put the tables into Microsoft Excel and use the formula below used in standard geology textbooks we can calculate dates from the undated isotopic ratios.

Table 5	SW USA	Mexico	Europe	Asia
Average	-90	-105	41	24
Maximum	1,336	431	1,168	1,130
Minimum	-754	-1,127	-1,386	-513
Difference	2,090	1,558	2,553	1,643
Model Age	1,550	1,750	1,620	1,580

The Origin Of Gold And Sulfides

These rocks from the Witwatersrand basin in South Africa were dated in 2000 by scientist from the University of Arizona and the CSIRO using the Rhenium/Osmium dating methods. [16] According to the article the true age is based on Rhenium depletion model is 3,300 million years old. "Rhenium depletion ages (TRD) range from 3.5 Ga to 2.9 Ga, with a median age of 3.3 Ga." [17] The article contains a table with Osmium 187/188 ratios that have no dates beside them. [18] If we put the tables into Microsoft Excel and use the formula below used in standard geology text books we can calculate dates from the undated isotopic ratios. There is a 12,766 million year discrepancy between the supposed true age [3,300 million years ago] and the Osmium isotope ratio age [9,466 million years future]. Column one has the Osmium isotope ratio age. Column two gives the percentage discordance between the model age [3,300] and column one. Column three gives the difference [million years] between the model age [3,300] and column one.

Table 6	Age (Ma)	% Discordance	Difference (Ma)
Average	-8,450	356	11,750
Maximum	-7,600	387	12,766
Minimum	-9,466	330	10,900
Difference	1,866	57	1,866

Diagram 1

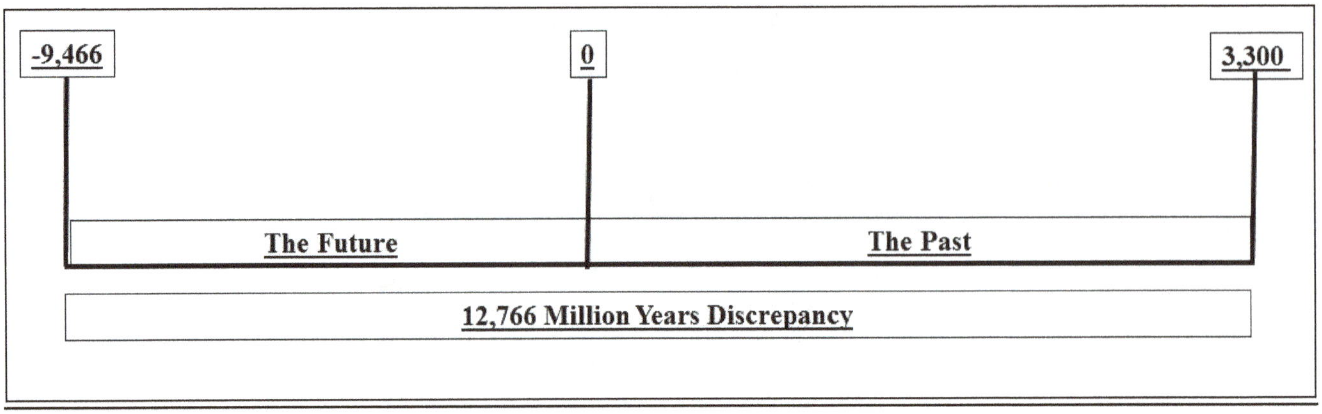

Rhenium-Osmium Systematics

These meteorites were dated in 2000 by scientist from the Department of Geology, University of Maryland using the Rhenium/Osmium dating methods. [19] According to the article the true age is based on Rhenium depletion model and 206Pb-207Pb method is 4,558 million years old. "An age of 4,558 Ma is assumed for the IIIA iron meteorites based on 53Mn-53Cr similarities between angrite meteorites and IIIA irons." [20] "The inferred IIIA age is only slightly younger than the oldest solar system objects known, Ca-Al-rich inclusions (CAIs) from the Allende meteorite, dated at 4,566 Ma, using the 206Pb-207Pb method." [20] The article contains a table with Osmium 187/188 ratios that have no dates beside them. [21] If we put the tables into Microsoft Excel and use the formula below used in standard geology text books we can calculate dates from the undated isotopic ratios. There is a 6,610 million year discrepancy between the supposed true age [4,558 million years ago] and the Osmium isotope ratio age [2,052 million years future]. The article claims that the Rhenium/Osmium dating method is 99.8% accurate: "The 187Re-187Os decay system potentially provides a unique chronometer to obtain absolute age constraints on processes that affected highly siderophile elements (HSE) during early solar system processing. Precise Re-Os ages (0.2– 0.6%, error) obtained on various groups of iron meteorites likely reflect system closure subsequent to metal crystallization in asteroidal cores." [20]

Table 7	Age (Ma)	% Discordance	Difference (Ma)
Average	-393	21,186	4,951
Maximum	624	696,126	6,610
Minimum	-2,052	322	3,934
Difference	2,676	695,804	2,676

Diagram 2

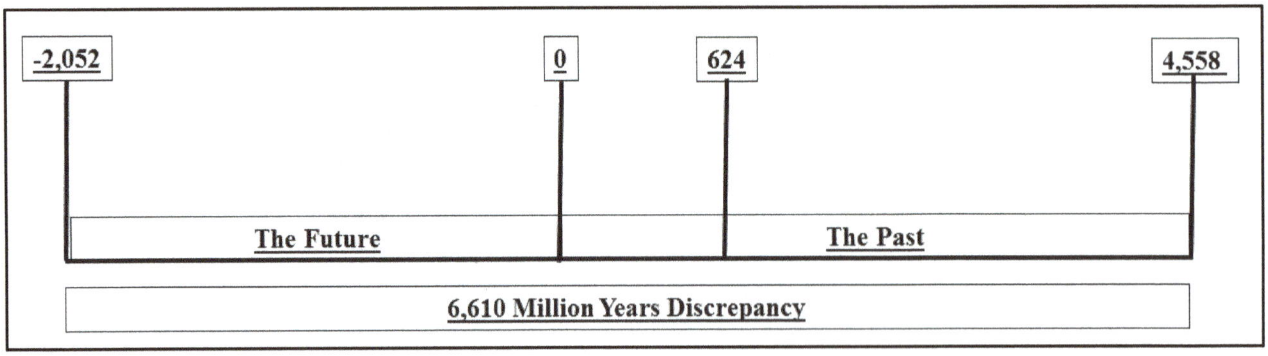

190Pt-186Os and 187Re-187Os Systematics

These sulphide ores from the Sudbury Igneous Complex, Ontario were dated in 2000 by scientist from the Colorado State University using the Rhenium/Osmium dating methods. [22] According to the article the true age is 1,850 million years old. "At McCreedy West and Falconbridge, the isochron Re–Os ages are 1835 Ma and 1827 Ma, and the initial 187Os/188Os ratios 0.514 and 0.550, respectively. The ages agree with the canonical value of 1850 Ma for the Sudbury Igneous Complex (SIC). For Hanging wall and Deep Zone ores at Strathcona, the age of 1780 Ma may reflect resetting by dyke activity." [22] The article contains a table with Osmium 188/187 ratios that have no dates beside them. [23] If we put the tables into Microsoft Excel and use the formula in standard geology textbooks we can calculate dates from the undated isotopic ratios. There is a 2.2 trillion-year discrepancy between the supposed true age [1,850 million years ago] and the Osmium isotope ratio age [2,257 million years future]. The article claims that the 1850 Ma date method is at most one million years in error [21] [99.95% accurate] but the error level obtained from the undated ratios gives an error level of 122,007,880%! This means that their calculation of the maximum error level is 1,220,689 times too small.

Table 8	Age (Ma)	% Discordance	Difference (Ma)
Average	-632,140	34,270	633,990
Maximum	-128,289	129,042	2,387,285
Minimum	-2,385,435	7,035	130,139
Difference	-2,257,146	122,008	2,257,146

Behavior of Re and Os

These soil samples from the Himalayas were dated in 2001 by scientist from the Centre for Geochemical Research in Notre-Dame, France using the Rhenium/Osmium dating methods. [24] According to the article the true age is based on the Rhenium/Osmium method is 840 million years old. [25] The author admits that many dates are impossible and the true age is just a guess:

"This apparent Re loss is confirmed by the impossibly high Re/Os model ages of nearly all of the soils, most of which exceed the age of the earth." [25]

"The median model age of the soils (10 Ga) is much higher than those of typical HHC and LH rocks, indicating that the soils have in general suffered much more extensive recent Re loss." [25]

"These soils display very radiogenic Os isotopic ratios that cannot be explained by their 187Re/188Os ratios, which imply impossible model ages (11–13.5 Ga for MO 601; 16–23 Ga for MO 602, and 6.1 Ga for saprolite MO 600). [25]

The article contains a table with Osmium 188/187 ratios that have no dates beside them. [26] If we put the tables into Microsoft Excel and use the formula below used in standard geology text books we can calculate dates from the undated isotopic ratios. There is a 1.86 trillion year discrepancy between the oldest model age [331,800 million years ago] and the Osmium isotope ratio age [1,528,332 million years future].

Table 9	Age (Ma)	% Discordance	Difference (Ma)
Average	-760,654	11,810	768,700
Maximum	-154,967	35,287	1,696,333
Minimum	-1,683,299	1,664	156,367
Difference	-1,528,332	33,623	1,539,967

Table 10

Model Age (Ma)	Model Age (Ma)	Model Age (Ma)	Model Age (Ma)
331,800	19,800	17,600	12,900
68,200	19,500	15,600	11,500
22,200	19,100	13,400	10,600

187Os Isotopic Constraints

These Lava flows from Belingwe, Zimbabwe were dated in 2001 by scientist from the University of Maryland and the University of London using the Rhenium/Osmium dating methods. [28] According to the article the true age is based on the Lead 207/206 and Neodymium/Samarium dating methods is 2,720 million years old. "Regression of the data for the mineral concentrates yields an age of 2.721 +- 21 Ga, which is consistent with Pb-Pb and Sm-Nd ages that have been previously reported for the komatiites, and an initial 187Os/188Os ratio of 0.11140" [28] The article contains a table with Osmium 188/187 ratios that have no dates beside them. [29] If we put the tables into Microsoft Excel and use the formula in standard geology text books we can calculate dates from the undated isotopic ratios. There is a 456,586 million year discrepancy between the supposed true age [2,720 million years ago] and the Osmium isotope ratio age [-456,934 million years future]. The article claims that the 2,721 Ma date is only has an error margin of 21 million years [99.559% accurate]. [28] Since there is a 456,586 million year discrepancy between dates the error margin is 21,742 times too small.

Table 11	Age (Ma)	% Discordance	Difference (Ma)
Average	-62,776	2,425	65,476
Maximum	2,372	17,023	459,634
Minimum	-456,934	12	328
Difference	459,306	17,011	459,306

Comparative 187Re-187Os Systematics Of Chondrites

These meteorites were dated in 2002 by scientist from the University of California using the 187Re/187Os dating methods. [30] According to the article the true age is based on 187Re/187Os method is 4,500 million years old. "Chondrites are among the most primitive of solar system materials. Assuming derivation from a reservoir with a uniform initial 187Os/188Os ratio, it would be expected that bulk chondrites should plot very close to the Re-Os isochron defined by the IIIAB irons, which are assumed to have crystallized within 10 to 20 Ma of the inception of the solar system." [31] The article contains a table [32] with Osmium 188/187 ratios that have no dates beside them. If we put the tables into Microsoft Excel and use the formula in standard geology text books we can calculate dates from the undated isotopic ratios. There is a 46,318 million year discrepancy between the supposed true age [4,500 million years ago] and the Osmium isotope ratio age [50,818 million years future].

Table 12	Age (Ma)	% Discordance	Difference (Ma)
Average	-1,422	3,207	5,980
Maximum	878	81,909	55,376
Minimum	-50,818	109	3,680
Difference	51,696	81,800	51,696

Conclusion

Evolutionists Schmitz and Bowring claim that Uranium/Lead dating is 99% accurate. [33] Looking at some of the dating it is obvious that precision is much lacking. The Bible believer who accepts the creation account literally has no problem with such unreliable dating methods. Much of the data used in this dating method is selectively taken to suit and ignores data to the contrary.

Yuri Amelin states in the journal Elements that radiometric dating is extremely accurate: "However, four 238U/235U-corrected CAI dates reported recently (Amelin et al. 2010; Connelly et al. 2012) show excellent agreement, with a total range for the ages of only 0.2 million years – from 4567.18 ± 0.50 Ma to 4567.38 ± 0.31 Ma." [34-36] To come within 0.2 million years out of 4,567.18 million years means an accuracy of 99.99562%. Looking at some of the dating it is obvious that precision is much lacking. The Bible believer who accepts the creation account literally has no problem with such unreliable dating methods. Much of the data in radiometric dating is selectively taken to suit and ignores data to the contrary.

Prominent evolutionist Brent Dalrymple states: "Several events in the formation of the Solar System can be dated with considerable precision." [37] Looking at some of the dating it is obvious that precision is much lacking. He then goes on: "Biblical chronologies are historically important, but their credibility began to erode in the eighteenth and nineteenth centuries when it became apparent to some that it would be more profitable to seek a realistic age for the Earth through observation of nature than through a literal interpretation of parables." [38] The Bible believer who accepts the creation account literally has no problem with such unreliable dating methods. Much of the data in Dalrymple's book is selectively taken to suit and ignores data to the contrary.

References

1 **Pt-Re-Os systematics, Geochimica et Cosmochimica Acta, 2004, Volume 68, Number 6, Pages 1413,**

2 **Reference 1, page 1416**

3 **Re-Os, and Mo isotope systematics, Geochimica et Cosmochimica Acta, 2005, Volume 69, Number 7, Pages 1787,**

4 **Reference 3, page 1792**

5 **Evolution of the South China block, Geochimica et Cosmochimica Acta, 2009, Volume 73, Pages 4531,**

6 **Reference 5, page 4537-4539**

7 http://www.creationismonline.com/Isoplot/Isoplot.html
https://www.bgc.org/isoplot

8 **Principles of Isotope Geology, Second Edition, By Gunter Faure, Published By John Wiley And Sons, New York, 1986. Pages 269.**

9 **Isotopes in the Earth Sciences, By H.G. Attendorn, R. Bowen Chapman And Hall Publishers, London, 1994. Page 289**
http://books.google.com.au/books?id=k90iAnFereYC&printsec=frontcover

10 **Introduction to Geochemistry: Principles and Applications, Page 241**
By Kula C. Misra, Wiley-Blackwell Publishers, 2012
http://books.google.com.au/books?id=ukOpssF7zrIC&printsec=frontcover

11 **Radioactive and Stable Isotope Geology, Issue 3, By H. G. Attendorn, Robert Bowen, Page 298**
Chapman and Hall Publishers, London, 1997
http://books.google.com.au/books?id=-bzb_XU7OdAC&printsec=frontcover

12 **http://www.geo.cornell.edu/geology/classes/Geo656/656notes03/656%2003Lecture11.pdf**

13 **Isotopic compositions of mantle xenoliths, Geochimica et Cosmochimica Acta, 2000, Volume 65, Number 8, Pages 1311–1323**

14 **Reference 13, page 1318**

15 **Reference 13, page 1312**

16 **The origin of gold and sulfides, Geochimica et Cosmochimica Acta, 2001, Volume 65, Number 13, Pages 2149–2159**

17 **Reference 16, page 2149**

18 **Reference 16, page 2153**

19 **Rhenium–osmium systematics, Geochimica et Cosmochimica Acta, 2001, Volume 65, Number 19, Pages 3379–3390**

20 **Reference 19, page 3379**

21 Reference 19, page 3382
22 190Pt–186Os and 187Re–187Os systematics, Geochimica et Cosmochimica Acta, 2002, Volume 66, Number 2, Pages 273–290
23 Reference 22, page 278, 279
24 Behavior of Re and Os, Geochimica et Cosmochimica Acta, 2002, Volume 66, Number 9, Pages 1539,
25 Reference 24, page 1545
26 Reference 24, page 1542
27 Reference 24, page 1542, 1545
28 187Os isotopic constraints, Geochimica et Cosmochimica Acta, 2002, Volume 66, Number 18, Pages 3317,
29 Reference 28, page 3318
30 Comparative 187Re-187Os systematics of chondrites, Geochimica et Cosmochimica Acta, 2002, Volume 66, Number 23, Pages 4187,
31 Reference 30, page 4192
32 Reference 30, page 4190, 4191
33 An assessment of high-precision U-Pb geochronology. Geochimica et Cosmochimica Acta, 2001, Volume 65, Pages 2571-2587
34 Dating the Oldest Rocks in the Solar System, Elements, 2013, Volume 9, Pages 39-44
35 Amelin, Earth and Planetary Science Letters, 2010, Volume 300, Pages 343-350
36 Connelly, Science, 2012, Volume 338, Pages 651-655
37 The Age Of The Earth, By G. Brent Dalrymple, 1991, Stanford University Press, Stanford, California, Page 10.
38 Reference 43, Page 23

Chapter 17

The Rubidium-Strontium Dating Method

By Paul Nethercott, October 2012

Origin Of The Indian Ocean-Type Isotopic Signature

According to the article [1] this rock formation the Philippine Sea plate was dated in 1998 by scientists from Department of Geology, Florida International University, Miami. According to the essay the true age is: "Spreading centers in three basins, the West Philippine Basin (37-60 Ma), the Parece Vela Basin (18-31 Ma), and the Shikoku Basin (17-25 Ma) are extinct, and one, the Mariana Trough (0-6 Ma), is active (Figure 1)." [1] Numerous table and charts affirm this as the true age. [2] Two tables [3] in the essay have isotopic ratios which can be calculated. As we can see below they are all at radical disagreement with each other. There is a spread of dates of almost 100 billion years! None of the Uranium/Lead based dating methods even come vaguely close to the so called true age. The oldest date is 3,971 times older than the youngest date.

Table 1

Dating Summary	Age 87Rb/86Sr	Age 147Sm/144Nd	Age 207Pb/206Pb	Age 206Pb/238U	Age 208Pb/232Th
Average	42	41	4,960	4,260	8,373
Maximum	55	54	4,989	7,093	13,430
Minimum	19	20	4,921	1,904	3,065
Difference	37	33	68	5,188	10,365

U–Th–Pb Dating Of Secondary Minerals

According to the article [4] this rock formation Yucca Mountain, Nevada was dated in 2008 by scientists from United States Geological Survey, Geological Survey of Canada, and the Australian National University. According to the essay the true age is unknown. [5] Other authors have affirmed the same problem. [6] Two tables [7] in the essay have isotopic ratios which can be calculated. As we can see below they are all at radical disagreement with each other. There is a spread of dates of almost 353 billion years! None of the Uranium/Lead based dating methods even come vaguely close to the so called true age. The oldest date is 350,000 times older than the youngest date.

Table 2

Dating Summary	207Pb/206Pb Age	206Pb/238U Age	208Pb/232Th Age	87Rb/86Sr Age
Average	3,459	4,891	9,984	12
Maximum	8,126	31,193	352,962	13
Minimum	-445	1	2	11
Difference	8,571	31,192	352,960	2

Another table [8] in the essay has a list of calculated dates As we can see below they are all at radical disagreement with each other. There is a spread of dates of 82 billion years! None of the Uranium/Lead based dating methods even come vaguely close to the so called true age. The oldest date is 82,000 times older than the youngest date.

Table 3

Dating Summary	206Pb/238U Age	207Pb/235U Age	208Pb/232Th Age	87Rb/86Sr Age
Average	1,540	46	7,687	12
Maximum	20,209	486	82,030	13
Minimum	1	0	3	11
Difference	20,208	486	82,027	2

If we run the isotopic ratios give in standard geology magazines through the computer program Isoplot we find that the Uranium/Thorium/Lead isotopic ratios in the rocks disagree radically with the Rubidium/Strontium ages. The U/Th/Pb ratios give ages older than the evolutionist age of the Earth, Solar System, Galaxy and Universe. How can Earth rocks be dated as being older than the Big Bang?

If we use isotopic formulas [8-11] given in standard geology text we can arrive at ages from the Rb/Sr and Nd/Sm ratios. The formula for Rb/Sr age is given as:

$$t = \frac{2.303}{\lambda}\log\left(\frac{(87Sr/86Sr)-(87Sr/86Sr)_0}{(87Rb/86Sr)}+1\right) \quad [1]$$

Where t equals the age in years. λ equals the decay constant. (87Sr/86Sr) = the current isotopic ratio. $(87Sr/86Sr)_0$ = the initial isotopic ratio. (87Rb/86Sr) = the current isotopic ratio. The same is true for the formula below.

$$t = \frac{2.303}{\lambda}\log\left(\frac{(143Nd/144Nd)-(143Nd/144Nd)_0}{(147Sm/144Nd)}+1\right) \quad [2]$$

Here are examples of isotopic ratios taken from several articles in major geology magazines which give absolutely absurd dates.

Early Archaean Rocks At Fyfe Hills

These early Archaean rocks from Fyfe Hills in Antarctica were dated in 1982 by scientists form the Australian Bureau of Mineral Resources, The University of Adelaide, Adelaide, and the University of Tasmania, Hobart. [12] Several isotopic samples [13] gave negative ages [-24 billion, -14 billion, -108 billion, -43 billion]. How can a rock that exists in the present and formed in the past have formed 108 billion years in the future?

Table 4

Average	-3,556
Maximum	4,925
Minimum	-108,362
Difference	113,287

The Uranium/Lead ratios [14] give uniform values of 2,500 million years old. The thirty 87Rb/86Sr ratios have nineteen that give ages much older [3,039 to 4,925 Million years] and seven [1,835 to -108,362 Million years] much younger. The author's choice of age is purely arbitrary.

Shock-Melted Antarctic LL-Chondrites

These meteorite samples were dated in 1990 by scientists from the Department of Earth Sciences, Kohe University, Japan. [15] According to the article [16] the meteorite is 4.55 billion years old. The article claims that the maximum range of model ages is 3.11 to 7.33 billion years. [17] If we run the isotopic ratios through Microsoft Excel we get ages from 4 to 21 billion years old. Thirty six dates are over 5 billion years. Nine are over 10 billion years. If the Solar System is less than 5 billion years old how can the meteorite be older than the assumed age of the galaxy [10 billion years]?

Table 5

Age Million Years	Age Million Years	Age Million Years
21,611	9,015	6,756
14,466	8,988	6,556
12,968	8,921	6,192
12,354	8,869	6,157
11,946	8,753	5,981
10,868	8,675	5,677
10,727	8,556	5,491
10,623	8,405	5,483
10,162	8,153	5,458
9,888	7,590	5,453
9,237	6,947	5,388
9,161	6,899	5,319

Table 6

Average	8,585
Maximum	21,611
Minimum	3,969
Difference	17,642

Diamonds And Mantle-Derived Xenoliths

These samples from South African diamond mines were dated in 1979 by scientist from the University of the Witwatersrand, Johannesburg, South Africa. According to the isochron diagrams [17] the age of the sample is 2.4 billion years. If we run the Lead isotope ratios [18] through Isoplot we get the following values:

Table 7

Average	4,995
Maximum	5,249
Minimum	4,885
Std Deviation	122

If we run the 87Rb/86Sr isotope ratios [18] through Microsoft Excel we get the following values:

Table 8

Average	28,429
Maximum	91,957
Minimum	3,257
Difference	88,700

There is almost a 90 billion years difference between the oldest and youngest dates. Below we can see some of the maximum ages and how stupid they are.

Table 9

Age	Age
Million Years	Million Years
91,957	18,139
53,584	17,036
51,582	15,716
43,201	15,340
33,542	13,633
24,366	12,202

87Rb/87Sr Isochron Of The Norton County Achondrite

This meteorite dating was done in 1967 by scientist [20] from the California Institute of Technology. In this article we will find that dating done 45 years later [2008] is giving just as absurd results. According to the Argon dating results [21] the meteorite is between 2.3 and 5.1 billion years old. If we run the 87Rb/86Sr isotope ratios [22] through Microsoft Excel we get the following values:

Table 10

Average	1,375
Maximum	4,871
Minimum	-16,277
Difference	21,149

Base and Precious Metal Veins

According to the article the dating [Coeur D'Alene Mining District, Idaho] was done in 2002 by scientists from the U.S. Geological Survey, California, the Department of Earth and Planetary Sciences, Washington University, Saint Louis, Missouri, the Lawrence Livermore National Laboratory, Livermore, California and the Sunshine Precious Metals Company, Idaho. [22] If we run the 87Rb/86Sr isotope ratios [23] from Table 1 in the article through Microsoft Excel we get the following values:

Table 11

Average	128,708
Maximum	508,074
Minimum	7,990
Difference	516,064

There is a 500 billion year difference between the youngest and oldest dates. The average age is over 120 billion years. Below we can see some of the maximum ages and how stupid they are.

Table 12

Age Million Years	Age Million Years	Age Million Years	Age Million Years
508,074	157,304	125,399	86,483
314,336	151,142	114,796	75,684
302,580	150,089	114,795	72,915
287,077	149,802	113,950	71,225
207,257	144,826	111,884	69,729
201,185	142,977	110,719	63,934
191,104	138,115	109,164	63,406
190,573	134,866	108,617	61,740
189,167	134,061	108,278	56,735
186,066	134,039	102,140	52,117
183,607	132,885	99,952	47,926
183,225	132,746	93,848	46,968
163,764	131,670	89,246	39,944
158,436	130,664	88,626	37,623
158,282	129,495	87,708	16,153

If we run the 87Rb/86Sr isotope ratios [24] from Table 2 in the article through Microsoft Excel we get the following values:

Table 13

Average	139,471
Maximum	508,074
Minimum	12,314
Difference	520,388

There is a 520 billion year difference between the youngest and oldest dates. The average age is almost 140 billion years. Below we can see some of the maximum ages and how stupid they are. The oldest dates is over half a trillion years old.

Table 14

Age Million Years	Age Million Years	Age Million Years
508,074	147,429	87,708
314,336	138,882	84,716
165,542	118,679	82,294
157,714	98,450	59,080
157,589	91,450	45,663
151,317	89,236	12,314

If we run the 87Rb/86Sr isotope ratios [25] from Table 4 in the article through Microsoft Excel we get the following values:

Table 15

Average	**88,571**
Maximum	**288,775**
Minimum	**-170,232**
Difference	**459,007**

There is a 560 billion year difference between the youngest and oldest dates. The average age is almost 90 billion years. Below we can see some of the maximum ages and how stupid they are. The oldest date is almost 300 billion years old. The youngest is negative 170 billion years old.

Table 16

Age Million Years	Age Million Years	Age Million Years	Age Million Years	Age Million Years	Age Million Years
288,775	97,242	94,819	93,079	90,891	85,924
102,716	97,117	94,465	92,995	90,700	85,805
101,380	97,033	94,453	92,972	90,536	85,263
100,277	96,792	94,431	92,967	90,367	84,990
99,779	96,687	94,408	92,963	90,127	83,914
99,683	96,655	94,397	92,915	90,089	83,584
99,369	96,602	94,345	92,878	90,018	82,639
99,238	96,293	94,339	92,863	89,838	80,962
99,177	96,252	94,249	92,829	89,736	80,214
98,948	96,236	94,235	92,634	89,466	79,082
98,765	96,043	94,139	92,630	89,236	78,053
98,736	95,981	94,100	92,374	89,171	76,750
98,685	95,894	93,928	92,315	88,932	76,256
98,591	95,761	93,841	92,309	88,876	76,178
98,436	95,711	93,766	92,205	88,540	75,048
98,285	95,609	93,730	92,140	88,295	72,004
98,243	95,522	93,582	92,108	87,585	70,479
97,979	95,510	93,574	91,906	87,359	69,790
97,830	95,388	93,504	91,674	87,260	55,157
97,628	95,218	93,401	91,650	86,826	53,568
97,604	95,197	93,394	91,435	86,691	51,934
97,545	95,185	93,271	91,238	86,474	-39,207
97,421	95,125	93,199	91,189	86,136	-89,656
97,402	94,994	93,124	91,005	86,050	-170,232

The Munchberg Massif, Southern Germany

According the article, this dating was done in 1990 by scientists from the Koln University, Germany and the Scripps Institution of Oceanography, La Jolla, California. [26] There is an 8 billion year difference between the youngest and oldest dates.

Table 17

Average	1,105
Maximum	7,834
Minimum	-296
Difference	8,130

Rocks of the Central Wyoming Province

These rock samples were dated in 2005 by scientists from the University of Wyoming. [27] If we run the Rubidium/Strontium and Neodymium/Samarium isotope ratios [28] from the article through Microsoft Excel we get the following values:

Table 18

Dating Summary	Age 87Rb/86Sr	Age 147Sm/144Nd	Age 207Pb/206Pb	Age 208Pb/232Th	Age 206Pb/238U
Average	2,863	2,869	5,123	17,899	11,906
Maximum	2,952	2,954	5,294	38,746	18,985
Minimum	2,630	2,631	4,662	6,650	7,294
Std Deviation	38	39	152	9,754	3,298

The Uranium/Lead dates [29] are up to sixteen billion years older than the Rubidium/Strontium and Neodymium/Samarium dates. The Thorium/Lead dates are up to thirty six billion years older. The so called true age is just a guess.

Basalts From Apollo 15

According the article, this Moon rock dating was done in 1972 by scientists from the California Institute of Technology, Pasadena, California. [30] According to the essay the rock is 3.4 billion years old. 31 If we run the 87Rb/86Sr isotope ratios [32] from Table 4 in the article through Microsoft Excel we get the following values:

Table 19

Average	3,045
Maximum	27,211
Minimum	-3,808
Difference	31,019

Of the 21 isotopic ratios, seven were below 500 million years old. Two were over six billion years old.

History Of The Pasamonte Achondrite

According to the article this meteorite specimen was dated in 1977 by scientists from the United States Geological Survey, Colorado and the Department of Chemistry and Geochemistry, Colorado School of Mines. [33] The article states that Rubidium/Strontium dating affirms that this material is 4.5 billion years old. [34] If we run the various isotope ratios [34] from two different tables in the article through Microsoft Excel we get the following values respectively:

Table 20

Summary	206Pb/238U	207Pb/235U	207Pb/206Pb	208Pb/232Th
Average	3,088	3,666	4,566	2,263
Maximum	5,694	5,032	4,963	14,800
Minimum	103	865	4,440	-10,700
Difference	5,591	4,167	523	25,500

If we run the 87Rb/86Sr isotope ratios [34] from the article through Microsoft Excel we get the following values:

Table 21

Average	4,403
Maximum	6,674
Minimum	2,412
Difference	4,262

The Thorium/Lead dates are up to twelve billion years older. The so called true age is just a guess.

Sr Isotopic Composition Of Afar Volcanics

According to the article [35] this specimen [basalts from the Afar depression in Ethiopia] was dated in 1977 by scientists from Italy and France. The article states that the formation is of the late Quaternary period and thus very young. If we run the 87Rb/86Sr isotope ratios [36] from the article through Microsoft Excel we get the following values:

Table 22

Average	183
Maximum	2,260
Minimum	-108
Difference	2,368

As far as the rocks being of a Quaternary age, the dates just don't line up.

Orogenic Lherzolite Complexes

According to the article [37] this specimen from Gibraltar was dated in 1979 by scientists from France. According to the article [38] the maximum age of the samples is 103 million years. If we run the 87Rb/86Sr isotope ratios [39] from the two different tables in the article [Tables 2 and 3] through Microsoft Excel we get the following values respectively:

Table 23

Summary	Table 2	Table 3
Average	-52,203	-29,099
Maximum	-2,229	-1,258
Minimum	-135,140	-102,498
Difference	132,911	101,240

The dates are light years different from what the essay claims. They are just absurd.

Isotopic Geochemistry (Os, Sr, Pb)

According to the article [40] this specimen [the Golda Zuelva and Mboutou anorogenic complexes, North Cameroun] was dated in 1982 by scientists from France. According to the article [40] the maximum age of the sample is 66 million years. If we run the 87Rb/86Sr isotope ratios [41] from the two different tables in the article [Tables 1and 2] through Microsoft Excel we get the following values respectively:

Table 24

Dating	87Rb/86Sr	87Rb/86Sr	Pb207/Pb206
Summary	Age	Age	Age
Average	321	57	4,982
Maximum	1,635	141	5,080
Minimum	52	0	4,932
Difference	1,687	141	10,012

If we run the 207Pb/206Pb isotope ratios [42] from the article [Table 3] through Microsoft Excel we get the following values respectively:

Table 25

Age	Age
5,080	4,964
5,048	4,958
4,990	4,957
4,984	4,938
4,980	4,932
4,975	

The so called true age is just a guess.

Cretaceous-Tertiary Boundary Sediments

According to the article [43] this specimen [from the Barranco del Gredero, Caravaca, Spain] was dated in 1983 by scientists from University of California, Los Angeles, the United States Geological Survey, and the Geological Institute, University of Amsterdam. According to the article [44] the maximum age of the sample is 65 million years. If we run the 87Rb/86Sr isotope ratios [44] from the article through Microsoft Excel we get the following values respectively:

Table 26

Average	740
Maximum	5,157
Minimum	-266
Difference	5,423

Out of the 16 dates derived from isotopic ratios, ten were over 100 million years old. Two were over 4 billion years old. One was negative 266 million years old. How can a rock that formed in the past have a negative age! The choice of 65 million years is just a guess.

Correlated N D, Sr And Pb Isotope Variation

According to the article [45] this specimen [Walvis Ridge, Walvis Bay] was dated in 1982 by scientists from the Massachusetts Institute of Technology, and the Department of Geochemistry, University of Cape Town, South Africa. According to the article [45] the age of the sample is 70 million years. If we run the various isotope ratios [46] from the article through Microsoft Excel we get the following values respectively:

Table 27

Summary	Pb207/Pb206	147Sm/144Nd	87Rb/86Sr
Average	5,033	70	64
Maximum	5,061	70	93
Minimum	5,004	69	0
Difference	57	140	93

A Depleted Mantle Source For Kimberlites

According to the article [47] this specimen [kimberlites from Zaire] was dated in 1984 by scientists from Belgium. According to the article [48] the age of the samples is 70 million years. If we run the various isotope ratios [49] from the article through Microsoft Excel we get the following values respectively:

Table 28

Summary	207Pb/206Pb	206Pb/238U	87Rb/86Sr	147Sm/144Nd
Average	4,977	4,810	86	72
Maximum	5,017	10,870	146	80
Minimum	4,909	1,391	50	63
Difference	108	9,478	196	17

The 207Pb/206Pb maximum age is 34 times older than the 87Rb/86Sr maximum age. The 206Pb/238U maximum age is 74 times older than the 147Sm/144Nd maximum age. There is a 10.8 billion year difference between the oldest and youngest age attained.

Sm-Nd Isotopic Systematics

According to the article [50] this specimen [Enderby Land, East Antarctic] was dated in 1984 by scientists from the Australian National University, Canberra, and the Bureau of Mineral Resources, Canberra. According to the article [50] the age of the sample is 3,000 million years. If we run the Rb/Sr isotope ratios [51] from the article through Microsoft Excel we get the following values respectively:

Table 29

Average	-873
Maximum	3,484
Minimum	-25,121
Difference	28,605

There is almost a 30 billion year difference between the oldest and youngest dates.

Strontium, Neodymium And Lead Compositions

According to the article [52] this specimen [Snake River Plain, Idaho] was dated in 1985 by scientists from the Geology Department, Rice University, Houston, Texas, the Earth Sciences Department, Open University, England and the Geology Department, Ricks College, Idaho. According to the article [52] the age of the sample is 3.4 billion years. If we run the various isotope ratios [53] from the article through Microsoft Excel we get the following values respectively:

Table 30

Summary	Pb207/Pb206	Pb207/Pb206	87Rb/86Sr
Average	5,143	5,138	40,052
Maximum	5,362	5,314	205,093
Minimum	4,698	4,940	1,443
Difference	664	374	203,650

The Lead isotope ratios from two different tables give dates 200 billion years younger than the Rb/Sr isotope ratios. The Average age of the Rb/Sr isotope ratios is 40 billion years. Below we can see some of the maximum ages and how stupid they are.

Table 31

Age Million Years	Age Million Years
205,093	11,974
189,521	11,908
188,777	9,960
95,450	9,101
52,643	7,124
13,119	6,022
12,220	5,089

Trace Element And Sr And Nd Isotope

According to the article [54] this specimen [West Germany] was dated in 1986 by scientists from Germany and California. According to the article [54] the age of the samples is 2 billion years. If we run the various isotope ratios [55] from the article through Microsoft Excel we get the following values respectively:

Table 32

Average	41,573
Maximum	175,289
Minimum	-30,734
Difference	206,022

Many of the Rb/Sr isotopic ratios would not produce proper ages. Those that did gave absurd values. Below are some dates taken from another table [56] in the original article.

Table 33

TABLE 5	Sm-Nd	Rb-Sr
Sample	Age	Age
Ib/K1	2,090	2,210
Ib/8	2,900	1,790
D1	1,450	1,660
Ib/5	1,100	1,430
D45	1,630	530
D58	3,200	1,930

The Southeast Australian Lithosphere Mantle

According to the article [57] this specimen was dated in 1987 by scientists from The Australian National University. According to the article [58] the age of the samples is 1.5 billion years. If we run the various isotope ratios [59] from two different tables in the article through Microsoft Excel we get the following values respectively:

Table 34

Average	1,905	42,639
Maximum	11,657	218,042
Minimum	134	-15,716
Difference	11,523	233,758

Below we can see the maximum ages obtained from the second table. The oldest age is 18 times older than the Big Bang explosion. It is sixty two times older than the so called age of the Earth.

Table 35

Age	Age
218,042	45,207
64,770	38,581
54,457	26,113
48,074	17,246
45,734	11,813

Strontium, Neodymium and Lead Isotopic

According to the article [60] this specimen was dated in 1988 by scientists from the Department of Terrestrial Magnetism. Carnegie Institution of Washington. Throughout the article the author admits that the dates are contradicting and unreliable: "For sample 7541. the apatite eclogite, the range observed in both Rh/Sr and Sm/Nd for the whole-rock and mineral separates is quite small resulting in very imprecise "ages" of 400 Ma for Rb-Sr and 1110 Ma for Sm-Nd." [61] If we run the Lead isotope ratios [62] from the article through Microsoft Excel we get the following values respectively:

Table 36

Age	Age
4,933	4,928
4,961	4,956
4,952	4,947
4,952	4,957
4,942	4,927
4,978	4,952
4,940	4,954
4,947	

Sr, Nd, and Os Isotope Geochemistry

According to the article [63] this specimen [Camp Creek area, Arizona] was dated in 1987 by scientists from The University of Tennessee, the University of Michigan, the University of California, Leeds University, and the University of Chicago. According to the article [64] the age of the samples is 120 million years. If we run the various isotope ratios [65] from two different tables in the article through Microsoft Excel we get the following values respectively:

Table 37

Summary	87Rb/86Sr	87Rb/86Sr	147Sm/144Nd	147Sm/144Nd
Average	310	103	120	159
Maximum	1,092	207	123	400
Minimum	0	0	120	119
Difference	1,092	207	3	281

The author's choice of 120 million years is just a guess.

Pb, Nd and Sr Isotopic Geochemistry

According to the article [66] this specimen [Bellsbank kimberlite, South Africa] was dated in 1991 by scientists from the University Of Rochester, New York, Guiyang University in China, and the United States Geological Survey, Colorado. According to the article [67] the age of the samples is just 1 million years. If we run the various isotope ratios [68] from two different tables in the article through Microsoft Excel we get the following values respectively:

Table 38

Table	207Pb/206Pb	206Pb/238U	208Pb/232Th	87Rb/86Sr
Summaries	Age	Age	Age	Age
Average	5,057	5,092	10,182	-1,502
Maximum	5,120	8,584	17,171	0
Minimum	5,002	0	0	-3,593
Difference	118	8,584	17,171	3,593

In tables 39 to 42 we can see some of the astounding spread of dates [million of years]. The oldest date is over 17 billion years old. The youngest is less than negative 3.5 billion years. The difference between the two is over 20 billion years. According to the article the true age of the rock is just one million years old!

Table 39

Age	Age	Age	Age
17,171	13,322	9,737	7,968
15,343	13,202	9,707	7,830
15,299	13,001	9,049	7,250
15,136	11,119	8,420	6,972
15,054	10,873	8,419	6,628
13,476	10,758	8,368	6,577

Table 40

Age	Age	Age
8,584	6,656	5,576
7,975	6,654	5,520
7,314	6,518	5,285
7,184	6,448	5,159
6,861	5,758	5,099

Table 41

Age	Age	Age	Age
5,120	5,067	5,060	5,049
5,109	5,066	5,059	5,045
5,097	5,066	5,051	5,044
5,077	5,065	5,050	5,044
5,067	5,062	5,050	5,033
5,067	5,060	5,050	5,022

Table 42

Age	Age	Age	Age
-3,593	-2,981	-1,917	-1,323
-3,231	-2,725	-1,611	-1,245
-3,089	-2,050	-1,499	-1,229
-3,067	-1,926	-1,370	-1,194

Sr, Nd, and Pb isotopes

According to the article [68] this specimen [eastern China] was dated in 1992 by scientists from the University Of Rochester, New York, Guiyang University in China, and the United States Geological Survey, Colorado. According to the article: "Observed high Th/U, Rb/Sr,

87Sr/86 Sr and Delta 208, low Sm/Nd ratios, and a large negative Nd in phlogopite pyroxenite with a depleted mantle model age of 2.9 Ga, support our contention that metasomatized continental lower mantle lithosphere is the source for the EMI component." [68] If we run the various isotope ratios [69] from two different tables in the article through Isoplot we get the following values respectively:

Table 43

Dating Summaries	232Th/208Pb Age	206Pb/238U Age	207Pb/206Pb Age
Average	14,198	7,366	5,014
Maximum	94,396	22,201	5,077
Minimum	79	1,117	4,945
Difference	94,317	21,083	131

If the true age is 2.9 billion years why so much discordance? In tables 41 to 43 we can see some of the astounding spread of dates [million of years]. The oldest date is over 94 billion years old. The youngest is 79 million years. The difference between the two is over 94 billion years. The oldest date is 1,194 times older than the youngest. According to the article the true age of the rock is 2.9 billion years old!

Table 44

Age	Age	Age	Age
94,396	39,267	10,595	8,171
90,683	26,266	10,284	7,789
74,639	18,334	9,328	7,638
58,153	16,357	8,821	7,375
55,324	14,250	8,771	7,317
45,242	11,215	8,403	5,759

Table 45

Age	Age	Age	Age
22,201	9,878	7,348	5,746
21,813	9,656	7,335	5,700
19,320	9,054	7,249	5,218
16,656	8,242	7,202	5,201
16,200	8,044	7,019	5,163
14,748	7,996	6,923	5,159
13,607	7,590	6,848	5,099
11,256	7,422	6,292	4,812

Production of Jurassic Rhyolite

According to the article [70] this specimen [Patagonia, South America] was dated in 1994 by scientists from the British Antarctic Survey, National University, Argentina. According to the article: "Primary magmas of andesitic composition were generated by partial melting of mafic" Grenvillian" lower crust, indentified by depleted-mantle model ages of 1150-1600 Ma." [70] If we run the various isotope ratios [71] from two different tables in the article through Microsoft Excel we get the following values respectively:

Table 46

Average	432
Maximum	17,387
Minimum	-4,633
Difference	22,020

Evolution of Reunion Hotspot Mantle

According to the article [72] this specimen [Reunion and Mauritius Islands] was dated in 1995 by scientists from the University of Hawaii. According to the article: "Whole-rock powder obtained from P. Krishnamurthy. (87Sr/86 Sr), and em(T) are age-corrected values; *T* = 66 Ma for the drill hole lavas." [73] If we run the various isotope ratios [74] from two different tables in the article through Isoplot we get the following values respectively:

Table 47

Table	232Th/208Pb	206Pb/238U	207Pb/206Pb
Summaries	Age	Age	Age
Average	8,079	4,449	4,976
Maximum	13,287	6,285	5,016
Minimum	5,641	3,010	4,953
Difference	7,646	3,276	63

Table 48

Age	Age	Age	Age
13,287	8,725	7,363	6,540
11,832	8,609	7,362	6,479
11,017	7,541	7,080	6,323
10,357	7,517	7,017	5,660
9,101	7,446	6,679	5,641

Table 49

Age	Age	Age	Age
6,285	4,903	4,141	3,875
6,165	4,633	4,133	3,647
5,767	4,342	4,011	3,548
5,553	4,258	4,001	3,369
5,152	4,220	3,973	3,010

According to dating charts in the article, the true age is just 66 million years old! [74]

An Extremely Low U/Pb Source

According to the article [75] this specimen [lunar meteorite] was dated in 1993 by scientists from the United States Geological Survey, Colorado, the United States Geological Survey, California and The National Institute of Polar Research, Tokyo. According to the article: "The Pb-Pb internal isochron obtained for acid leached residues of separated mineral fractions yields an age of 3940 ± 28 Ma, which is similar to the U-Pb (3850 ± 150 Ma) and Th-Pb (3820 ± 290 Ma) internal isochron ages. The Sm-Nd data for the mineral separates yield an internal isochron age of 3871 ± 57 Ma and an initial 143Nd/I44Nd value of 0.50797 ± 10. The Rb-Sr data yield an internal isochron age of 3840 ± 32 Ma." [75]

Table 50

Average	3,619
Maximum	5,385
Minimum	721
Difference	4,664

Table 51

Table Summaries	207Pb/206Pb Age	206Pb/238U Age	208Pb/232Th Age	207Pb/235U Age
Average	4,673	8,035	10,148	4,546
Maximum	5,018	56,923	65,286	8,128
Minimum	3,961	1,477	2,542	2,784
Difference	1,057	55,445	62,744	5,344

The article claims that the Rb/Sr age is 3.8 billion years for this meteorite. If that is the true age why are all the Uranium/Thorium/Lead dates [76] so stupid? Or are they right and the Rb/Sr is wrong?

Table 52

Age	Age	Age	Age
65,286	14,430	9,094	5,401
33,898	14,410	6,520	5,396
25,013	13,107	6,166	5,365
22,178	12,738	6,121	5,098
21,204	11,641	5,671	5,035
17,611	11,174	5,408	4,678

Table 53

Age	Age	Age	Age
56,923	10,895	6,764	5,777
27,313	10,278	6,670	5,625
17,873	9,653	6,449	5,602
13,680	8,009	6,436	5,278
13,623	7,395	6,070	5,147

The 72 Ma Geochemical Evolution

According to the article [77] this specimen [Madeira Archipelago] was dated in 2000 by scientists from Germany. The average Lead date is 705 times older than the average Rubidium date. The true age is claimed to be 430 million years old. [77] If we run the various isotope ratios [78] from two different tables in the article through Isoplot we get the following values respectively:

Table 54

Table	207Pb/206Pb	87Rb/86Sr	147Sm/144Nd
Summaries	Age	Age	Age
Average	4,938	7	10
Maximum	5,199	55	164
Minimum	4,898	-4	0
Difference	302	59	164

If the true age is 430 million years than none of the dating methods are even vaguely close. The oldest date is 731 times older than the youngest.

The Himalayan Collision Zone

According to the article [79] this specimen [East Tibet] was dated in 2000 by scientists from Germany. As far as the age goes the author states: "Partial melting of the mantle source was most likely triggered by a Cenozoic asthenospheric mantle diapir related to Indian–Asian continent collision at 65–45Ma. Rising and emplacement of carbonatitic magmas with coeval potassium-rich magmas took place in the tectonic regime of the transition from transpression to transtension at Eocene/Oligocene boundary in the EIACZ." [80] He also states: "The initial "Nd values and 87Sr / 86Sr ratios were calculated at *t*=35Ma." [81] If we run the various isotope ratios [82] from two different tables in the article through Isoplot we get the following values respectively:

Table 55

Dating	207Pb/206Pb	87Rb/86Sr
Summary	Age	Age
Average	5,015	0
Maximum	5,023	0
Minimum	4,976	0
Difference	47	0

If the specimen is of the Eocene era [Less than 100 million years old] how can the Lead/Lead dating produce such rubbish? If we run the Rb/Sr ratios through Microsoft Excel we get zero ages!

Evidence for a Non Magmatic component

According to the article [83] this specimen [Yukon, Canada] was dated in 2001 by Canadian scientists from the University of Alberta, and Dalhousie University, Halifax. According to Argon dating the age of the material is 70 million years. [84] If we run the various isotope ratios [85] from two different tables in the article through Isoplot we get the following values respectively:

Table 56

Table Summaries	207Pb/206Pb Age	87Rb/86Sr Age
Average	4,955	71
Maximum	5,214	101
Minimum	4,918	60
Difference	296	41

If we look at the average ages we see that there is a 7 thousand percent difference between them! If we compare the youngest and oldest dates we see that there is an 8,540 percent difference between them.

The Origin Of Geochemical Diversity

According to the article [86] this specimen [lunar basalt] was dated in 2007 by scientists from New Mexico University. According to Rb/Sr isochron diagram the age of the material is 3.678 billion years. [87] If we run the various isotope ratios [88] from two different tables in the article through Isoplot we get the following values respectively:

Table 57

Table Summaries	207Pb/206Pb Age	206Pb/238U Age	87Rb/86Sr Age
Average	4,635	6,565	4,672
Maximum	5,111	18,213	7,094
Minimum	4,028	3,706	3,476
Difference	1,082	14,506	3,618

The dating methods all disagree with each other. There is a wide spread of dates which are just random.

Mechanisms For Incompatible-Element Enrichment

According to the article [89] this specimen [meteorite Northwest Africa] was dated in 2009 by scientists from Lawrence Livermore National Laboratory, University of New Mexico, the University of California, Berkeley, and Arizona State University. The author states: "Rubidium–Strontium isotopic analyses yield an age of 2,947 ± 16 Ma" If we run the various isotope ratios [90] from a table in the article through Microsoft Excel we get the following values respectively:

Table 58

Average	5,483
Maximum	13,497
Minimum	1,917
Difference	11,579

Out of the eleven isotope ratios, two returned dates over ten billion years old.

Constraints On Martian Differentiation Processes

According to the article [91] this specimen [Martian meteorite] was dated in 1997 by scientists from the NASA Johnson Space Centre, Houston, Texas, the University of Tennessee, and Lockheed Martin, Houston, Texas. According to the article [91] the age range is: "The neodymium isotopic systematics of QUE 94201 are not consistent with significant melting between 4.525 Ga and 327 Ma." If we run the various isotope ratios [92] from two different tables [1 and 4] in the article through Microsoft Excel we get the following values respectively:

Table 59

Summary	Table 1	Table 4
Average	618	-34,834
Maximum	1,765	4,642
Minimum	-98	-118,922
Difference	1,668	123,564

Instead of having a 4.2 billion year spread we have a 123 billion year spread of dates. Both tables in the article give dates way off the so called true age.

Geochemistry of the Volcan de l'Androy

According to the article [93] this specimen from the Androy massif in south eastern Madagascar was dated in 2008 by scientists from the University Of Hawaii. According to the article Argon and Rubidium dating defined the so called true ages as: "The R2 rhyolites define a whole-rock Rb/Sr isochron of 84 Ma, the same, within error, as an 40Ar/39Ar sanidine age reported by earlier workers." [93] If we run the various isotope ratios [94] from a table in the article through Isoplot we get the following values respectively:

Table 60

Average	5,004	4,999
Maximum	5,048	5,029
Minimum	4,980	4,984
Difference	67	18

The Lead dating give ages that are sixty times older than the Rb/Sr dates.

Continental Lithospheric Contribution

According to the article [95] this specimen from southern Portugal was dated in 1997 by scientists from France. According to the article Argon and Rubidium dating defined the so called true ages as: "The age of the intrusion and crystallization of the alkaline rocks of the Serra de Monchique is 72 Ma, based on Rb/Sr and K/Ar dating." [96] If we run the various isotope ratios [97] from a table in the article through Isoplot we get the following values respectively:

Table 61

Table	207Pb/206Pb	208Pb/232Th	206Pb/238U	87Rb/86Sr
Summaries	Age	Age	Age	Age
Average	4,920	6,126	4,539	-62
Maximum	4,949	10,084	7,723	-50
Minimum	4,894	2,616	2,306	-75
Difference	55	7,467	5,417	25

The date of 72 million years is just a guess. The Thorium/Lead method gives dates 140 times older. The Uranium/Lead methods give dates 107 times older. Below we can see the maximum ages [million years] calculated form isotope ratios. Compare these with the so called true age!

Table 62

208Pb/232Th	206Pb/238U
10,084	7,723
9,320	7,060
8,101	6,507
7,502	6,387
7,080	6,206
6,891	5,143
6,655	4,734
6,313	4,186
5,830	3,768
5,755	3,761
5,029	3,487

Garnet Granulite Xenoliths

According to the article [98] this specimen from the northern Baltic shield was dated in 2001 by scientists from England, USA and Russia. According to the article Argon dating defined the so called true ages as 400 to 2200 million years. [99] If we run the various isotope ratios [100] from table 4 in the article through Isoplot we get the following values respectively:

Table 63

Table	206Pb/238U	207Pb/206Pb
Summaries	Age	Age
Average	17,002	5,046
Maximum	40,059	5,295
Minimum	1,608	3,908
Difference	38,452	1,387

Below are the maximum ages calculated from isotope ratios in tables 4 and 5 in the article:

Table 64

206Pb/238U	206Pb/238U	206Pb/238U	206Pb/238U
Age	Age	Age	Age
40,059	28,118	21,092	13,724
35,742	27,127	16,026	13,404
34,459	25,884	14,371	12,747
33,978	21,209	14,272	10,956

Table 65

206Pb/238U	206Pb/238U	206Pb/238U
Age	Age	Age
20,648	13,724	10,956
17,527	13,404	10,049

16,336	12,622	6,792
15,626	12,165	6,265
15,018	11,432	5,865

If we run more ratios form and online supplement we get ages uniformly 5 billion years old. Compare these with the so called true age!

The Isotope and Trace Element Budget

According to the article [102] this specimen from the Devil River Arc System, New Zealand was dated in 2000 by scientists from Germany. According to the article, the so called true ages is Cambrian. [102] If we run the various isotope ratios [103] from table 4 in the article through Isoplot we get the following values respectively:

Table 66

Table	207Pb/206Pb	206Pb/238U	87Rb/86Sr
Summaries	Age	Age	Age
Average	4,970	19,143	500
Maximum	4,986	21,761	501
Minimum	4,932	15,150	495
Difference	54	6,611	6

The Lead/Lead dates are ten times too old and the Uranium/Lead dates are 40 times too old!

Fluid Flow and Diffusion

According to the article [104] this specimen from the Waterville Formation in south–central Maine, USA, was dated in 1997 by scientists from England and USA. According to the article, the so called true age is: "the 376±6 Ma Rb–Sr whole-rock age of the syn-metamorphic Hallowell pluton." [104] According to isochron diagrams in the article [105] the model age is between 342 to 391 million years. The article has an age range diagram [106] which claims that the maximum age is 425 million years. If we run the various isotope ratios [107] from table 4 in the article through Isoplot we get the following values respectively:

Table 67

Average	746
Maximum	2,063
Minimum	316
Difference	1,747

Out of the 150 isotopic ratios in the essay, 134 gave ages greater than the so called maximum age limit. Twenty six gave ages that were more than twice the maximum limit.

Temporal Evolution of the Lithospheric Mantle

According to the article [108] this specimen from the Eastern North China Craton was dated in 2009 by scientists from China, USA and Australia. Various tables [109] in the essay have either calculated dates or ratios which can be calculated. As we can see below they are all at strong disagreement with each other. There is a spread of dates over a 32 billion year range.

Table 68

Table Summaries	147Sm/144Nd Age	176Lu/176Hf Age	187Re/188Os Age	87Rb/86Sr Age
Average	291	-220	1,048	9
Maximum	3,079	4,192	20,710	22
Minimum	-3,742	-9,369	-11,060	0
Difference	6,821	13,561	31,770	22

Petrogenesis and Origins of Mid-Cretaceous

According to the article [110] this specimen from the Intraplate Volcanism in Marlborough, New Zealand was dated in 2010 by scientists from New Zealand. According to the essay: "the intraplate basalts in New Zealand that have been erupted intermittently over the last c. 100 Myr" [111] Various tables [112] in the essay have isotopic ratios which can be calculated. As we can see below they are all at strong disagreement with each other. There is a spread of dates over a 10 billion year range. None of the Lead based dating methods even come vaguely close to a Cretaceous age.

Table 69

Table Summaries	207Pb/206Pb Age	207Pb/235U Age	87Rb/86Sr Age	208Pb/232Th Age	206Pb/238U Age
Average	4,876	4,416	59	6,333	3,515
Maximum	4,945	5,159	85	10,716	5,717
Minimum	4,836	4,088	15	4,785	2,712
Difference	109	1,071	70	5,931	3,005

The Petrogenetic Association of Carbonatite

According to the article [113] this specimen from the Spitskop Complex, South Africa was dated in 1999 by scientists from South Africa. According to the essay: "The 1,341 Ma old Spitskop Complex in South Africa is one of a series of intrusions of alkaline affinity." [113] Various tables [114] in the essay have isotopic ratios which can be calculated. As we can see below they are all at strong disagreement with each other.

Table 70

Dating Summary	87Rb/86Sr Age	207Pb/206Pb Age
Average	-6,012	5,056
Maximum	2,762	5,126
Minimum	-66,499	4,649
Difference	69,262	477

Nine of the twenty six Rb/Sr dates are over three billion years in error. Seven are over eleven billion years in error. The thirteen Lead 206/207 dates are all totally way off.

Geochemistry Of The Jurassic Oceanic Crust

According to the article [115] this specimen from the Canary Islands was dated in 1998 by scientists from Germany. According to the essay: "An Sm–Nd isochron gives an age of 178 ± 17 Ma, which agrees with the age predicted from paleomagnetic data."[115] The article places the age in the late Cretaceous period. Various tables [116] in the essay have isotopic ratios which can be calculated. As we can see below they are all at strong disagreement with each other. There is a spread of dates over a 350 billion year range! None of the Lead or Rubidium based dating methods even come vaguely close to a Jurassic age.

Table 71

Dating	87Rb/86Sr	207Pb/206Pb
Summary	Age	Age
Average	-149,488	4,974
Maximum	51,967	5,024
Minimum	-299,346	4,845
Difference	351,313	179

The Age Of Dar Al Gani 476

According to the article [117] this Martian meteorite was dated in 2003 by scientists from the University of New Mexico, NASA Johnson Space Centre, Lockheed Engineering and Science Company. According to the essay: "In either case, the fact that the Martian meteorites define a whole rock Rb-Sr isochron with an age of 4.5 Ga require these reservoirs to have formed near the time of planet formation." [117] A table [118] in the essay has isotopic ratios which can be calculated. As we can see below they are all at strong disagreement with the assumed age. There is a spread of dates of almost 18 billion year range! None of the Rubidium based dating methods even come vaguely close to the so called true age.

Table 72

Average	-9,398
Maximum	-2,142
Minimum	-20,004
Difference	17,862

Petrogenesis Of The Flood Basalts

According to the article [119] this basalt form the Northern Kerguelen Archipelago was dated in 1998 by scientists from the Massachusetts Institute Of Technology, University of Brussels, Belgium and the San Diego State University. According to the essay: "The dominance of this isotopic signature in archipelago lavas for 30 my and its presence in ~40 Ma gabbros is consistent with the previous interpretation that these are isotopic characteristics of the Kerguelen Plume." [119] Various tables [120] in the essay have isotopic ratios which can be calculated. As we can see below they are all at strong disagreement with each other. There is a spread of dates of over a 44 billion year range! None of the Uranium/Lead based dating methods even come vaguely close to the so called true age.

Table 73

Mt Rabouillere	Age	Age	Age	Age	Age
Summary	87Rb/86Sr	207Pb/206Pb	206Pb/238U	207Pb/235U	208Pb/232Th
Average	21	5,008	4,903	4,975	6,142
Maximum	30	5,019	5,355	5,100	7,788
Minimum	-7	5,000	4,305	4,793	2,799
Difference	38	20	1,050	307	4,989

Table 74

Mount Bureau Summary	Age 87Rb/86Sr	Age 207Pb/206Pb	Age 206Pb/238U	Age 207Pb/235U	Age 208Pb/232Th
Average	27	5,006	5,924	5,161	8,410
Maximum	30	5,020	23,366	8,496	44,378
Minimum	24	4,994	3,335	4,454	2,650
Difference	6	26	20,031	4,042	41,728

Nature Of The Source Regions

According to the article [121] this lava from southern Tibet was dated in 2004 by scientists from the Open University in Milton Keynes, the University of Bristol and Cardiff University. According to the essay: "Most samples are Miocene in age, ranging from 10 to 25Ma in the south and 19Ma to the present day in northern Tibet" [122] Various tables [123] in the essay have isotopic ratios which can be calculated. As we can see below they are all at strong disagreement with each other. There is a spread of dates of over a 88 billion year range! None of the Uranium/Lead based dating methods even come vaguely close to the so called true age.

Table 75

North Tibet Summary	208Pb/232Th Million Years	207Pb/235U Million Years	207Pb/206Pb Million Years	206Pb/238U Million Years
	11,420	5,136	4,980	7,783
87Rb/86Sr	11,350	5,138	4,980	8,023
Model Age	13,475	5,135	4,987	8,305
13 Million Years	11,504	5,140	4,989	7,349
	81,614	7,470	4,987	33,751
	88,294	7,471	4,991	33,742

Table 76

South Tibet Summary	208Pb/232Th Million Years	207Pb/235U Million Years	207Pb/206Pb Million Years	206Pb/238U Million Years
	11,102	313	4,982	6,331
	6,092	946	4,919	5,799
87Rb/86Sr	9,265	266	4,980	6,682
Model Age	4,826	238	4,992	4,086
13 Million Years	8,205	294	4,980	5,567
	25,015	447	4,994	13,328
	33,191	482	4,992	15,053

Generation Of Palaeocene Adakitic Andesites

According to the article [124] this rock formation from North Eastern China was dated in 2007 by scientists from China and Japan. According to the essay the true age is: "Palaeocene (c. 55-58Ma) adakitic andesites from the Yanji area." [124] Numerous table and charts affirm this as the true age. [125] A table [126] in the essay have isotopic ratios which can be calculated. As we can see below they are all at radical disagreement with each other. There is a spread of dates of over 10 billion years! None of the Uranium/Lead based dating methods even come vaguely close to the so called true age.

Table 77

Dating	87Rb/86Sr	207Pb/206Pb	208Pb/232Th	206Pb/238U	207Pb/235U
Summary	Age	Age	Age	Age	Age
Average	51	5,022	8,941	8,754	5,908
Maximum	66	5,024	10,518	9,669	6,052
Minimum	40	5,020	7,800	7,403	5,641
Difference	26	3	2,718	2,266	411

Evidence For A Widespread Tethyan

According to the article [127] this rock formation from North Eastern China was dated in 2007 by scientists from China and Japan. According to the essay the true age is: "Here, we report age-corrected Nd–Pb–Sr isotope data for 100–350 Ma basalt, diabase, and gabbro from widely separated Tethyan locations in Tibet, Iran, Albania, the eastern Himalayan syntaxis, and the seafloor off NW Australia (Fig. 1)." [128] The author concludes that the rocks are from the Cretaceous and Jurassic time periods: "We collected Early Jurassic to Early Cretaceous Neotethyan magmatic rocks in 1998 from outcrops along 1300 km of the Indus–Yarlung suture zone." [129] Several tables [130] in the essay have isotopic ratios which can be calculated. As we can see below they are all at radical disagreement with each other. There is a spread of dates of almost 60 billion years! None of the Uranium/Lead based dating methods even come vaguely close to the so called true age.

Table 78

Dating	87Rb/86Sr	207Pb/206Pb	208Pb/232Th	206Pb/238U
Summary	Age	Age	Age	Age
Average	168	4,999	22,356	7,014
Maximum	1,739	5,236	58,796	15,747
Minimum	0	4,982	10,699	5,042
Difference	1,739	254	48,096	10,705

Table 79

208Pb/232Th	208Pb/232Th	208Pb/232Th	208Pb/232Th
58,796	29,705	18,607	11,427
54,206	27,710	18,121	11,377
48,252	27,422	17,797	11,366
47,976	26,674	17,787	11,241
46,117	26,369	17,591	10,718
42,203	25,972	17,536	10,699
42,192	25,590	17,054	10,699
41,604	25,096	16,053	10,300
41,343	24,010	15,299	9,357
41,231	22,718	14,340	8,632
39,637	22,307	13,845	8,486

38,125	22,228	13,772	8,057
37,115	21,827	13,652	6,497
35,012	21,560	13,404	5,573
33,584	19,910	13,403	5,425
31,556	19,594	13,006	4,869
31,286	19,148	12,171	
30,740	18,765	11,540	

Table 80

206Pb/238U	206Pb/238U	206Pb/238U	206Pb/238U	206Pb/238U
15,747	11,309	8,770	6,602	5,724
15,067	11,248	8,508	6,589	5,720
14,363	10,360	8,315	6,421	5,601
13,580	9,643	8,314	6,398	5,599
13,204	9,427	8,072	6,369	5,573
12,780	9,300	8,024	6,357	5,515
11,757	9,123	7,604	6,219	5,462
11,659	9,014	7,504	5,863	5,311
11,537	8,996	7,056	5,861	5,286
11,313	8,954	7,002	5,807	5,120

Post-Collisional Potassic And Ultrapotassic

According to the article [131] this rock formation from south west Tibet was dated in 1999 by scientists from Austria. According to the essay the true age is: "Volcanic rocks from SW Tibet, with 40Ar/39Ar ages in the range 17–25 Ma." [131] Numerous table and charts affirm this as the true age. [132] Two tables [133] in the essay have isotopic ratios which can be calculated. As we can see below they are all at radical disagreement with each other. There is a spread of dates of almost 100 billion years! None of the Uranium/Lead based dating methods even come vaguely close to the so called true age. The oldest date is 3,971 times older than the youngest date.

Table 81

87Rb/86Sr	207Pb/206Pb	208Pb/232Th	206Pb/238U
Maximum Age	Age	Age	Age
25	5,007	99,275	6,944
25	5,007	95,541	5,560
25	5,001	71,706	5,013
25	5,000	70,277	4,715
25	4,997	68,343	3,745
25	4,988	67,704	2,646

Conclusion

Brent Dalrymple states in his anti creationist book The Age of the Earth: "Several events in the formation of the Solar System can be dated with considerable precision." [134]

Looking at some of the dating it is obvious that precision is much lacking. He then goes on: "Biblical chronologies are historically important, but their credibility began to erode in the eighteenth and nineteenth centuries when it became apparent to some that it would be more profitable to seek a realistic age for the Earth through observation of nature than through a literal interpretation of parables." [135]

In his book he gives a table [136] with radiometric dates of twenty meteorites. If you run the figures through Microsoft Excel, you will find that they are 98.7% in agreement. There is only a seven percent difference between the ratio of the smallest and oldest dates. As we have seen in this essay, such a perfect fit is attained by selecting data and ignoring other data. A careful study of the latest research shows that such perfection is illusionary at best. The Bible believer who accepts the creation account literally has no problem with such unreliable dating methods. Much of the data in Dalrymple's book is selectively taken to suit and ignores data to the contrary.

References

1 **Origin of the Indian Ocean-type isotopic signature, Journal Of Geophysical Research, 1998, Volume 103, Number B9, Pages 20,963**
2 **Reference 1, Pages 20965, 20969**
3 **Reference 1, Pages 20968, 20969**
4 **U–Th–Pb Dating Of Secondary Minerals, Geochimica et Cosmochimica Acta, 2008, Volume 72, Pages 2067**
5 **Reference 4, Pages 2067, 2068**
6 **Reference 4, Pages 2072-2073, 2074**
7 **Reference 4, Pages 2080, 2081**
8 **Radioactive and Stable Isotope Geology, By H.G. Attendon, Chapman And Hall Publishers, 1997. Page 73 [Rb/Sr], 195 [K/Ar], 295 [Re/OS], 305 [Nd/Nd].**
9 **Principles of Isotope Geology, Second Edition, By Gunter Faure, Published By John Wiley And Sons, New York, 1986. Pages 120 [Rb/Sr], 205 [Nd/Sm], 252 [Lu/Hf], 266 [Re/OS], 269 [Os/OS].**
10 **Absolute Age Determination, Mebus A. Geyh, Springer-Verlag Publishers, Berlin, 1990. Pages 80 [Rb/Sr], 98 [Nd/Sm], 108 [Lu/Hf], 112 [Re/OS].**
11 **Radiogenic Isotope Geology, Second Edition, By Alan P. Dickin, Cambridge University Press, 2005. Pages 43 [Rb/Sr], 70 [Nd/Sm], 205 [Re/OS], 208 [Pt/OS], 232 [Lu/Hf].**
12 **Early Archaean Rocks At Fyfe Hills, Precambrian Research, Volume 21, 1983, Pages 197**
13 **Reference 12, Page 211**
14 **Reference 12, Page 215**
15 **Shock-Melted Antarctic LL-Chondrites, Geochimica et Cosmochimica Acta, 1990, Voume 54, Pages 3509**
16 **Reference 15, Page 3517**
17 **Diamonds And Mantle-Derived Xenoliths, Earth and Planetary Science Letters, Volume 42, 1979, Pages 58**
18 **Reference 17, Page 66**
19 **Reference 17, Page 64**
20 **87Rb-87Sr Isochron Of The Norton County Achondrite, Earth And Planetary Science Letters, Volume 3, 1967, Pages 179**
21 **Reference 20, Page 182**
22 **Base and Precious Metal Veins, Economic Geology, Volume 97, 2002, Pages 23**
23 **Reference 22, Page 27, 28**
24 **Reference 22, Page 29**
25 **Reference 22, Page 34-37**
26 **The Munchberg Massif, Southern Germany, Earth and Planetary Science Letters, Volume 99, 1990, Pages 230**
27 **Rocks of the Central Wyoming Province, Canadian Journal Of Earth Science, 2006, Volume 43, Pages 1419**
28 **Reference 27, Page 1436-1437**
29 **Reference 27, Page 1439**
30 **Basalts From Apollo 15, Earth and Planetary Science Letters, Volume 17, 1973, Pages 324**

31 Reference 30, Page 334
32 Reference 30, Page 332
33 History Of The Pasamonte Achondrite, Earth and Planetary Science Letters, Volume 37, 1977, Pages 1
34 Reference 33, Pages 3, 9
35 Sr Isotopic Composition Of Afar Volcanics, Earth and Planetary Science Letters, Volume 50, 1980, Pages 247
36 Reference 35, Page 249
37 Reference 35, Page 250, 251
38 Orogenic Lherzolite Complexes, Earth and Planetary Science Letters, Volume 51, 1980, Pages 71
39 Reference 37, Page 72
40 Reference 37, Pages 78-80
40 Isotopic Geochemistry (O, Sr, Pb), Earth and Planetary Science Letters, Volume 61, 1982, Pages 97
41 Reference 40, Pages 101, 102
42 Reference 40, Pages 104
43 Cretaceous-Tertiary Boundary Sediments, Earth and Planetary Science Letters, Volume 64, 1983, Pages 356
44 Reference 43, Pages 361
45 Correlated N D, Sr And Pb Isotope Variation, Earth and Planetary Science Letters, Volume 59, 1982, Pages 327
46 Reference 45, Pages 330, 331
47 A Depleted Mantle Source For Kimberlites, Earth and Planetary Science Letters, Volume 73, 1985, Pages 269
48 Reference 47, Pages 270
49 Reference 47, Pages 271, 273
50 Sm-Nd Isotopic Systematics, Earth and Planetary Science Letters, Volume 71, 1984, Pages 46
51 Reference 50, Pages 49
52 Strontium, Neodymium And Lead Compositions, Earth and Planetary Science Letters, Volume 75, 1985, Pages 354-368
53 Reference 52, Pages 356, 363
54 Trace Element And Sr And Nd Isotope, Earth and Planetary Science Letters, Volume 80, 1986, Pages 281-298
55 Reference 54, Pages 287
56 Reference 54, Pages 289
57 The southeast Australian Lithosphere Mantle, Earth and Planetary Science Letters, Volume 86, 1987, Pages 327
58 Reference 57, Pages 332
59 Reference 57, Pages 330, 332
60 Strontium, neodymium and lead isotopic, Earth and Planetary Science Letters, Volume 90, 1988, Pages 26-40
61 Reference 60, Pages 35
62 Reference 60, Pages 31
63 Sr, Nd, and Os isotope geochemistry, Earth and Planetary Science Letters, Volume 99, 1990, Pages 362
64 Reference 63, Pages 364
65 Reference 63, Pages 365, 368
66 Pb, Nd and Sr isotopic geochemistry, Earth and Planetary Science Letters, Volume 105, 1991, Pages 149
67 Reference 66, Pages 154, 160
67 Reference 66, Pages 156, 157
68 Sr, Nd, and Pb isotopes, Earth and Planetary Science Letters, Volume 113, 1992, Pages 107
69 Reference 68, Pages 110
70 Production of Jurassic Rhyolite, Earth and Planetary Science Letters, Volume 134, 1995, Pages 23-36
71 Reference 70, Pages 25
72 Evolution of Reunion Hotspot Mantle, Earth and Planetary Science Letters, Volume 134, 1995, Pages 169-185
73 Reference 72, Pages 173
73 Reference 72, Pages 174
74 Reference 72, Pages 180
75 An extremely low U/Pb source, Geochimica et Cosmochimica Acta, 1993, Volume 57, Pages 4687-4702
76 Reference 75, Pages 4690, 4691
77 The 72 Ma Geochemical Evolution, Earth and Planetary Science Letters, Volume 183, 2000, Pages 73
78 Reference 77, Pages 76-79
79 The Himalayan collision zone, Earth and Planetary Science Letters, Volume 244, 2006, Pages 234

80 Reference 79, Pages 234, 235
81 Reference 79, Pages 238
82 Reference 79, Pages 242
83 Evidence for a Non Magmatic Component, Geochimica et Cosmochimica Acta, 2001, Volume 65, Number 4, Pages 571
84 Reference 83, Pages 581
85 Reference 83, Pages 576, 577
86 The Origin of Geochemical Diversity, Geochimica et Cosmochimica Acta, Volume 71, 2007, Pages 3656
87 Reference 86, Pages 3661
88 Reference 86, Pages 3660
89 Mechanisms for Incompatible-Element Enrichment, Geochimica et Cosmochimica Acta, Volume 73, 2009, Pages 3963
90 Reference 89, Pages 3967
91 Constraints on Martian Differentiation Processes, Geochimica et Cosmochimica Acta, 1997, Volume 61, Number 22, Pages 4915
92 Reference 91, Pages 4918, 4924
93 Geochemistry of the Volcan de l'Androy, Journal Of Petrology, 2008, Volume 49, Number 6, Pages 1069
94 Reference 93, Pages 1078
95 Continental Lithospheric Contribution, Journal Of Petrology, 1997, Volume 38, Number 1, Pages 115
96 Reference 95, Pages 119
97 Reference 95, Pages 124
98 Garnet Granulite Xenoliths, Journal Of Petrology, 2001, Volume 42, Number 4, Pages 731
99 Reference 98, Pages 742, 743
100 Reference 98, Pages 737-740
101 http://petrology.oxfordjournals.org/content/suppl/2001/04/27/42.4.731.DC1/ege033SUPPLEM.csv
102 The Isotope and Trace Element Budget, Journal Of Petrology, 2000, Volume 41, Number 6, Pages 759
103 Reference 102, Pages 772-774
104 Fluid Flow and Diffusion, Journal Of Petrology, 1997, Volume 38, Number 11, Pages 1489
105 Reference 104, Pages 1497
106 Reference 104, Pages 1498
107 Reference 104, Pages 1492-1495
108 Temporal Evolution of the Lithospheric Mantle, Journal Of Petrology, 2009, Volume 50, Number 10, Pages 1857
109 Reference 108, Pages 1873, 1874, 1877, 1879, 1880
110 Petrogenesis and Origins of Mid-Cretaceous, Journal Of Petrology, 2010, Volume 51, Number 10, Pages 2003-2045
111 Reference 110, Pages 2038
112 Reference 110, Pages 2024-2026
113 The Petrogenetic Association of Carbonatite, Journal Of Petrology, 1999, Volume 40, Number 4, Pages 525
114 Reference 113, Pages 534, 535
115 Geochemistry of Jurassic Oceanic Crust, Journal Of Petrology, 1998, Volume 39, Number 5, Pages 859–880
116 Reference 115, Pages 867, 868
117 The age of Dar al Gani 476, Geochimica Et Cosmochimica Acta, 2003, Volume 67, Number 18, Pages 3519–3536
118 Reference 117, Pages 3523
119 Petrogenesis of the Flood Basalts, Journal Of Petrology, 1998, Volume 39, Number 4, Pages 711–748
120 Reference 119, Pages 729, 730
121 Nature of the Source Regions, Journal Of Petrology, 2004, Volume 45, Number 3, Pages 555
122 Reference 121, Pages 556
123 Reference 121, Pages 566, 575, 576
124 Generation of Palaeocene Adakitic Andesites, Journal Of Petrology, 2007, Volume 48, Number 4, Pages 661
125 Reference 124, Pages 676-678
126 Reference 124, Pages 684
127 Evidence for a Widespread Tethyan, Journal Of Petrology, 2005, Volume 46, Number 4, Pages 829-858
128 Reference 127, Pages 831
129 Reference 127, Pages 840
130 Reference 127, Pages 832-837

131 Post-Collisional Potassic and Ultrapotassic , Journal Of Petrology, 1999, Volume 40, Number 9, Pages 1399-1424
132 Reference 131, Pages 1403, 1405, 1406
133 Reference 131, Pages 1414, 1415
132 The Age Of The Earth, By G. Brent Dalrymple, 1991, Stanford University Press, Stanford, California, Page 10.
133 Reference 142, Page 23
134 Reference 142, Page 287

Chapter 18

The Thorium Lead Dating Method

Sr, Nd, and Pb isotopes

According to the article the rock formation is 2,900 million years. [1] If we run the Pb/Th ratios [1] through Isoplot we see that the dates vary from 79 to over 94,000 million years old [Table 1]. In Table 2 we can see the maximum ages for the Thorium/Lead dating method.

Table 1: Uranium/Thorium/Lead - Ages Summary

Dating	232Th/208Pb	206Pb/238U	207Pb/206Pb
Summaries	Age	Age	Age
Average	14,198	7,366	5,014
Maximum	94,396	22,201	5,077
Minimum	79	1,117	4,945
Difference	94,317	21,083	131

Table 2: Thorium/Lead – Maximum Ages

Age	Age	Age	Age
94,396	39,267	10,595	8,171
90,683	26,266	10,284	7,789
74,639	18,334	9,328	7,638
58,153	16,357	8,821	7,375
55,324	14,250	8,771	7,317
45,242	11,215	8,403	5,759

An Extremely low U/Pb Source

According to the article: "The Rb-Sr data yield an internal isochron age of 3,840 ± 32 Ma." [2] If we run the Pb/Th ratios [3] through Isoplot we see that the dates vary from 5,000 to over 13,000 million years old. In Table 3 we can see the maximum ages for the Thorium/Lead dating method.

Table 3: Uranium/Thorium/Lead - Ages Summary

Table	207Pb/206Pb	206Pb/238U	208Pb/232Th	207Pb/235U	87Rb/86Sr
Summaries	Age	Age	Age	Age	Age
Average	4,673	8,035	10,148	4,546	3,619
Maximum	5,018	56,923	65,286	8,128	5,385
Minimum	3,961	1,477	2,542	2,784	721
Difference	1,057	55,445	62,744	5,344	4,664

Table 4: Thorium/Lead – Maximum Ages

Age	Age	Age	Age
65,286	14,430	9,094	5,401
33,898	14,410	6,520	5,396
25,013	13,107	6,166	5,365
22,178	12,738	6,121	5,098
21,204	11,641	5,671	5,035
17,611	11,174	5,408	4,678

Petrogenesis and Origins of Mid-Cretaceous

According to the article: "The basal lava flow displays a sharp contact with the underlying terrestrial sediments, which in turn rest on shallow marine sediments of Ngaterian age (100.2-95.2Ma)." [4] If we run the Rb/Sr ratios [5] through Microsoft Excel we see that the dates vary from 15 to 85 million years old. If we run the Pb/Th ratios [6] through Isoplot we see that the dates vary from 4,000 to over 10,000 million years old. In Table 5 we can see the maximum ages for the Thorium/Lead dating method.

Table 5: Dating Methods - Ages Summary

Table	207Pb/206Pb	207Pb/235U	87Rb/86Sr	208Pb/232Th	206Pb/238U
Summaries	Age	Age	Age	Age	Age
Average	4,876	4,416	59	6,333	3,515
Maximum	4,945	5,159	85	10,716	5,717
Minimum	4,836	4,088	15	4,785	2,712
Difference	109	1,071	70	5,931	3,005

Table 6: Thorium/Lead – Maximum Ages

Age	Age	Age
10,716	6,355	5,655
7,520	6,354	5,598
7,259	6,138	5,519
7,145	6,032	5,515
6,559	5,972	5,505
6,511	5,972	5,210

Tracing the Indian Ocean Mantle

These samples were dated in 1998 by scientists from the School Of Ocean And Earth Science And Technology, University Of Hawaii, Honolulu. According to this article the samples were taken from volcanic material that is only 100 million years old.[7] If we put isotopic ratios [8] into Microsoft Excel and run the through Isoplot [9] we find the average age is almost 17 billion years old. In Table 8 we see some fantastic dates.

Table 7

Average	16,890
Maximum	82,561
Minimum	1,139
Difference	81,422

Table 8: Thorium/Lead – Maximum Ages

$^{206}Pb/^{238}U$	$^{206}Pb/^{238}U$	$^{208}Pb/^{232}Th$	$^{208}Pb/^{232}Th$	$^{208}Pb/^{232}Th$	$^{207}Pb/^{206}Pb$	$^{207}Pb/^{206}Pb$	$^{207}Pb/^{206}Pb$
15,747	7,504	58,796	22,718	11,427	5,236	5,013	4,960
15,067	7,056	54,206	22,307	11,377	5,062	5,012	4,959
14,363	7,002	48,252	22,228	11,366	5,052	5,012	4,959
13,580	6,602	47,976	21,827	11,241	5,052	5,012	4,959
13,204	6,589	46,117	21,560	10,718	5,051	5,011	4,953
12,780	6,421	42,203	19,910	10,699	5,051	5,010	4,951
11,757	6,398	42,192	19,594	10,699	5,049	5,010	4,951
11,659	6,369	41,604	19,148	10,300	5,044	5,008	4,947
11,537	6,357	41,343	18,765	9,357	5,043	5,006	4,947
11,313	6,219	41,231	18,607	8,632	5,043	5,006	4,942
11,309	5,863	39,637	18,121	8,486	5,042	5,005	4,939
11,248	5,861	38,125	17,797	8,057	5,041	5,005	4,930
10,360	5,807	37,115	17,787	6,497	5,041	5,004	4,928
9,643	5,724	35,012	17,591	5,573	5,038	5,004	4,901
9,427	5,720	33,584	17,536	5,425	5,029	5,002	4,873
9,300	5,601	31,556	17,054	4,869	5,027	5,001	4,822
9,123	5,599	31,286	16,053		5,027	4,998	
9,014	5,573	30,740	15,299		5,025	4,995	
8,996	5,515	29,705	14,340		5,021	4,990	
8,954	5,462	27,710	13,845		5,020	4,989	
8,770	5,311	27,422	13,772		5,018	4,988	
8,508	5,286	26,674	13,652		5,016	4,982	
8,315	5,120	26,369	13,404		5,015	4,981	
8,314	5,115	25,972	13,403		5,014	4,967	
8,072	5,042	25,590	13,006		5,014	4,964	
8,024	4,749	25,096	12,171		5,013	4,962	
7,604	4,702	24,010	11,540		5,013	4,960	

Petro Genesis of the Flood Basalts

These samples were dated in 1998 by scientists from the Department Of Earth, Atmospheric And Planetary Sciences, Massachusetts Institute of Technology. According to this article the samples were taken from the volcanic crust of the Kerguelen Archipelago that is only 30 million years old.[10] If we put isotopic ratios [11] into Microsoft Excel and run the through Isoplot we find the average age of Mount Bureau is over 5 billion years old. In Table 9 we see some fantastic dates for both mountains.

Table 9: Mount Bureau – Maximum Ages

$^{207}Pb/^{206}Pb$	$^{206}Pb/^{238}U$	$^{207}Pb/^{235}U$	$^{208}Pb/^{232}Th$
5,020	23,366	8,496	44,378
5,019	6,607	5,428	9,092
5,018	6,329	5,365	8,651
5,017	6,194	5,330	8,624
5,015	5,726	5,213	8,144

5,013	5,355	5,100	8,142
5,012	5,260	5,082	8,023
5,012	5,198	5,062	7,788
5,012	5,163	5,055	7,518
5,011	5,158	5,052	7,507
5,010	5,132	5,048	7,416
5,009	5,127	5,037	7,245
5,008	5,117	5,035	7,046
5,007	5,094	5,034	6,961
5,006	5,094	5,034	6,560

Table 10: Mount Rabouillere – Maximum Ages

$^{207}Pb/^{206}Pb$	$^{206}Pb/^{238}U$	$^{207}Pb/^{235}U$	$^{208}Pb/^{232}Th$
5,006	5,084	5,030	6,548
5,006	4,894	4,964	6,422
5,005		4,945	6,328
5,004		4,938	6,216
5,004		4,915	5,966
5,004		4,913	5,787
5,004		4,879	5,773
5,003		4,815	5,639
5,001			5,613
5,001			5,107
5,000			
5,000			
4,994			
4,994			
4,994			

Nature of the Source Regions

These samples were dated in 2004 by scientists from the Department Of Earth Sciences, The Open University, England. According to the article: "Most samples are Miocene in age, ranging from 10 to 25Ma in the south and 19Ma to the present day in northern Tibet."[12, 13] If we run the 87Rb/86Sr ratios [14] in the essay through Isoplot we get dates between 1 and 24 million years. If we run the Uranium/Lead ratios [15] in the essay through Isoplot we get unbelievable dates as listed below in Table 11.

Table 11: North Tibet – Maximum Ages

$^{208}Pb/^{232}Th$	$^{206}Pb/^{238}U$	$^{207}Pb/^{235}U$	$^{207}Pb/^{206}Pb$
11,420	7,783	5,136	4980
11,350	8,023	5,138	4980
13,475	8,305	5,135	4987
11,504	7,349	5,140	4989
81,614	33,751	7,470	4987
88,294	33,742	7,471	4991

Table 12: South Tibet – Maximum Ages

$^{208}Pb/^{232}Th$	$^{206}Pb/^{238}U$	$^{207}Pb/^{235}U$	$^{207}Pb/^{206}Pb$
11,102	6,331	313	4982
6,092	5,799	946	4919
9,265	6,682	266	4980
4,826	4,086	238	4992
8,205	5,567	294	4980
25,015	13,328	447	4994
33,191	15,053	482	4992

Generation of Palaeocene Adakitic Andesites

These samples were dated in 2007 by scientists from the Chinese Academy Of Sciences, Wushan, Guangzhou. According to the article: "The initial Sr, Nd and Pb isotopic ratios were corrected using the Ar/Ar age of 55Ma."[16, 17] If we run the Uranium/Lead ratios [18] in the essay through Isoplot we get unbelievable dates as listed below in Table 13.

Table 13: Thorium/Lead – Maximum Ages

$^{207}Pb/^{206}Pb$	$^{208}Pb/^{232}Th$	$^{206}Pb/^{238}U$	$^{207}Pb/^{235}U$
5,024	10,518	9,669	6,052
5,023	10,277	9,552	6,051
5,023	8,529	9,526	6,051
5,023	8,360	8,443	5,828
5,021	8,165	7,929	5,826
5,020	7,800	7,403	5,641

Evidence for a Widespread Tethyan Upper Mantle

In 2005 scientists from the School of Ocean and Earth Science and Technology, University of Hawaii, Honolulu dated these rocks. According to the article: "Isotopic data for such sites show that mantle similar to that beneath the modern Indian Ocean was present, at least in places, as long ago as 140 Ma, the age of the oldest true Indian Ocean crust yet sampled." [19, 20] If we run the Rb/Sr ratios [21] through Isoplot we see that the average age is 168 million years. [Table 14]

Table 14: Rb/Sr Ages Summary

Average	168
Maximum	1,739
Minimum	0
Difference	1,739

If we run the Pb/Th ratios [22] through Isoplot we see that the average age is 22,675 million years. [Table 15]

Table 15: Pb/Th Ages Summary

$^{206}Pb/^{238}U$	$^{206}Pb/^{238}U$	$^{208}Pb/^{232}Th$	$^{208}Pb/^{232}Th$	$^{208}Pb/^{232}Th$	$^{207}Pb/^{206}Pb$	$^{207}Pb/^{206}Pb$	$^{207}Pb/^{206}Pb$
15,747	7,504	58,796	22,718	11,427	5,236	5,013	4,960
15,067	7,056	54,206	22,307	11,377	5,062	5,012	4,959
14,363	7,002	48,252	22,228	11,366	5,052	5,012	4,959
13,580	6,602	47,976	21,827	11,241	5,052	5,012	4,959
13,204	6,589	46,117	21,560	10,718	5,051	5,011	4,953
12,780	6,421	42,203	19,910	10,699	5,051	5,010	4,951
11,757	6,398	42,192	19,594	10,699	5,049	5,010	4,951
11,659	6,369	41,604	19,148	10,300	5,044	5,008	4,947
11,537	6,357	41,343	18,765	9,357	5,043	5,006	4,947
11,313	6,219	41,231	18,607	8,632	5,043	5,006	4,942
11,309	5,863	39,637	18,121	8,486	5,042	5,005	4,939
11,248	5,861	38,125	17,797	8,057	5,041	5,005	4,930
10,360	5,807	37,115	17,787	6,497	5,041	5,004	4,928
9,643	5,724	35,012	17,591	5,573	5,038	5,004	4,901
9,427	5,720	33,584	17,536	5,425	5,029	5,002	4,873
9,300	5,601	31,556	17,054	4,869	5,027	5,001	4,822
9,123	5,599	31,286	16,053		5,027	4,998	
9,014	5,573	30,740	15,299		5,025	4,995	
8,996	5,515	29,705	14,340		5,021	4,990	
8,954	5,462	27,710	13,845		5,020	4,989	
8,770	5,311	27,422	13,772		5,018	4,988	
8,508	5,286	26,674	13,652		5,016	4,982	
8,315	5,120	26,369	13,404		5,015	4,981	
8,314	5,115	25,972	13,403		5,014	4,967	
8,072	5,042	25,590	13,006		5,014	4,964	
8,024	4,749	25,096	12,171		5,013	4,962	
7,604	4,702	24,010	11,540		5,013	4,960	

Post-Collisional Potassic and Ultrapotassic

According to the article: "Major and trace element, Sr–Nd–Pb–O isotope and mineral chemical data are presented for post-collisional ultrapotassic, silicic and high-K calc-alkaline volcanic rocks from SW Tibet, with 40Ar/39Ar ages in the range 17–25 Ma." [23, 24] If we run the Rb/Sr ratios [25] through Isoplot we see that the average age is 43 million years. [Table 16]

Table 16: Rb/Sr Ages Summary

Average	43
Maximum	1,258
Minimum	-1,439
Difference	2,697

If we run the Pb/Th ratios [26] through Isoplot we see that the average age is 78,808 million years. [Table 17]

Table 17: Pb/Th Ages Summary

$^{207}Pb/^{206}Pb$	$^{208}Pb/^{232}Th$	$^{206}Pb/^{238}U$	Model Age
5,000	68,343	2,646	3,600
5,001	67,704	5,013	3,600
5,007	70,277	3,745	3,600
5,007	71,706	4,715	3,600
4,988	95,541	5,560	3,600
4,997	99,275	6,944	3,600

In Table 17 we see a comparison between the model age ["True Age"] and the isotopic age derived from atomic ratios. We can see how far in error the Thorium dating system is.

Continental Lithospheric Contribution to Alkaline

According to the article: "These two genetically related alkaline complexes were emplaced at the east Atlantic continent-ocean boundary during the Upper Cretaceous, i.e. 66-72 m. y. ago" [27] If we run the Rb/Sr ratios [28] through Isoplot we see that the average age is 65 million years. [Table 18]

Table 18: Rb/Sr Ages Summary

Average	65
Maximum	74
Minimum	4
Difference	78

If we run the Pb/Th ratios [28] through Isoplot we see that the average age is 6,126 million years. [Table 19]

Table 19: Pb/Th Ages Summary

$^{207}Pb/^{206}Pb$	$^{208}Pb/^{232}Th$	$^{206}Pb/^{238}U$
4,888	2,616	2,306
4,906	4,005	2,397
4,907	4,188	2,852
4,908	4,237	2,965
4,913	4,416	3,150
4,914	5,029	3,487
4,917	5,755	3,761
4,918	5,830	3,768
4,924	6,313	4,186
4,925	6,655	4,734
4,925	6,891	5,143
4,926	7,080	6,206
4,926	7,502	6,387
4,930	8,101	6,507
4,932	9,320	7,060
4,949	10,084	7,743

Pin Pricking The Elephant

According to tables [29] in the article, the rock formation is only 120 million years old. If we run the $^{207}Pb/^{206}Pb$ ratios [30] through Isoplot we get an average age of 5,000 million years. If we run the Pb/Th ratios [31] through Isoplot we see in Table 20 that the age is between 12 billion and 14 billion years old.

Table 20

208Pb/232Th	206Pb/238U	$^{207}Pb/^{206}Pb$
12,090	4,579	5,379
12,845	5,498	5,385
14,459	6,936	5,000

Chronology And Geochemistry Of Lavas

According to the article: "New $^{40}Ar/^{39}Ar$ incremental heating age determinations for dredged rocks from volcanoes east of Salas y Gomez Island show that, with very few exceptions, ages increase steadily to the east from 1.4 to 30 Ma" [32] Tables [33] in the article affirms this as the true age of the geological formation. [33] If we run the Pb/Th ratios [34] through Isoplot we see that the average age is 8,325 million years. In Table 21 we see some of the incredible dates all the way from 5 billion to almost 24 billion years old.

Table 21: Pb/Th Ages Summary

$^{207}Pb/^{206}Pb$	206Pb/238U	208Pb/232Th
4,881	1,166	4,129
4,886	2,380	4,462
4,896	2,801	5,446
4,897	2,828	5,453
4,899	2,915	5,754
4,906	2,965	5,848
4,907	2,996	5,896
4,907	3,063	6,245
4,908	3,170	6,298
4,908	3,175	6,396
4,909	3,249	6,421
4,910	3,266	6,498
4,918	3,273	6,607
4,922	3,350	7,054
4,922	3,358	7,101
4,923	3,421	7,599
4,926	3,433	7,654
4,927	3,511	8,393
4,930	3,723	9,061
4,953	3,818	13,004
4,964	6,315	15,364
4,971	7,140	16,942
4,971	9,645	23,850

Ion Microprobe U-Th-Pb Dating

According to the article: "The formation age of this meteorite is 1.53 ± 0.46 Ga. On the other hand, the data of nine apatite grains from Lafayette are well represented by planar regression rather than linear regression, indicating that its formation age is 1.15 ± 0.34 Ga" [35] If we run the Pb/Th ratios [36] through Isoplot we see that the age is up to 23,800 million years. [Table 22]

Table 22: Uranium/Thorium/Lead - Ages Summary

U238/Pb206	Th232/Pb208	$^{207}Pb/^{206}Pb$
11,214	23,837	5,348
8,087	17,095	5,103
7,820	16,633	5,093
7,810	15,784	5,019
6,568	14,905	5,000
6,330	14,091	4,922
5,552	12,754	4,905
4,901	11,781	4,893
4,699	11,359	4,835
4,681	10,694	4,776
4,427	10,423	4,756
4,062	8,734	4,750
4,041	7,573	4,710
4,005	7,013	4,654
3,959	6,444	4,637
3,549	5,612	4,625
2,813	5,493	4,581
2,050	3,184	4,087
1,842	2,963	3,897

U-Th-Pb Dating Of Secondary Minerals

This dating was done in 2008 on minerals from Yucca Mountain, Nevada. It was done by scientists from the U.S. Geological Survey, Denver, Colorado, the Geological Survey of Canada, Ottawa, Ontario and the Research School of Earth Sciences and Planetary Science Institute, The Australian National University. According to the article: "Most $^{206}Pb/^{238}U$ ages determined for the calcite subsamples are much older than the 12.8-Ma age of the host tuff (Fig. 5) and thus unreasonable." [37] If we run the Pb/Th ratios [38] through Isoplot we see that the average age is 10,000 million years. The Rb/Sr ratios [39] gave a uniform result of 11 to 13 million years old [Table 23].

Table 23: 208Pb/232Th Ages Versus Rb/Sr Ages

Summary	$^{207}Pb/^{206}Pb$	206Pb/238U	208Pb/232Th	87Rb/86Sr
Average	3,459	4,891	9,984	12
Maximum	8,126	31,193	352,962	13
Minimum	-445	1	2	11
Difference	8,571	31,192	352,960	2

Another set of dates [40] in the essay [Table 24] give dates as high as 352 billion years old.

Table 24: Uranium/Thorium/Lead - Ages Summary

$^{207}Pb/^{206}Pb$	$^{207}Pb/^{206}Pb$	206Pb/238U	206Pb/238U	206Pb/238U	208Pb/232Th	208Pb/232Th	207Pb/235U
8,126	3,092	31,193	2,173	301	352,962	161	486
5,024	2,876	28,058	2,023	280	82,030	159	285
5,023	2,759	23,318	1,978	249	57,900	112	277
5,020	2,639	21,152	1,872	185	28,600	96	162
5,019	2,621	20,209	1,738	176	12,900	88	81
5,000	2,258	16,434	1,711	163	5,389	75	64
4,999	2,214	15,648	1,473	149	3,225	69	49
4,989	2,167	15,434	1,455	146	1,951	66	29
4,965	1,820	13,503	1,404	145	1,864	63	23
4,959	1,384	12,301	1,304	143	1,839	59	19
4,950	1,241	11,023	1,044	119	1,557	58	16
4,949	1,103	8,442	997	104	1,555	54	15
4,943	882	8,010	921	81	1,300	53	12
4,904	472	7,940	830	78	1,120	49	11
4,883	282	7,556	798	67	938	43	10
4,845	-54	7,504	776	42	649	42	9
4,837	-332	6,416	745	38	607	37	6
4,671	-445	6,372	716	18	529	35	5
4,654		6,372	696	12	489	28	4
4,566		6,292	631	11	454	26	3
4,333		5,612	506	10	445	25	
4,122		4,844	475	8	276	22	
4,073		4,795	457	7	250	19	
4,012		4,511	442	6	236	14	
3,884		4,423	437	3	216	11	
3,818		3,557	420	1	208	9	
3,541		3,353	408		177	5	
3,365		3,270	399		167	3	
3,141		2,734	365		166	2	

The Influence of High U-Th Inclusions

This dating was done in 1998 by scientists from Zurich, Switzerland. According to the article: "The U-Th-Pb data from the bulk dissolutions are highly complex and yield apparent ages ranging from 1000 Ma to 30 Ma." [41]

If we run the Pb/Th ratios [42,43] through Isoplot we see that the dates vary from 300 to over 14,000 million years old [Table 25].

Table 25: Uranium/Thorium/Lead - Ages Summary

206Pb/238U	208-Pb/232-Th	207Pb/206Pb
270	288	4,720
277	307	4,721
321	354	4,813
398	467	4,835
660	1,333	4,882
695	1,816	4,897

765	1,944	4,905
797	1,957	4,961
881	3,061	4,981
7,832	13,524	4,981
8,025	14,316	4,986
19,484		4,987
29,040		5,000
		5,042

U, Th And Pb Isotope Compositions

These samples were dated in 2009 by scientists from the Arthur Holmes Isotope Geology Laboratory, Department of Earth Sciences, Durham University. [44] According to the article: "Detailed petrographic and geochemical descriptions of the samples presented here can be found elsewhere" [45] If we examine what these other people [46-49] have said about the same rock formation the consensus is that it is three million years old. If we run the Pb/Th ratios [50] through Isoplot we see that the dates vary from 2,000 to over 92,000 million years old [Table 26].

Table 26: Uranium/Thorium/Lead - Ages Summary

207Pb/206Pb	207Pb/206Pb	206Pb/238U	206Pb/238U	208Pb/232Th	208Pb/232Th
4,871	4,964	1,437	3,693	1,939	10,022
4,873	4,984	1,457	3,767	2,018	10,621
4,874	4,996	1,492	4,508	2,051	10,956
4,877	5,000	1,524	5,625	2,058	13,018
4,881	5,001	1,527	9,291	2,109	29,253
4,882	5,008	1,577	10,312	2,238	42,038
4,883	5,016	1,679	10,772	2,394	73,503
4,890	5,021	1,966	15,307	2,598	92,495
4,893	5,025	1,980	18,639	2,799	
4,923	5,068	2,131		3,133	
4,943	5,085	2,259		4,429	
4,949	5,086	2,921		4,956	
4,959	5,090	3,348		5,204	

U-Th-Pb Isotope Data

According to the article: "In contrast to the apparent ^{207}Pb–^{206}Pb ages, the minimum depositional age of the Warrawoona Group is 3,426Ma based on a U–Pb zircon age from the Panorama Formation." [51] If we run the Pb/Th ratios [52] through Isoplot we see that the dates vary from 25,000 to over 100,000 million years old. In Table 27 we can see the maximum ages for each dating method.

Table 27: Uranium/Thorium/Lead - Ages Summary

207Pb/206Pb	206Pb/238U	208Pb/232Th	206Pb/235U
5,403	31,005	100,601	63,067
5,395	20,343	84,457	51,813
5,390	19,584	73,968	50,973
5,351	17,306	67,423	48,585
5,339	17,088	58,353	48,346
5,332	13,410	57,116	44,333

5,328	13,022	55,311	43,879
5,315	11,479	51,607	42,056
5,298	11,353	44,439	41,899
5,296	10,652	39,090	41,043
5,289	9,926	26,361	40,146
5,269	7,138	24,980	36,327

Evolution Of Reunion Hotspot Mantle

According to the article: "In the same context, the Trend 1 data imply that (1) the isotopic composition of the Reunion end-member has changed relatively little in the last 66 m.y." [53] If we run the Pb/Th ratios [54] through Isoplot we see that the dates vary from 5,000 to over 13,000 million years old. In Table 28 we can see the maximum ages for the Thorium/Lead dating method.

Table 28: Thorium/Lead – Maximum Ages

238U/206Pb	232Th/208Pb	207Pb/206Pb
3,010	5,641	4,953
3,369	5,660	4,954
3,548	6,323	4,957
3,647	6,479	4,962
3,875	6,540	4,963
3,973	6,679	4,964
4,001	7,017	4,965
4,011	7,080	4,967
4,133	7,362	4,969
4,141	7,363	4,969
4,220	7,446	4,970
4,258	7,517	4,971
4,342	7,541	4,973
4,633	8,609	4,989
4,903	8,725	4,991
5,152	9,101	4,995
5,553	10,357	4,999
5,767	11,017	4,999
6,165	11,832	5,000
6,285	13,287	5,016

Continental Growth 3.2 Gyr Ago

According to the article the rock formation is 3,200 million years old. [55] If we run the Pb/Th ratios [55] through Isoplot we see that the dates vary from negative 24,000 to over 11,000 million years old [Table 29]. In Table 30 we can see the maximum ages for the Thorium/Lead dating method.

Table 29: Uranium/Thorium/Lead - Ages Summary

Summary	208Pb/232Th	238U/206Pb	207Pb/206Pb
Average	3,273	3,300	3,296
Maximum	11,517	4,463	3,897
Minimum	-24,295	1,560	2,667
Difference	35,813	2,902	1,229

Table 30:cThorium/Lead – Maximum Ages

Age	Age	Age	Age	Age
11,517	5,322	5,083	4,668	4,601
6,027	5,289	4,776	4,662	-366
5,806	5,130	4,709	4,638	-2,485
5,704	5,095	4,704	4,614	-24,295
5,568	5,085	4,690	4,610	-24,295

Uranium-Lead Zircon Ages

If we run the Pb/Th ratios [56] through Isoplot we see that the dates vary from 6,000 to over 55,000 million years old [Table 31]. In Table 32 we can see the maximum ages for each dating method.

Table 31: Uranium/Thorium/Lead - Ages Summary

Dating Summary	206Pb/238U Age	208Pb/232Th Age	207Pb/206Pb Age
Average	11,159	17,193	4,933
Maximum	23,421	55,110	4,997
Minimum	3,108	6,130	4,799
Std Deviation	6,223	13,524	59

Table 32: Uranium/Thorium/Lead – Maximum Ages

206Pb/238U Age	208Pb/232Th Age	207Pb/206Pb Age
23,421	55,110	4,997
20,387	29,742	4,991
18,909	27,889	4,981
17,143	27,051	4,976
16,784	21,318	4,972
15,320	19,224	4,969
12,851	18,091	4,965
12,012	17,944	4,957
10,579	16,474	4,953
9,677	15,059	4,949
9,424	14,779	4,947
9,099	13,374	4,945
9,044	11,951	4,925
8,094	10,783	4,921

6,776	9,336	4,915
5,719	8,644	4,910
5,500	8,058	4,892

Table 33: Thorium/Lead – Maximum Ages

Age	Age	Age	Age
55,110	19,224	14,779	8,644
29,742	18,091	13,374	8,058
27,889	17,944	11,951	6,721
27,051	16,474	10,783	6,185
21,318	15,059	9,336	6,130

The Pilbara Craton in Western Australia

According to the article the rock formation is 3,200 million years old. [57] If we run the Pb/Th ratios [58] through Isoplot we see that the dates vary from 2,000 to over 8,000 million years old [Table 34]. In Table 35 we can see the maximum ages for the Thorium/Lead dating method.

Table 34: Thorium/Lead - Ages Summary

Average	4,853
Maximum	8,728
Minimum	2,792
Std Deviation	1,040

Table 35: Thorium/Lead – Maximum Ages

Age	Age	Age	Age	Age
8,728	6,241	5,721	5,430	5,058
8,296	6,191	5,643	5,417	5,042
7,017	6,076	5,578	5,288	5,032
6,433	5,786	5,533	5,171	5,027
6,431	5,759	5,522	5,138	4,999

If we run another set of Pb/Th ratios [59] through Isoplot we see that the dates vary from 500 to over 17,000 million years old [Table 36]. In Table 37 we can see the maximum ages for the Thorium/Lead dating method.

Table 36: Uranium/Thorium/Lead - Ages Summary

Dating	207Pb/235U	206Pb/238U	208Pb/232Th
Summary	Age	Age	Age
Average	2,955	2,956	6,286
Maximum	4,220	8,073	17,500
Minimum	1,921	1,074	535
Std Deviation	392	1,019	3,196

Table 37: Thorium/Lead – Maximum Ages

Age	Age	Age	Age
17,500	8,891	7,493	5,743
13,259	8,768	7,443	5,594
13,100	8,689	7,368	5,512
12,821	8,343	7,343	5,512
12,662	8,320	7,240	5,455
12,212	8,247	7,192	5,432
11,163	8,232	7,148	5,255
10,959	8,197	7,047	5,253
10,783	8,064	6,478	5,229
10,668	8,013	6,270	5,154
10,384	7,949	6,199	5,148
9,945	7,947	6,152	5,135
9,580	7,861	6,083	5,115
9,124	7,702	6,052	5,047
8,908	7,692	5,885	5,033
8,905	7,612	5,803	4,889

Timing of Sedimentation, Metamorphism, and Plutonism

According to the article the rock formation is 478 million years old. [60] If we run the Pb/Th ratios [61] through Isoplot we see that the dates vary from 500 to over 80,000 million years old [Table 39]. In Table 40 we can see the maximum ages for the Thorium/Lead dating method.

Table 39: Thorium/Lead - Ages Summary

Average	19,539
Maximum	80,532
Minimum	489
Std Deviation	27,260

Table 40: Thorium/Lead – Maximum Ages

Age	Age	Age	Age
80,532	66,448	51,879	24,604
74,016	65,076	51,751	16,809
70,713	65,000	51,545	15,748
69,057	61,342	34,766	15,365
68,831	60,335	31,045	13,384
68,503	58,364	28,397	11,945
67,672	56,792	24,733	9,477

U-Th and U-Pb Systematics in Zircons

According to the article: "At Taupo, the zircon model ages range from <20 ka to >500 Ma." [62] If we run the Pb/Th ratios [63] through Isoplot we see that the dates vary from 11,000 to over 41,000 million years old [Table 41]. In Table 42 we can see the maximum ages for the Thorium/Lead dating method.

Table 41: Thorium/Lead - Ages Summary

Average	22,847
Maximum	41,460
Minimum	11,390
Std Deviation	6,191

Table 42: Thorium/Lead – Maximum Ages

Age	Age	Age	Age	Age
41,460	26,447	23,441	21,348	18,534
34,824	25,988	23,025	20,730	18,140
33,392	25,525	22,704	19,977	17,701
29,182	24,858	22,560	19,950	17,357
29,126	24,325	22,493	19,738	16,455
28,671	24,160	22,138	19,422	16,221
27,733	23,992	21,885	19,360	15,726
27,587	23,665	21,877	19,307	15,301
26,533	23,448	21,390	19,024	11,390

Hydrothermal Zebra Dolomite

According to the article the rock formation is 416 million years old. [64] If we run the Pb/Th ratios [65] through Isoplot we see that the dates vary from 6,000 to over 55,000 million years old [Table 43]. In Table 44 we can see the maximum ages for the Thorium/Lead dating method.

Table 43: Uranium/Thorium/Lead - Ages Summary

Dating	Pb206/U238	Pb208/Th232	Pb207/Pb206
Summary	Age	Age	Age
Average	11,353	17,193	4,933
Maximum	23,421	55,110	4,997
Minimum	1,715	6,130	4,799
Std Deviation	5,055	11,459	53

Table 44: Thorium/Lead – Maximum Ages

Age	Age
55,110	14,779
29,742	13,374
27,889	11,951
27,051	10,783
21,318	9,336
19,224	8,644
18,091	8,058
17,944	6,721
16,474	6,185
15,059	6,130

If we run the Pb/Th ratios [65] in the second spreadsheet table through Isoplot we see that the dates vary from 6,000 to over 270,000 million years old [Table 45]. In Table 46 we can see the maximum ages for the Thorium/Lead dating method.

Table 45: Thorium/Lead - Ages Summary

Average	90,690
Maximum	277,727
Minimum	6,643
Std Deviation	47,209

Table 46: Thorium/Lead – Maximum Ages

Billion Years	Quantity	Billion Years	Quantity
0 To 20	2	130 To 140	6
20 To 30	1	140 To 150	6
30 To 40	22	150 To 160	2
40 To 50	19	160 To 170	6
50 To 60	33	170 To 180	1
60 To 70	17	180 To 190	5
70 To 80	23	190 To 200	1
80 To 90	18	200 To 210	3
90 To 100	14	210 To 220	1
100 To 110	18	220 To 230	2
110 To 120	21	240 To 250	1
120 To 130	13	270 To 280	2

Origin of Indian Ocean Seamount Province

According to the article the rock formation is 6 million years old. [66] If we run the Pb/Th ratios [67] through Isoplot we see that the dates vary from 2,000 to over 28,000 million years old [Table 47]. In Table 48 we can see the maximum ages for the Thorium/Lead dating method.

Table 47: Uranium/Thorium/Lead - Ages Summary

Dating	207Pb/206Pb	206Pb/238U	208Pb/232Th
Summary	Age	Age	Age
Average	5,015	5,191	7,740
Maximum	5,087	18,210	28,677
Minimum	4,921	890	1,943
Std Deviation	48	3,634	4,590

Table 48: Thorium/Lead – Maximum Ages

Age	Age	Age	Age	Age
28,677	10,719	9,515	7,923	6,512
12,829	10,626	9,506	7,669	6,333
12,028	10,425	9,146	7,407	6,199
11,798	10,378	9,073	7,380	6,198
11,552	10,240	9,019	7,380	6,085
11,317	10,201	8,916	7,367	6,051
11,113	10,082	8,298	7,030	5,999
10,773	10,055	8,111	6,910	5,493
10,725	9,678	8,001	6,651	5,418

Geochemistry Geophysics Geosystems

According to the article the rock formation is 100 million years old. [68] If we run the Pb/Th ratios [68] through Isoplot we see that the dates vary from 5,000 to over 82,000 million years old [Table 48]. In Table 49 we can see the maximum ages for the Thorium/Lead dating method.

Table 49: Uranium/Thorium/Lead - Ages Summary

Dating	206Pb/238U	207Pb/235U	207Pb/206Pb	208Pb/232Th
Summary	Age	Age	Age	Age
Average	15,345	7,019	4,936	39,068
Maximum	38,340	10,872	5,043	82,865
Minimum	3,125	4,385	4,760	5,577
Std Deviation	9,657	1,750	63	27,390

Table 50: Thorium/Lead – Maximum Ages

Age	Age	Age
82,865	51,821	16,417
81,065	45,608	7,512
75,644	45,035	6,840
72,833	42,233	6,626
64,393	39,019	6,322
58,240	27,562	5,579
57,334	23,571	5,577
56,640	19,834	

Continental Lithospheric Contribution

According to the article the rock formation is 72 million years old. [69] If we run the Pb/Th ratios [69] through Isoplot we see that the dates vary from 5,000 to over 82,000 million years old [Table 50]. In Table 51 we can see the maximum ages for the Thorium/Lead dating method.

Table 51: Dating Methods - Ages Summary

Dating	207Pb/206Pb	208Pb/232Th	206Pb/238U	87Rb/86Sr
Summaries	Age	Age	Age	Age
Average	4,920	6,126	4,539	-47
Maximum	4,949	10,084	7,723	0
Minimum	4,894	2,616	2,306	-75
Difference	55	7,467	5,417	75

Table 52: Thorium/Lead – Maximum Ages

Age
10,084
9,320
8,101
7,502
7,080

6,891
6,655
6,313
5,830
5,755
5,029

Cenozoic Volcanic Rocks of Eastern China

According to the article the rock formation is Quaternary in age. [70] If we run the Pb/Th ratios [71] through Isoplot we see that the dates vary from 4,000 to over 17,000 million years old [Table 52]. In Table 53 we can see the maximum ages for the Thorium/Lead dating method.

Table 53: Dating Methods - Ages Summary

Table	207Pb/206Pb	206Pb/238U	208Pb/232Th	87Rb/86Sr
Summaries	Age	Age	Age	Age
Average	5,057	5,296	10,589	-1,502
Maximum	5,120	8,584	17,171	0
Minimum	5,002	1,136	4,042	-3,593
Difference	118	7,448	13,129	3,593

Table 54: Thorium/Lead – Maximum Ages

Age	Age	Age	Age
17,171	13,322	9,737	7,968
15,343	13,202	9,707	7,830
15,299	13,001	9,049	7,250
15,136	11,119	8,420	6,972
15,054	10,873	8,419	6,628
13,476	10,758	8,368	6,577

Conclusion

If we use the standard formula for calculating Rb/Sr ages we find on many occasions that the Uranium/Thorium/Lead dates are all wrong! Evolutionist Brent Dalrymple states:

"Several events in the formation of the Solar System can be dated with considerable precision." [72]

Looking at some of the dating it is obvious that precision is much lacking. He then goes on:

"Biblical chronologies are historically important, but their credibility began to erode in the eighteenth and nineteenth centuries when it became apparent to some that it would be more profitable to seek a realistic age for

the Earth through observation of nature than through a literal interpretation of parables." [73]

In his book he gives a table [74] with radiometric dates of twenty meteorites. If you run the figures through Microsoft Excel, you will find that they are 98.7% in agreement. There is only a seven percent difference between the ratio of the smallest and oldest dates. As we have seen in this essay, such a perfect fit is attained by selecting data and ignoring other data. A careful study of the latest research shows that such perfection is illusionary at best.

Much of the data in Dalrymple's book is selectively taken to suit and ignores data to the contrary. The Bible believer who accepts the creation account literally has no problem with such unreliable dating methods. Much of the data in Dalrymple's book is selectively taken to suit and ignores data to the contrary.

References

1 Sr, Nd, and Pb isotopes, Earth and Planetary Science Letters, Volume 113 (1992), Pages 107, 110
2 An extremely low U/Pb source, Geochimica et Cosmochimica Acta, 1993, Volume 57, Pages 4687
3 Reference 2, Pages 4690, 4691
4 Petrogenesis and Origins of Mid-Cretaceous, Journal Of Petrology, 2010, Volume 51, Number 10, Pages 2005
5 Reference 4, Page 2024
6 Reference 4, Page 2025
7 Tracing the Indian Ocean Mantle, Journal Of Petrology, 1998, Volume 39, Number 7, Pages 1288
8 Reference 8, Page 1292-1294
9 http://www.creationismonline.com/Isoplot/Isoplot.html
https://www.bgc.org/isoplot
10 Petrogenesis of the Flood Basalts, Journal Of Petrology, 1998, Volume 39, Number 4, Pages 711
11 Reference 10, Page 729-730
12 Nature of the Source Regions, Journal Of Petrology, 2004, Volume 45, Number 3, Pages 556
13 Reference 12, Table 1, Page 558
14 Reference 12, Page 566
15 Reference 12, Page 575-576
16 Generation of Palaeocene Adakitic Andesites, Journal Of Petrology, 2007, Volume 48, Number 4, Pages 667
17 Reference 16, Table 5, Page 676, 677
18 Reference 16, Table 9, Page 684
19 Evidence for a Widespread Tethyan Upper Mantle, Journal Of Petrology, 2005, Volume 46, Number 4, Pages 830
20 Reference 19, Charts, Pages 843, 844, 845, 849
21 Reference 19, Pages 832-834
22 Reference 19, Pages 835-837
23 Post-Collisional Potassic and Ultrapotassic Magmatism, Journal Of Petrology, 1999, Volume 40, Number 9, Pages 1399
24 Reference 23, Page 1403
25 Reference 23, Page 1414
26 Reference 23, Page 1415
27 Continental Lithospheric Contribution to Alkaline, Journal Of Petrology, 1997, Volume 38, Number 1, Pages 115
28 Reference 27, Pages 124, 125
29 Pin Pricking The Elephant, Geological Society Of London, Special Publications, 2004, Volume 229, Pages 139, 140, 144
30 Reference 29, Pages 138, 143
31 Reference 29, Pages 143
32 Chronology And Geochemistry Of Lavas, Journal Of Petrology, April 11, 2012, Pages 1
33 Reference 32, Pages 5, 6, 7, 14
34 Reference 32, Pages 12
35 Ion microprobe U-Th-Pb dating, Meteoritics & Planetary Science, 2004, Volume 39, Number 12, Pages 2033
36 Reference 35, Pages 2036
37 U–Th–Pb Dating Of Secondary Minerals, Geochimica et Cosmochimica Acta, 2008, Volume 72, Pages 2078
38 Reference 37, Pages 2072, 2073
39 Reference 37, Pages 2074
40 Reference 37, Pages 2080, 2081
41 The Influence of High U-Th, Geochimica et Cosmochimica Acta, 1998, Volume 62, Numbers 21/22, Pages 3527
42 Reference 41, Pages 3529

43 Reference 41, Pages 3531
44 U, Th And Pb Isotope Compositions, Geochimica et Cosmochimica Acta, 2009, Volume 73, Pages 469
45 Reference 44, Pages 471
46 Earth Planetary Science Letters, 1987, Volume 82, Pages 121–135.
47 Chemical Geology, 2003, Volume 200, Pages 71–87.
48 Journal Petrology, 1993, Volume 34, Pages 125–172.
49 Geochimica et Cosmochimica Acta, 2007, Volume 71, Pages 1290–1311
50 Reference 44, Pages 475, 476
51 U–Th–Pb Isotope Data, Earth and Planetary Science Letters, 2012, Volume 319-320, Pages 200
52 Reference 51, Pages 199
53 Evolution Of Reunion Hotspot Mantle, Earth and Planetary Science Letters, 1995, Volume 134, Pages 169
54 Reference 53, Page 174
55 Continental Growth 3.2 Gyr Ago, Nature, 2012, Volume 485, Pages 627–630, http://www.nature.com/nature/journal/v485/n7400/extref/nature11140-s3.xls
56 Uranium-Lead Zircon Ages, http://pubs.usgs.gov/of/2008/1142/tables/table04.xls
57 Pilbara Craton in Western Australia, http://www.geo.uu.nl/~kikeb/thesis/thesis/thesis_frame.html
58 http://www.geo.uu.nl/~kikeb/thesis/database/upb/KB746.xls
59 www.geo.uu.nl/~kikeb/thesis/database/upb/KB770.xls
60 Timing of sedimentation, metamorphism, and plutonism, http://geosphere.gsapubs.org/content/3/6/683.abstract
61 http://geosphere.gsapubs.org/content/suppl/2009/02/18/3.6.683.DC1/00138_App3.xls
62 U–Th and U–Pb Systematics in Zircons, Journal Petrology, January 1, 2005, volume 46, Number 1, Pages 27
63 http://petrology.oxfordjournals.org/content/suppl/2004/09/24/egh060.DC1/Table_3.xls
64 Hydrothermal Zebra Dolomite, Geosphere, October 2010, Volume 6, Number 5, Pages 663-690,
65 http://geosphere.gsapubs.org/content/suppl/2010/09/29/6.5.663.DC1/530_suppl.xls
66 Origin of Indian Ocean Seamount Province, Nature Geoscience, 2011, Volume 4, Pages 883-887
67 www.nature.com/ngeo/journal/v4/n12/extref/ngeo1331-s2.xls
68 Geochemistry Geophysics Geosystems, 2003, Volume 4, Page 1089, http://earthref.org/ERDA/download:147/
69 Continental Lithospheric Contribution, Journal Of Petrology, 1997, Volume 38, Number 1, Pages 124
70 Cenozoic Volcanic Rocks of Eastern China, Earth and Planetary Science Letters, Volume 105 (1991), Pages 154
71 Reference 70, Pages 156, 157
72 The Age Of The Earth, By G. Brent Dalrymple, 1991, Stanford University Press, Stanford, California, Page 10.
73 Reference 72, Page 23
74 Reference 72, Page 287

Chapter 19

The Uranium 235 Dating Method

By Paul Nethercott, August 2013

Petrogenesis and Origins of Mid-Cretaceous

According to the article [1] this specimen from the Intraplate Volcanism in Marlborough, New Zealand was dated in 2010 by scientists from New Zealand. According to the essay: "the intraplate basalts in New Zealand that have been erupted intermittently over the last c. 100 Myr." [2] Various tables [3] in the essay have isotopic ratios which can be calculated. As we can see below they are all at strong disagreement with each other. There is a spread of dates over a 10 billion year range. None of the Lead based dating methods even come vaguely close to a Cretaceous age.

Table 1

Table	207Pb/206Pb	207Pb/235U	87Rb/86Sr	208Pb/232Th	206Pb/238U
Summaries	Age	Age	Age	Age	Age
Average	4,876	4,416	59	6,333	3,515
Maximum	4,945	5,159	85	10,716	5,717
Minimum	4,836	4,088	15	4,785	2,712
Difference	109	1,071	70	5,931	3,005

U–Th–Pb Dating Of Secondary Minerals

According to the article [4] this rock formation Yucca Mountain, Nevada was dated in 2008 by scientists from United States Geological Survey, Geological Survey of Canada, and the Australian National University. According to the essay the true age is unknown: "The U–Pb system in opal and chalcedony allows dating in the age range from 50 ka to millions of years and older (Ludwig et al., 1980; Neymark et al., 2000, 2002). Recently, the reliability of U–Pb dating of opal was questioned." [5] Other authors have affirmed the same problem. [5] Two tables [6] in the essay have isotopic ratios which can be calculated. As we can see below they are all at radical disagreement with each other. There is a spread of dates of almost 353 billion years! None of the Uranium/Lead based dating methods even come vaguely close to the so called true age. The oldest date is 350,000 times older than the youngest date.

Table 2

Dating	207Pb/206Pb	206Pb/238U	208Pb/232Th	87Rb/86Sr
Summary	Age	Age	Age	Age
Average	3,459	4,891	9,984	12
Maximum	8,126	31,193	352,962	13
Minimum	-445	1	2	11
Difference	8,571	31,192	352,960	2

Another table [6] in the essay has a list of calculated dates. As we can see below they are all at radical disagreement with each other. There is a spread of dates of 82 billion years! None of the Uranium/Lead based dating methods even come vaguely close to the so called true age. The oldest date is 82,000 times older than the youngest date.

Table 3

Dating	206Pb/238U	207Pb/235U	208Pb/232Th	87Rb/86Sr
Summary	Age	Age	Age	Age
Average	1,540	46	7,687	12
Maximum	20,209	486	82,030	13
Minimum	1	0	3	11
Difference	20,208	486	82,027	2

Here are more examples of isotopic ratios taken from several articles in major geology magazines which give absolutely absurd dates.

History Of The Pasamonte Achondrite

According to the article this meteorite specimen was dated in 1977 by scientists from the United States Geological Survey, Colorado and the Department of Chemistry and Geochemistry, Colorado School of Mines. [7] The article states that Rubidium/Strontium dating affirms that this material is 4.5 billion years old. [8] If we run the various isotope ratios [8] from two different tables in the article through Microsoft Excel we get the following values respectively:

Table 4

Summary	206Pb/238U	207Pb/235U	207Pb/206Pb	208Pb/232Th
Average	3,088	3,666	4,566	2,263
Maximum	5,694	5,032	4,963	14,800
Minimum	103	865	4,440	-10,700
Difference	5,591	4,167	523	25,500

If we run the 87Rb/86Sr isotope ratios [8] from the article through Microsoft Excel we get the following values:

Table 5: Rb/Sr Age Dating Summary

Average	4,403
Maximum	6,674
Minimum	2,412
Difference	4,262

The Thorium/Lead dates are up to twelve billion years older. The so called true age is just a guess.

An Extremely Low U/Pb Source

According to the article [9] this specimen [lunar meteorite] was dated in 1993 by scientists from the United States Geological Survey, Colorado, the United States Geological Survey, California and The National Institute of Polar Research, Tokyo. According to the article: "The Pb-Pb internal isochron obtained for acid leached residues of separated mineral fractions yields an age of 3940 ± 28 Ma, which is similar to the U-Pb (3850 ± 150 Ma) and Th-Pb (3820 ± 290 Ma) internal isochron ages. The Sm-Nd data for the mineral separates yield an internal isochron age of 3871 ± 57 Ma and an initial 143Nd/I44Nd value of 0.50797 ± 10. The Rb-Sr data yield an internal isochron age of 3840 ± 32 Ma." [9]

Table 6: Rb/Sr Age Dating Summary

Average	3,619
Maximum	5,385
Minimum	721
Difference	4,664

Table 7: Uranium Age Dating Summary

Table Summaries	207Pb/206Pb Age	206Pb/238U Age	208Pb/232Th Age	207Pb/235U Age
Average	4,673	8,035	10,148	4,546
Maximum	5,018	56,923	65,286	8,128
Minimum	3,961	1,477	2,542	2,784
Difference	1,057	55,445	62,744	5,344

The article claims that the Rb/Sr age is 3.8 billion years for this meteorite. If that is the true age why are all the Uranium/Thorium/Lead dates [10] so stupid? Or are they right and the Rb/Sr [11] is wrong?

Table 8: 208Pb/232Th, Maximum Ages

Age	Age	Age	Age
65,286	14,430	9,094	5,401
33,898	14,410	6,520	5,396
25,013	13,107	6,166	5,365
22,178	12,738	6,121	5,098
21,204	11,641	5,671	5,035
17,611	11,174	5,408	4,678

Table 9: 206Pb/238U, Maximum Ages

Age	Age	Age	Age
56,923	10,895	6,764	5,777
27,313	10,278	6,670	5,625
17,873	9,653	6,449	5,602
13,680	8,009	6,436	5,278
13,623	7,395	6,070	5,147

Petrogenesis of the Flood Basalts

According to the article [12] this basalt form the Northern Kerguelen Archipelago was dated in 1998 by scientists from the Massachusetts Institute Of Technology, University of Brussels, Belgium and the San Diego State University. According to the essay: "The dominance of this isotopic signature in archipelago lavas for 30 my and its presence in ~40 Ma gabbros is consistent with the previous interpretation that these are isotopic characteristics of the Kerguelen Plume." [12] Various tables [13] in the essay have isotopic ratios which can be calculated. As we can see below they are all at strong disagreement with each other. There is a spread of dates of over a 44 billion year range! None of the Uranium/Lead based dating methods even come vaguely close to the so called true age.

Table 10

Mount Bureau Summary	Age 207Pb/206Pb	Age 206Pb/238U	Age 207Pb/235U	Age 208Pb/232Th
Average	5,006	5,924	5,161	8,410
Maximum	5,020	23,366	8,496	44,378
Minimum	4,994	3,335	4,454	2,650
Difference	26	20,031	4,042	41,728

Table 11

Mt. Rabouillere Summary	Age 207Pb/206Pb	Age 206Pb/238U	Age 207Pb/235U	Age 208Pb/232Th
Average	5,008	4,903	4,975	6,142
Maximum	5,019	5,355	5,100	7,788
Minimum	5,000	4,305	4,793	2,799
Difference	20	1,050	307	4,989

Nature of the Source Regions

According to the article [14] this lava from southern Tibet was dated in 2004 by scientists from the Open University in Milton Keynes, the University of Bristol and Cardiff University. According to the essay: "Most samples are Miocene in age, ranging from 10 to 25Ma in the south and 19Ma to the present day in northern Tibet." [15] Various tables [16] in the essay have isotopic ratios which can be calculated. As we can see below they are all at strong disagreement with each other. There is a spread of dates of over an 88 billion year range! None of the Uranium/Lead based dating methods even come vaguely close to the so called true age.

Table 12

207Pb/235U Age	Model Age	Ratio	Percentage
5,136	0.5	10,273	10,272,962
5,138	0.5	10,275	10,275,154
5,135	13	395	395,000
5,140	18.5	278	277,839
7,470	13	575	574,597
7,471	12.5	598	597,649

Table 13

207Pb/235U Age	Model Age	Ratio	Percentage
313	24.0	13	13,026
946	13.8	69	68,534
266	13.8	19	19,267
238	13.8	17	17,265
294	13.3	22	22,095
447	18.8	24	23,757
482	17.3	28	27,878

Table 14

Statistics	Maximum	Minimum	Difference
208Pb/232Th	88,294	4,826	83,469
206Pb/238U	33,751	4,086	29,665
207Pb/235U	7,471	238	7,232
207Pb/206Pb	4,994	4,919	75

Table 15: North Tibet Summary. Rb/Sr model age = 13 million years

208Pb/232Th	207Pb/235U	207Pb/206Pb	206Pb/238U
Million Years	Million Years	Million Years	Million Years
11,420	5,136	4,980	7,783
11,350	5,138	4,980	8,023
13,475	5,135	4,987	8,305
11,504	5,140	4,989	7,349
81,614	7,470	4,987	33,751
88,294	7,471	4,991	33,742

Table 16: South Tibet Summary. Rb/Sr model age = 13 million years

208Pb/232Th	207Pb/235U	207Pb/206Pb	206Pb/238U
Million Years	Million Years	Million Years	Million Years
11,102	313	4,982	6,331
6,092	946	4,919	5,799
9,265	266	4,980	6,682
4,826	238	4,992	4,086
8,205	294	4,980	5,567
25,015	447	4,994	13,328
33,191	482	4,992	15,053

Generation of Palaeocene Adakitic Andesites

According to the article [17] this rock formation from North Eastern China was dated in 2007 by scientists from China and Japan. According to the essay the true age is: "Palaeocene (c. 55-58Ma) adakitic andesites from the Yanji area." [17] Numerous table and charts affirm this as the true age. [18] A table [19] in the essay have isotopic ratios which can be calculated. As we can see below they are all at radical disagreement with each other. There is a spread of dates of over 10 billion years! None of the Uranium/Lead based dating methods even come vaguely close to the so called true age.

Table 17

207Pb/206Pb	208Pb/232Th	206Pb/238U	207Pb/235U
Age	Age	Age	Age
5,024	10,518	9,669	6,052
5,023	10,277	9,552	6,051
5,023	8,529	9,526	6,051
5,023	8,360	8,443	5,828
5,021	8,165	7,929	5,826
5,020	7,800	7,403	5,641

Ivisaartoq Greenstone Belt

According to the article [20] this rock formation from southern West Greenland was dated in 2007 by scientists from Canada, Denmark, USA and Austria. According to the essay the true age is: "The Mesoarchean (ca. 3075Ma) Ivisaartoq greenstone belt in southern West Greenland."

[20] A table [21] in the essay have isotopic ratios which can be calculated. As we can see below they are all at radical disagreement with each other. There is a spread of dates of over 3 billion years!

Table 18

207Pb/235U Age	208Pb/232Th Age	206Pb/238U Age	207Pb/206Pb Age
5,288	2,671	2876	3082
5,162	2,860	2712	2998
5,299	2,586	2955	3046
5,407	2,305	3195	3059
5,302	2,726	2930	3067

References

1 Petrogenesis and Origins of Mid-Cretaceous, Journal Of Petrology, 2010, Volume 51, Number 10, Pages 2003-2045
2 Reference 1, Pages 2038
3 Reference 1, Pages 2024-2026
4 U–Th–Pb Dating Of Secondary Minerals, Geochimica et Cosmochimica Acta, 2008, Volume 72, Pages 2067
5 Reference 4, Pages 2068
6 Reference 4, Pages 2072-2074, 2080-2081
7 History Of The Pasamonte Achondrite, Earth and Planetary Science Letters, Volume 37, 1977, Pages 1
8 Reference 7, Pages 3, 9
9 An extremely low U/Pb source, Geochimica et Cosmochimica Acta, 1993, Volume 57, Pages 4687-4702
10 Reference 9, Pages 4690, 4691
11 Reference 9, Pages 4696
12 Petrogenesis of the Flood Basalts, Journal Of Petrology, 1998, Volume 39, Number 4, Pages 711–748
13 Reference 119, Pages 729, 730
14 Nature of the Source Regions, Journal Of Petrology, 2004, Volume 45, Number 3, Pages 555
15 Reference 121, Pages 556
16 Reference 121, Pages 566, 575, 576
17 Generation of Palaeocene Adakitic Andesites, Journal Of Petrology, 2007, Volume 48, Number 4, Pages 661
18 Reference 124, Pages 676-678
19 Reference 124, Pages 684
20 Ivisaartoq Greenstone Belt, Gondwana Research, Volume 11 (2007) Page 69
21 Reference 20, Pages 86

Chapter 20

The Uranium 238 Dating Method

By Paul Nethercott, July 2013

Origin Of The Indian Ocean-Type Isotopic Signature

According to the article [1] this rock formation the Philippine Sea plate was dated in 1998 by scientists from Department of Geology, Florida International University, Miami. According to the essay the true age is: "Spreading centers in three basins, the West Philippine Basin (37-60 Ma), the Parece Vela Basin (18-31 Ma), and the Shikoku Basin (17-25 Ma) are extinct, and one, the Mariana Trough (0-6 Ma), is active (Figure 1)." [1] Numerous table and charts affirm this as the true age. [2] Two tables [3] in the essay have isotopic ratios which can be calculated. As we can see below they are all at radical disagreement with each other. There is a spread of dates of almost 100 billion years! None of the Uranium/Lead based dating methods even come vaguely close to the so called true age. The oldest date is 3,971 times older than the youngest date.

Table 1

Dating Summary	Age 87Rb/86Sr	Age 147Sm/144Nd	Age 207Pb/206Pb	Age 206Pb/238U	Age 208Pb/232Th
Average	42	41	4,960	4,260	8,373
Maximum	55	54	4,989	7,093	13,430
Minimum	19	20	4,921	1,904	3,065
Difference	37	33	68	5,188	10,365

U-Th-Pb Dating Of Secondary Minerals

According to the article [4] this rock formation Yucca Mountain, Nevada was dated in 2008 by scientists from United States Geological Survey, Geological Survey of Canada, and the Australian National University. According to the essay the true age is unknown. [5] Other authors have affirmed the same problem. [6] Two tables [7] in the essay have isotopic ratios which can be calculated. As we can see below they are all at radical disagreement with each other. There is a spread of dates of almost 353 billion years! None of the Uranium/Lead based dating methods even come vaguely close to the so called true age. The oldest date is 350,000 times older than the youngest date.

Table 2

Dating Summary	207Pb/206Pb Age	206Pb/238U Age	208Pb/232Th Age	87Rb/86Sr Age
Average	3,459	4,891	9,984	12
Maximum	8,126	31,193	352,962	13
Minimum	-445	1	2	11
Difference	8,571	31,192	352,960	2

Another table [7] in the essay has a list of calculated dates. As we can see below they are all at radical disagreement with each other. There is a spread of dates of 82 billion years! None of the Uranium/Lead based dating methods even come vaguely close to the so called true age. The oldest date is 82,000 times older than the youngest date.

Table 3

Dating	206Pb/238U	207Pb/235U	208Pb/232Th	87Rb/86Sr
Summary	Age	Age	Age	Age
Average	1,540	46	7,687	12
Maximum	20,209	486	82,030	13
Minimum	1	0	3	11
Difference	20,208	486	82,027	2

Rocks Of The Central Wyoming Province

These rock samples were dated in 2005 by scientists from the University of Wyoming. [8] If we run the Rubidium/Strontium and Neodymium/Samarium isotope ratios [9] from the article through Microsoft Excel and use the formulas listed in Gunter Faure's book [10] we get the following values:

$$t = \frac{2.303}{(0.693 \div h)} \log\left(\frac{(143Nd/144Nd) - (143Nd/144Nd)_0}{(144Sm/147Nd)} + 1 \right)$$

h = Half life, 106 billion years

$$t = \frac{2.303}{(0.693 \div h)} \log\left(\frac{(87Sr/86Sr) - (87Sr/86Sr)_0}{(87Rb/86Sr)} + 1 \right)$$

h = Half life, 48.8 billion years

Where t equals the age in years. (87Sr/86Sr) = the current isotopic ratio. $(87Sr/86Sr)_0$ = the initial isotopic ratio. (87Rb/86Sr) = the current isotopic ratio. The same is true for the formula below

Table 4

Dating	Age	Age	Age	Age	Age
Summary	87Rb/86Sr	147Sm/144Nd	207Pb/206Pb	208Pb/232Th	206Pb/238U
Average	2,863	2,869	5,123	17,899	11,906
Maximum	2,952	2,954	5,294	38,746	18,985
Minimum	2,630	2,631	4,662	6,650	7,294
Std Deviation	38	39	152	9,754	3,298

The Uranium/Lead dates [11] are up to sixteen billion years older than the Rubidium/Strontium and Neodymium/Samarium dates. The Thorium/Lead dates are up to thirty six billion years older. The so called true age is just a guess.

History Of The Pasamonte Achondrite

According to the article this meteorite specimen was dated in 1977 by scientists from the United States Geological Survey, Colorado and the Department of Chemistry and Geochemistry, Colorado School of Mines. [12] The article states that Rubidium/Strontium dating affirms that this material is 4.5 billion years old. [34] If we run the various isotope ratios [13] from two different tables in the article through Microsoft Excel we get the following values respectively:

Table 5

Summary	206Pb/238U	207Pb/235U	207Pb/206Pb	208Pb/232Th
Average	3,088	3,666	4,566	2,263
Maximum	5,694	5,032	4,963	14,800
Minimum	103	865	4,440	-10,700
Difference	5,591	4,167	523	25,500

If we run the 87Rb/86Sr isotope ratios [13] from the article through Microsoft Excel we get the following values:

Table 6: Rb/Sr Age Dating Summary

Average	4,403
Maximum	6,674
Minimum	2,412
Difference	4,262

The Thorium/Lead dates are up to twelve billion years older. The so called true age is just a guess.

A Depleted Mantle Source For Kimberlites

According to the article [14] this specimen [kimberlites from Zaire] was dated in 1984 by scientists from Belgium. According to the article [15] the age of the samples is 70 million years. If we run the various isotope ratios [16] from the article through Microsoft Excel we get the following values respectively:

Table 7

Summary	207Pb/206Pb	206Pb/238U	87Rb/86Sr	147Sm/144Nd
Average	4,977	4,810	86	72
Maximum	5,017	10,870	146	80
Minimum	4,909	1,391	50	63
Difference	108	9,478	196	17

The 207Pb/206Pb maximum age is 34 times older than the 87Rb/86Sr maximum age. The 206Pb/238U maximum age is 74 times older than the 147Sm/144Nd maximum age. There is a 10.8 billion year difference between the oldest and youngest age attained.

Pb, Nd And Sr Isotopic Geochemistry

According to the article [17] this specimen [Bellsbank kimberlite, South Africa] was dated in 1991 by scientists from the University Of Rochester, New York, Guiyang University in China, and the United States Geological Survey, Colorado. According to the article [18] the age of the samples is just 1 million years. If we run the various isotope ratios [19] from two different tables in the article through Microsoft Excel we get the following values respectively:

Table 8

Table Summaries	207Pb/206Pb Age	206Pb/238U Age	208Pb/232Th Age	87Rb/86Sr Age
Average	5,057	5,092	10,182	-1,502
Maximum	5,120	8,584	17,171	0
Minimum	5,002	0	0	-3,593
Difference	118	8,584	17,171	3,593

In tables 9 to 12 we can see some of the astounding spread of dates [million of years]. The oldest date is over 17 billion years old. The youngest is less than negative 3.5 billion years. The difference between the two is over 20 billion years. According to the article the true age of the rock is just one million years old!

Table 9: 208Pb/232Th, Maximum Ages

Age	Age	Age	Age
17,171	13,322	9,737	7,968
15,343	13,202	9,707	7,830
15,299	13,001	9,049	7,250
15,136	11,119	8,420	6,972
15,054	10,873	8,419	6,628
13,476	10,758	8,368	6,577

Table 10: 206Pb/238U, Maximum Ages

Age	Age	Age
8,584	6,656	5,576
7,975	6,654	5,520
7,314	6,518	5,285
7,184	6,448	5,159
6,861	5,758	5,099

Table 11: Pb 207/206, Maximum Ages

Age	Age	Age	Age
5,120	5,067	5,060	5,049
5,109	5,066	5,059	5,045
5,097	5,066	5,051	5,044
5,077	5,065	5,050	5,044
5,067	5,062	5,050	5,033
5,067	5,060	5,050	5,022

Table 12: 87Rb/86Sr, Minimum Ages

Age	Age	Age	Age
-3,593	-2,981	-1,917	-1,323
-3,231	-2,725	-1,611	-1,245
-3,089	-2,050	-1,499	-1,229
-3,067	-1,926	-1,370	-1,194

Sr, Nd, And Pb Isotopes

According to the article [20] this specimen [eastern China] was dated in 1992 by scientists from the University Of Rochester, New York, Guiyang University in China, and the United States Geological Survey, Colorado. According to the article: "Observed high Th/U, Rb/Sr, 87Sr/86 Sr and Delta 208, low Sm/Nd ratios, and a large negative Nd in phlogopite pyroxenite with a depleted mantle model age of 2.9 Ga, support our contention that metasomatized continental lower mantle lithosphere is the source for the EMI component." [20] If we run the various isotope ratios [21] from two different tables in the article through Isoplot we get the following values respectively:

Table 13

Dating	232Th/208Pb	206Pb/238U	207Pb/206Pb
Summaries	Age	Age	Age
Average	14,198	7,366	5,014
Maximum	94,396	22,201	5,077
Minimum	79	1,117	4,945
Difference	94,317	21,083	131

If the true age is 2.9 billion years, why so much discordance? In tables 14 to 15 we can see some of the astounding spread of dates [millions of years]. The oldest date is over 94 billion years old. The youngest is 79 million years. The difference between the two is over 94 billion years. The oldest date is 1,194 times older than the youngest. According to the article the true age of the rock is 2.9 billion years old!

Table 14: 208Pb/232Th, Maximum Ages

Age	Age	Age	Age
94,396	39,267	10,595	8,171
90,683	26,266	10,284	7,789
74,639	18,334	9,328	7,638
58,153	16,357	8,821	7,375
55,324	14,250	8,771	7,317
45,242	11,215	8,403	5,759

Table 15: 206Pb/238U, Maximum Ages

Age	Age	Age	Age
22,201	9,878	7,348	5,746
21,813	9,656	7,335	5,700
19,320	9,054	7,249	5,218
16,656	8,242	7,202	5,201
16,200	8,044	7,019	5,163
14,748	7,996	6,923	5,159
13,607	7,590	6,848	5,099
11,256	7,422	6,292	4,812

Evolution Of Reunion Hotspot Mantle

According to the article [22] this specimen [Reunion and Mauritius Islands] was dated in 1995 by scientists from the University of Hawaii. According to the article: "Whole-rock powder obtained from P. Krishnamurthy. (87Sr/86 Sr), and em(T) are age-corrected values; *T* = 66 Ma for the drill hole lavas." [23] If we run the various isotope ratios [24] from two different tables in the article through Isoplot we get the following values respectively:

Table 16

Table Summaries	232Th/208Pb Age	206Pb/238U Age	207Pb/206Pb Age
Average	8,079	4,449	4,976
Maximum	13,287	6,285	5,016
Minimum	5,641	3,010	4,953
Difference	7,646	3,276	63

Table 17: 208Pb/232Th, Maximum Ages

Age	Age	Age	Age
13,287	8,725	7,363	6,540
11,832	8,609	7,362	6,479
11,017	7,541	7,080	6,323
10,357	7,517	7,017	5,660
9,101	7,446	6,679	5,641

Table 18: 206Pb/238U, Maximum Ages

Age	Age	Age	Age
6,285	4,903	4,141	3,875
6,165	4,633	4,133	3,647
5,767	4,342	4,011	3,548
5,553	4,258	4,001	3,369
5,152	4,220	3,973	3,010

According to dating charts in the article, the true age is just 66 million years old! [25]

An Extremely Low U/Pb Source

According to the article [26] this specimen [lunar meteorite] was dated in 1993 by scientists from the United States Geological Survey, Colorado, the United States Geological Survey, California and The National Institute of Polar Research, Tokyo. According to the article: "The Pb-Pb internal isochron obtained for acid leached residues of separated mineral fractions yields an age of 3940 ± 28 Ma, which is similar to the U-Pb (3850 ± 150 Ma) and Th-Pb (3820 ± 290 Ma) internal isochron ages. The Sm-Nd data for the mineral separates yield an internal isochron age of 3871 ± 57 Ma and an initial 143Nd/I44Nd value of 0.50797 ± 10. The Rb-Sr data yield an internal isochron age of 3840 ± 32 Ma." [26]

Table 19: Rb/Sr Age Dating Summary

Average	3,619
Maximum	5,385
Minimum	721
Difference	4,664

Table 20: Uranium Age Dating Summary

Table Summaries	207Pb/206Pb Age	206Pb/238U Age	208Pb/232Th Age	207Pb/235U Age
Average	4,673	8,035	10,148	4,546
Maximum	5,018	56,923	65,286	8,128
Minimum	3,961	1,477	2,542	2,784
Difference	1,057	55,445	62,744	5,344

The article claims that the Rb/Sr age is 3.8 billion years for this meteorite. If that is the true age why are all the Uranium/Thorium/Lead dates [27] so stupid? Or are they right and the Rb/Sr is wrong?

Table 21: 208Pb/232Th, Maximum Ages

Age	Age	Age	Age
65,286	14,430	9,094	5,401
33,898	14,410	6,520	5,396
25,013	13,107	6,166	5,365
22,178	12,738	6,121	5,098
21,204	11,641	5,671	5,035
17,611	11,174	5,408	4,678

Table 22: 206Pb/238U, Maximum Ages

Age	Age	Age	Age
56,923	10,895	6,764	5,777
27,313	10,278	6,670	5,625
17,873	9,653	6,449	5,602
13,680	8,009	6,436	5,278
13,623	7,395	6,070	5,147

The Origin Of Geochemical Diversity

According to the article [28] this specimen [lunar basalt] was dated in 2007 by scientists from New Mexico University. According to Rb/Sr isochron diagram the age of the material is 3.678 billion years. [29] If we run the various isotope ratios [30] from two different tables in the article through Isoplot we get the following values respectively:

Table 23

Table Summaries	207Pb/206Pb Age	206Pb/238U Age	87Rb/86Sr Age
Average	4,635	6,565	4,672
Maximum	5,111	18,213	7,094
Minimum	4,028	3,706	3,476
Difference	1,082	14,506	3,618

The dating methods all disagree with each other. There is a wide spread of dates which are just random.

Continental Lithospheric Contribution

According to the article [31] this specimen from southern Portugal was dated in 1997 by scientists from France. According to the article Argon and Rubidium dating defined the so called true ages as: "The age of the intrusion and crystallization of the alkaline rocks of the Serra de Monchique is 72 Ma, based on Rb/Sr and K/Ar dating." [32] If we run the various isotope ratios [33] from a table in the article through Isoplot we get the following values respectively:

Table 24

Table Summaries	207Pb/206Pb Age	208Pb/232Th Age	206Pb/238U Age	87Rb/86Sr Age
Average	4,920	6,126	4,539	-62
Maximum	4,949	10,084	7,723	-50
Minimum	4,894	2,616	2,306	-75
Difference	55	7,467	5,417	25

The date of 72 million years is just a guess. The Thorium/Lead method gives dates 140 times older. The Uranium/Lead methods give dates 107 times older. Below we can see the maximum ages [million years] calculated form isotope ratios. Compare these with the so called true age!

Table 25: Maximum Ages

208Pb/232Th	206Pb/238U
10,084	7,723
9,320	7,060
8,101	6,507
7,502	6,387
7,080	6,206
6,891	5,143
6,655	4,734
6,313	4,186
5,830	3,768
5,755	3,761
5,029	3,487

Garnet Granulite Xenoliths

According to the article [34] this specimen from the northern Baltic shield was dated in 2001 by scientists from England, USA and Russia. According to the article Argon dating defined the so called true ages as 400 to 2200 million years. [35] If we run the various isotope ratios [36] from table 4 in the article through Isoplot we get the following values respectively:

Table 26

Table	206Pb/238U	207Pb/206Pb
Summaries	Age	Age
Average	17,002	5,046
Maximum	40,059	5,295
Minimum	1,608	3,908
Difference	38,452	1,387

Below are the maximum ages calculated from isotope ratios in tables 4 and 5 in the article:

Table 27: 206Pb/238U, Maximum Ages

206Pb/238U	206Pb/238U	206Pb/238U	206Pb/238U
Age	Age	Age	Age
40,059	28,118	21,092	13,724
35,742	27,127	16,026	13,404
34,459	25,884	14,371	12,747
33,978	21,209	14,272	10,956

Table 28: 206Pb/238U, Maximum Ages

206Pb/238U	206Pb/238U	206Pb/238U
Age	Age	Age
20,648	13,724	10,956
17,527	13,404	10,049
16,336	12,622	6,792
15,626	12,165	6,265
15,018	11,432	5,865

If we run more ratios form and online supplement [37] we get ages uniformly 5 billion years old. Compare these with the so called true age!

The Isotope And Trace Element Budget

According to the article [38] this specimen from the Devil River Arc System, New Zealand was dated in 2000 by scientists from Germany. According to the article, the so called true ages is Cambrian. [102] If we run the various isotope ratios [39] from table 4 in the article through Isoplot we get the following values respectively:

Table 28

Table Summaries	207Pb/206Pb Age	206Pb/238U Age	87Rb/86Sr Age
Average	4,970	19,143	500
Maximum	4,986	21,761	501
Minimum	4,932	15,150	495
Difference	54	6,611	6

The Lead/Lead dates are ten times too old and the Uranium/Lead dates are 40 times too old!

Petrogenesis And Origins Of Mid-Cretaceous

According to the article [40] this specimen from the Intraplate Volcanism in Marlborough, New Zealand was dated in 2010 by scientists from New Zealand. According to the essay "the intraplate basalts in New Zealand that have been erupted intermittently over the last c. 100 Myr." [41] Various tables [42] in the essay have isotopic ratios which can be calculated. As we can see below they are all at strong disagreement with each other. There is a spread of dates over a 10 billion year range. None of the Lead based dating methods even come vaguely close to a Cretaceous age.

Table 29

Table Summaries	207Pb/206Pb Age	207Pb/235U Age	87Rb/86Sr Age	208Pb/232Th Age	206Pb/238U Age
Average	4,876	4,416	59	6,333	3,515
Maximum	4,945	5,159	85	10,716	5,717
Minimum	4,836	4,088	15	4,785	2,712
Difference	109	1,071	70	5,931	3,005

Petrogenesis Of The Flood Basalts

According to the article [43] this basalt form the Northern Kerguelen Archipelago was dated in 1998 by scientists from the Massachusetts Institute Of Technology, University of Brussels, Belgium and the San Diego State University. According to the essay: "The dominance of this isotopic signature in archipelago lavas for 30 my and its presence in ~40 Ma gabbros is consistent with the previous interpretation that these are isotopic characteristics of the Kerguelen Plume." [43] Various tables [44] in the essay have isotopic ratios which can be calculated. As we can see below they are all at strong disagreement with each other. There is a spread of dates of over a 44 billion year range! None of the Uranium/Lead based dating methods even come vaguely close to the so called true age.

Table 30

Mt Rabouillere Summary	Age 87Rb/86Sr	Age 207Pb/206Pb	Age 206Pb/238U	Age 207Pb/235U	Age 208Pb/232Th
Average	21	5,008	4,903	4,975	6,142
Maximum	30	5,019	5,355	5,100	7,788
Minimum	-7	5,000	4,305	4,793	2,799
Difference	38	20	1,050	307	4,989

Table 31

Mount Bureau Summary	Age 87Rb/86Sr	Age 207Pb/206Pb	Age 206Pb/238U	Age 207Pb/235U	Age 208Pb/232Th
Average	27	5,006	5,924	5,161	8,410
Maximum	30	5,020	23,366	8,496	44,378
Minimum	24	4,994	3,335	4,454	2,650
Difference	6	26	20,031	4,042	41,728

Nature Of The Source Regions

According to the article [45] this lava from southern Tibet was dated in 2004 by scientists from the Open University in Milton Keynes, the University of Bristol and Cardiff University. According to the essay: "Most samples are Miocene in age, ranging from 10 to 25Ma in the south and 19Ma to the present day in northern Tibet." [46] Various tables [47] in the essay have isotopic ratios which can be calculated. As we can see below they are all at strong disagreement with each other. There is a spread of dates of over an 88 billion year range! None of the Uranium/Lead based dating methods even come vaguely close to the so called true age.

Table 32

North Tibet Summary	208Pb/232Th Million Years	207Pb/235U Million Years	207Pb/206Pb Million Years	206Pb/238U Million Years
	11,420	5,136	4,980	7,783
87Rb/86Sr	11,350	5,138	4,980	8,023
Model Age	13,475	5,135	4,987	8,305
13 Million Years	11,504	5,140	4,989	7,349
	81,614	7,470	4,987	33,751
	88,294	7,471	4,991	33,742

Table 33

South Tibet Summary	208Pb/232Th Million Years	207Pb/235U Million Years	207Pb/206Pb Million Years	206Pb/238U Million Years
	11,102	313	4,982	6,331
	6,092	946	4,919	5,799
87Rb/86Sr	9,265	266	4,980	6,682
Model Age	4,826	238	4,992	4,086
13 Million Years	8,205	294	4,980	5,567
	25,015	447	4,994	13,328
	33,191	482	4,992	15,053

Generation Of Palaeocene Adakitic Andesites

According to the article [48] this rock formation from North Eastern China was dated in 2007 by scientists from China and Japan. According to the essay the true age is: "Palaeocene (c. 55-58Ma) adakitic andesites from the Yanji area." [48] Numerous table and charts affirm this as the true age. [49] A table [50] in the essay have isotopic ratios which can be calculated. As we can see below they are all at radical disagreement with each other. There is a spread of dates of over 10 billion years! None of the Uranium/Lead based dating methods even come vaguely close to the so called true age.

Table 34

Dating Summary	87Rb/86Sr Age	207Pb/206Pb Age	208Pb/232Th Age	206Pb/238U Age	207Pb/235U Age
Average	51	5,022	8,941	8,754	5,908
Maximum	66	5,024	10,518	9,669	6,052
Minimum	40	5,020	7,800	7,403	5,641
Difference	26	3	2,718	2,266	411

Evidence For A Widespread Tethyan

According to the article [51] this rock formation from North Eastern China was dated in 2007 by scientists from China and Japan. According to the essay the true age is: "Here, we report age-corrected Nd–Pb–Sr isotope data for 100–350 Ma basalt, diabase, and gabbro from widely separated Tethyan locations in Tibet, Iran, Albania, the eastern Himalayan syntaxis, and the seafloor off NW Australia (Fig. 1)." [52] The author concludes that the rocks are from the Cretaceous and Jurassic time periods: "We collected Early Jurassic to Early Cretaceous Neotethyan magmatic rocks in 1998 from outcrops along 1300 km of the Indus–Yarlung suture zone." [53] Several tables [54] in the essay have isotopic ratios which can be calculated. As we can see below they are all at radical disagreement with each other. There is a spread of dates of almost 60 billion years! None of the Uranium/Lead based dating methods even come vaguely close to the so called true age.

Table 35

Dating Summary	87Rb/86Sr Age	207Pb/206Pb Age	208Pb/232Th Age	206Pb/238U Age
Average	168	4,999	22,356	7,014
Maximum	1,739	5,236	58,796	15,747
Minimum	0	4,982	10,699	5,042
Difference	1,739	254	48,096	10,705

Table 36: 208Pb/232Th, Maximum Ages

208Pb/232Th Age	208Pb/232Th Age	208Pb/232Th Age	208Pb/232Th Age
58,796	29,705	18,607	11,427
54,206	27,710	18,121	11,377
48,252	27,422	17,797	11,366
47,976	26,674	17,787	11,241
46,117	26,369	17,591	10,718
42,203	25,972	17,536	10,699
42,192	25,590	17,054	10,699
41,604	25,096	16,053	10,300
41,343	24,010	15,299	9,357
41,231	22,718	14,340	8,632
39,637	22,307	13,845	8,486
38,125	22,228	13,772	8,057
37,115	21,827	13,652	6,497
35,012	21,560	13,404	5,573

33,584	19,910	13,403	5,425
31,556	19,594	13,006	4,869
31,286	19,148	12,171	
30,740	18,765	11,540	

Table 37: 206Pb/238U, Maximum Ages

206Pb/238U	206Pb/238U	206Pb/238U	206Pb/238U	206Pb/238U
Age	Age	Age	Age	Age
15,747	11,309	8,770	6,602	5,724
15,067	11,248	8,508	6,589	5,720
14,363	10,360	8,315	6,421	5,601
13,580	9,643	8,314	6,398	5,599
13,204	9,427	8,072	6,369	5,573
12,780	9,300	8,024	6,357	5,515
11,757	9,123	7,604	6,219	5,462
11,659	9,014	7,504	5,863	5,311
11,537	8,996	7,056	5,861	5,286
11,313	8,954	7,002	5,807	5,120

Conclusion

Evolutionists **Schmitz and Bowring** claim that Uranium/Lead dating is 99% accurate. [55] Looking at some of the dating it is obvious that precision is much lacking. The Bible believer who accepts the creation account literally has no problem with such unreliable dating methods. Much of the data used in this dating method is selectively taken to suit and ignores data to the contrary.

References

1 Origin of the Indian Ocean-type isotopic signature, Journal Of Geophysical Research, 1998, Volume 103, Number B9, Pages 20,963

2 Reference 2, Pages 20965, 20969

3 Reference 2, Pages 20968, 20969

4 U–Th–Pb Dating Of Secondary Minerals, Geochimica et Cosmochimica Acta, 2008, Volume 72, Pages 2067

5 Reference 4, Pages 2067, 2068

6 Reference 4, Pages 2072-2073, 2074

7 Reference 4, Pages 2080, 2081

8 Rocks of the Central Wyoming Province, Canadian Journal Of Earth Science, 2006, Volume 43, Pages 1419

9 Reference 27, Page 1436-1437

10 Principles of Isotopic Geology, Gunter Faure, John Wiley Publishers. New York, 1986, Pages 120, 205

11 Reference 27, Page 1439

12 History Of The Pasamonte Achondrite, Earth and Planetary Science Letters, Volume 37, 1977, Pages 1

13 Reference 33, Pages 3, 9

14 A Depleted Mantle Source For Kimberlites, Earth and Planetary Science Letters, Volume 73, 1985, Pages 269

15 Reference 47, Pages 270

16 Reference 47, Pages 271, 273

17 Pb, Nd and Sr isotopic geochemistry, Earth and Planetary Science Letters, Volume 105, 1991, Pages 149

18 Reference 66, Pages 154, 160

19 Reference 66, Pages 156, 157

20 Sr, Nd, and Pb isotopes, Earth and Planetary Science Letters, Volume 113, 1992, Pages 107
21 Reference 68, Pages 110
22 Evolution of Reunion Hotspot Mantle, Earth and Planetary Science Letters, Volume 134, 1995, Pages 169-185
23 Reference 72, Pages 173
24 Reference 72, Pages 174
25 Reference 72, Pages 180
26 An extremely low U/Pb source, Geochimica et Cosmochimica Acta, 1993, Volume 57, Pages 4687-4702
27 Reference 75, Pages 4690, 4691
28 The Origin of Geochemical Diversity, Geochimica et Cosmochimica Acta, Volume 71, 2007, Pages 3656
29 Reference 86, Pages 3661
30 Reference 86, Pages 3660
31 Continental Lithospheric Contribution, Journal Of Petrology, 1997, Volume 38, Number 1, Pages 115
32 Reference 95, Pages 119
33 Reference 95, Pages 124
34 Garnet Granulite Xenoliths, Journal Of Petrology, 2001, Volume 42, Number 4, Pages 731
35 Reference 98, Pages 742, 743
36 Reference 98, Pages 737-740
37 http://petrology.oxfordjournals.org/content/suppl/2001/04/27/42.4.731.DC1/ege033SUPPLEM.csv
38 The Isotope and Trace Element Budget, Journal Of Petrology, 2000, Volume 41, Number 6, Pages 759
39 Reference 102, Pages 772-774
40 Petrogenesis and Origins of Mid-Cretaceous, Journal Of Petrology, 2010, Volume 51, Number 10, Pages 2003-2045
41 Reference 110, Pages 2038
42 Reference 110, Pages 2024-2026
43 Petrogenesis of the Flood Basalts, Journal Of Petrology, 1998, Volume 39, Number 4, Pages 711–748
44 Reference 119, Pages 729, 730
45 Nature of the Source Regions, Journal Of Petrology, 2004, Volume 45, Number 3, Pages 555
46 Reference 121, Pages 556
47 Reference 121, Pages 566, 575, 576
48 Generation of Palaeocene Adakitic Andesites, Journal Of Petrology, 2007, Volume 48, Number 4, Pages 661
49 Reference 124, Pages 676-678
50 Reference 124, Pages 684
51 Evidence for a Widespread Tethyan, Journal Of Petrology, 2005, Volume 46, Number 4, Pages 829-858
52 Reference 127, Pages 831
53 Reference 127, Pages 840
54 Reference 127, Pages 832-837
55 Schmitz MD, Bowring SA. An assessment of high-precision U-Pb geochronology. Geochimica et Cosmochimica Acta, 2001, Volume 65, Pages 2571-2587

Chapter 21

Very Old Rocks

By Paul Nethercott

August 2012

Comparison of African and Canadian Diamonds

Table 1

Congo Million Years	Leslie Million Years	Grizzly Million Years	Fox Million Years	Koala Million Years	Jwaneng Million Years
5,500	7,500	7,500	6,500	6,500	5,000
5,500	7,500	7,500	7,500	7,000	5,000
5,500	8,000		8,300	7,500	5,000
6,500					5,000
6,500					
6,500					

These samples were dated in the year 2000 [1] by scientists from the University of Manchester, University College London and the University of Glasgow in Scotland. Samples were taken from Canada (Fox, Grizzly, Leslie and Koala), the Democratic Republic of Congo and from Botswana (Jwaneng). The article states that "apparent ages for most diamonds are greater than the age of the Earth." [2] Twenty one dates in this table [2] are indeed older than the theory of evolution would allow. Fourteen are over six billion years old. The article admits that many dates are meaningless: "all apparent ages are higher than the host kimberlite eruption ages and most are higher than the 4.5 Ga geochron."[3]

Standard evolutionist geology views the Earth as being 4.5 billion years old. Here are some quotes from popular text: "The age of the Earth is 4.54 ± 0.05 billion years." [4] "The Solar System, formed between 4.53 and 4.58 billion years ago." [5] "The age of 4.54 billion years found for the Solar System and Earth." [5] "A valid age for the Earth of 4.55 billion years." [6, 7]

The Archaean Barberton Greenstone Belt

In 1998 diamond samples were dated by scientist from the Johannes Gutenberg University, Mainz, Germany, the Max-Planck Institute Chemistry, and the Centre Geochemistry, Strasbourg, France. [4] According to the author the true ages is 2.7 billion years: "All three isotopic systems of whole rocks indicate ages of ~2.7 Ga, much younger than the depositional age of the successions." [5] "By treating the primary isochron slope of the Pb-isotopic data of sample OG 1 as a secondary isochron, an additional recalculation of the 208Pb/204Pb isotopic values indicates that the 232Th/238U (k) isotopic ratio of sample OG 1 has had a value of 4.78 from~2.7 Ga, which is slightly higher than the typical k value of ~4 (Taylor and McLennan, 1985)." [6] When we run the 207Pb/206Pb ratios listed [7] in the essay through Isoplot we get dates almost 2 billion years older. A radically different answer!

Table 2

Sample Number	207Pb/206Pb Million Years	Sample Number	207Pb/206Pb Million Years
OG-1-a	4,557	OG-1-x	4,557
OG-1-b	4,544	OG-1-y	4,544
OG-1-c	4,554	OG-1-z	4,554
OG-1-d	4,476	OG-1-aa	4,476
OG-1-e	4,596	OG-1-1a	4,596
OG-1-f	4,560	OG-1-1b	4,560
OG-1-g	4,566	OG-1-2a	4,566
OG-1-h	4,499	OG-1-2b	4,499
OG-1-i	4,495	OG-1-3a	4,495
OG-1-j	4,507	OG-1-3b	4,507
OG-1-k	4,514	OG-1-7a	4,514
OG-1-l	4,518	OG-1-7b	4,518
OG-1-m	4,454	OG-1-8a	4,454
OG-1-n	4,570	OG-1-8b	4,570
OG-1-o	4,477	OG-1-9a	4,477
OG-1-p	4,517	OG-1-9b	4,517
OG-1-q	4,534	OG-1-12a	4,534
OG-1-r	4,563	OG-1-12b	4,563
OG-1-s	4,510	OG-1-13a	4,510
OG-1-t	4,535	OG-1-13b	4,535
OG-1-u	4,458	OG-1-14a	4,458
OG-1-v	4,587	OG-1-14b	4,587
OG	4,488		

Laser Argon-40/Argon-39 Age Determinations

This dating on Moon rocks was done in 1998 by scientists from the University of Manchester in England. "The Luna 24 mission returned 160 cm of core (0.17 kg) from the south eastern rim of Mare Crisium in August 1976." [8] Nineteen samples from this Russian space probe were dates by Argon dating as being older than the evolutionist age of the Moon. [9] "The presence of trapped Ar components is evident from the anomalously high apparent ages determined from the measured 40Ar/39Ar values for the initial 30-40% of K release." [10] "Interpretation of the apparent ages is problematic because neither the clast composition nor the proportions of clast and matrix in the analysed splits could be determined." [11] The current consensus among evolutionists is that the true age of the Moon is 4.5 billion years old. [12]

Table 3

Sample Number	Age, Million Years
lc_1	5,700
3_1	4,810
5_1	5,760
5_2	5,320
5_3	5,060

7a_1	6,930
7a_2	6,240
7a_3	5,760
7a_4	5,180
7a_7	4,810
7a_8	5,250
7a_9	4,880
7a_14	5,180
7b_1	5,400
7b_2	5,110
7c_1	6,080
7c_2	5,330
7c_4	4,990
7c_5	4,770

Meteorite: Northwest Africa 482

"Northwest Africa 482 (NWA 482) is the second largest lunar meteorite and the fifth found in the Sahara. The complete stone had a mass of 1.015 kg before cutting." [13] In 2002 it was dated by scientists from the Lunar and Planetary Laboratory, University of Arizona. The results of the dating [14] are summarized below in table 4.

Table 4

Bulk Sample	Age, Million Years
	9,670
	8,560
	8,127
	6,256
Glass Sample	**Age, Million Years**
	9,905
	7,388
	5,708

The author of the article explains why he thinks that the ages are so absurd: "We believe that this ^{40}Ar is probably dominated by terrestrial contamination." [15]

Rhenium-Osmium Isotopic Composition in Diamonds

These rock samples from the King Leopold ranges in Western Australia were dated in 2010 by scientists from the Department of Geological Sciences, University of Cape Town, South Africa and the Department of Terrestrial Magnetism, Carnegie Institution of Washington. [16] The difference between the oldest and youngest dates [17] as shown in table 5 is 16,254 million years. The author of the article explains why he thinks that the ages are so absurd: "The chalcopyrite inclusion from EL57 gives a model age older than the age of the Earth, evidence, perhaps, that this sulphide has suffered Re loss." [18]

Table 5

Sample Name	Age, Million Years
EL10	1,658
EL26	430
EL57	7,457
EL61	847
EL23	1,264
EL50	1,171
EL54_1	-8,281
EL54_3	-362
EL55_1	7,973
EL55_2	-104
EL65	-5,773

K-Ar Dating of Diamonds

This dating was done in 1983 by scientists from the Geophysical Institute, University ofTokyo, Tokyo. [19] Eight dates are older than the evolutionist age of the Earth. [20] The author blames Argon contamination for the bizarre dates that were obtained: "Because of the extremely small amount of argon, the hot blank corrections were similar to or even larger than the argon in the diamonds, resulting in a large uncertainty in the experimental results." [20] The author admits that the dates are absolutely meaningless: "The apparent K-Ar ages range from 150 million to nine billion years, indicating that the non radiogenic ^{40}Ar is significant. Since we have no way to make a correction for the non-radiogenic 40Ar, the apparent K-Ar age does not offer useful information on the age of the diamonds." [21] Whichever date the author accepts is simply an arbitrary choice. Any date is just as good as any other date.

Table 6

Sample Number	Million Years
Premier Mine	
82701N	5,800
827021	5,200
82703A	8,200
8270413	3,300
Unidentified Origin	
821104N	4,800
821105H	5,700
821106N	4,400
821107N	5,000
8211083	4,500
8211091	9,100
82111ON	6,600
821111N	150

Isotopic And Petrographic Evidence

This dating was done in 2008 by scientists from the Department of Earth & Atmospheric Sciences, University of Alberta, Canada and from the Department of Earth Sciences, The Open University, England. [22] Two meteorites (Allan Hills and Northwest Africa) were dated and fourteen dates are older than the evolutionist age of the Earth. [23] The article admits that the dates are meaningless: "The most striking observation is that all of

NWA 1950 shock melt data, and more than half of the ages derived from ALH 77005 shock melts, are impossibly ancient, older than the Solar System itself (4.567 Ga; Fig. 6). Moreover, ancient ages (>4.567 Ga) from shock melts are known in meteorites, in articular the Peace River L6 chondrite, studied by Ar–Ar stepped heating and localized outgassing by a laser probe (McConville et al., 1988)." [24] The article concludes with the following remarks: "Our Ar–Ar results for shock melts—ages in >4.567 Ga and 40Ar/36Ar ratios that overlap with previous measurements of the Martian atmosphere—indicate that shock melt 'ages' are meaningless in terms of any real event." [25]

Table 7

Sample Number	Age Million Years
1	8,064
2	7,192
3	7,064
4	6,872
5	6,679
6	6,423
7	6,205
8	6,179
9	6,103
10	5,346
11	5,103
12	5,103
13	5,026
14	4,654

Rhenium–Osmium Systematics Of Diamond-Bearing Eclogites

Scientists from the Department of Geological Sciences, University of Cape Town, South Africa and the Department of Terrestrial Magnetism, Carnegie Institution of Washington, preformed this dating in 2003. [26] There is a 31,600 million years between the oldest and youngest dates. [27] "Thus, the Re–Os model ages, when calculated relative to a mantle undergoing chondritic Os isotopic evolution, are considerably older, varying from 3.1 to 18.5Ga (for calculation parameters). Model ages older than the age of the Earth are a clear indication that at least some of the samples have not experienced the simple single-stage Re–Os evolution required by the model age calculation. The unrealistically old Re–Os model ages reflect Re/Os ratios too low to account for the high measured $^{187}Os/^{188}Os$." [28] The author concluded the article with the following remarks: "The scatter in Re–Os systematics reflects a complex history for these eclogites that makes it impossible to define a precise age." [29]

Table 8

Sample Name	Age, Billion Years
AHM-C5	-13.1
AHM-K1/1	5.86
AHM-K4/2	4.24
AHM-K5/2	4.47
AHM-K6/1	5.12
AHM-K6/2	5.14
AHM-K13	18.5
AHM-K14	4.09
AHM-K15	13.8

A Study Of Northern Canadian Cordillera Xenoliths

These samples were dated in the year 2000 by Geologists from the University Of Montreal, Canada and from the Earth and Planetary Sciences Department, McGill University, Canada. [30] The samples were taken from mountain ranges near the Canadian/Alaskan border. [31] The data [32] in table 9 contrasts model age versus minimum age. "The decoupling of $^{187}Re/^{188}Os$ and $^{187}Os/^{188}Os$ observed in the Canadian Cordillera xenolith data also affects the calculation of Os model ages, and leads to "future" ages or ages older than the Earth." [33] Because the data is so bad the author admits: "Because of the apparent perturbation of the Re/Os ratios, age information cannot be obtained from an isochron diagram." [33] How can a rock that exists in the present have formed million of years in the future? Such a proposition is illogical.

Table 9

Sample	Model Age	Minimum Age
Name	Billion Years	Billion Years
AL-42	Less Than Zero	0.46
AL-46		Less Than Zero
AL-75	Less Than Zero	0.43
AL-76	Less Than Zero	0.10
AL-86	Less Than Zero	0.52
AL-88	0.32	Less Than Zero
AL-41	Less Than Zero	0.48
AL-52	Less Than Zero	0.22
XLG-29A	Less Than Zero	0.92
XLG-12A	Less Than Zero	Less Than Zero
XLG-25A	0.54	Less Than Zero
KLX-47	Less Than Zero	0.33
BTX-26	Less Than Zero	Less Than Zero

Ar-Ar Chronology Of The Martian Meteorite

The Department of Earth Sciences, University of Manchester, dated these meteorite samples in 1997. [34] The samples are believed to be material ejected from the surface of Mars billion so years ago. [34] If we look at the data in table 10 we see that there is a 24,648 million difference between the oldest and youngest date. [35] If we look at the dates and error margins in Table 2 in the original article we see that the maximum age is 6,047 million years and the minimum is 257 million years. [36]

Table 10

Sample	Minimum Age	Maximum Age
Number	Million Years	Million Years
ALH84001,110		
1,300	4,626	5,236
1,450	4,345	5,013
ALH84001,111		
1,200	5,138	7,980
1,300	3,904	5,694
1,450	4,151	6,373

ALH84001,127		
400	2,660	5,062
450	4,106	5,018
500	4,012	4,550
550	4,442	4,614
700	4,036	4,942
800	4,179	4,847
1,200	-3,171	21,477
1,400	4,920	7,354

The Slave Craton, Canada

These samples from Canada were dated in 2010 by scientists from the Earth & Atmospheric Sciences, University of Alberta, Edmonton, Canada. [37] Some of the specimens were dated to be over 5.5 billion years old. [38] The author tells how the isochron gave absurd ages: "In contrast, the most radiogenic sulphides in sample 1636 plot about an impossible 5 Ga model isochron." [39] The admission is that the dates are impossible and meaningless: "The Re–Os isotope systematics of sulphides in sample 1636 are disturbed (Fig. 6e), with three of four samples falling on an impossible 5 Ga model isochron." [40]

U-Th-Pb Systematics In Lunar Highland Samples

California Institute of Technology, (Pasadena, California) dated these Lunar rocks in 1972. [41] Eighty one dates are older than the evolutionist age of the Solar System. Sixty three are over five billion years old. Seven are over six billion years old. [42]

Table 11

Space Probe/Sample	^{207}Pb	^{206}Pb	^{207}Pb	^{208}Pb
Luna 20	^{206}Pb	^{238}U	^{235}U	^{232}Th
22001, 1 A-2	4.94	5.83	5.19	5.87
	5.00	5.20	5.06	5.01
	4.92	6.09	5.24	6.24
22001, 1A-2	4.96	5.78	5.19	6.08
	5.01	5.25	5.08	5.30
	4.95	5.83	5.20	6.14
67481, 26	4.92	5.49	5.08	5.80
	4.94	5.29	5.04	5.52
	4.92	5.51	5.09	5.84
64421, 29	4.91	5.41	5.05	5.47
	4.94	5.00	4.96	4.91
	4.90	5.43	5.06	5.50
60501, 31	4.98	5.35	5.08	5.26
	4.99	5.23	5.06	5.10
	4.97	5.36	5.09	5.28
68501, 52	5.05	5.61	5.21	5.55
	5.06	5.48	5.18	5.37
	5.05	5.62	5.21	5.56
60025, 65	4.64	6.64	5.18	5.64
	4.75	3.75	4.42	2.51
	4.62	7.83	5.45	7.21

If we run the Lead 207/206 ratios [43] through Isoplot we get the following ages as listed in Table 12:

Table 12

Lead 207/206 Ratio	Million Years
0.8166	4,951
0.8196	4,956
0.8189	4,955
0.8190	4,955
0.7804	4,886
0.7800	4,886
0.7883	4,901
0.7886	4,901
0.8006	4,923
0.8008	4,923
0.8417	4,994
0.8417	4,994
0.7989	4,920
0.8015	4,924

The author comments on the major problems with dating these samples: "The data for all highland soils analyzed here are shown in fig. 4. All five data points lie far above the concordia curve and give ages for a single stage model which are in excess of 4.6 AE. The 206Pb-238U ages range up to 5.83 AE. The 207Pb-206Pb ages are also very high." [44] His calculations confirm the wrong ages radiometric dating gives: "Inspection of rows D and E shows the extreme limits of the207Pb-206Pb ages. All highland soils analyzed have 207Pb-206Pb model ages in excess of 4.90 AE. These are the highest values observed so far for samples of 'total lunar soil'." [45]

A 40Ar/39Ar Geochronological Study

Rock samples from the Lower Onverwacht Volcanics in Barberton Mountain Land, South Africa were dated in 1992 by geologists from the Department of Physics, University of Toronto, and the Department of Geological Sciences, Queen's University, Kingston, Ontario, Canada. [46] The youngest date was -4.5×10^{10} million years. [47] How can a rock that exists in the present have formed 45,000 trillion years in the future? Such a proposition is illogical.

Table 13

Sample Number	Age, Million Years
B40-A, Third Run	-45,000,000,000 -310,000
B40-E	-56,112 386 2,663 2,667 2,672 2,943 3,321 3,313 3,299

Kt-17b, First Run	**6,555** **6,296** **4,969** **5,117** **6,164** **5,228**
Kt-17b, Second Run	**6,848** **6,479** **5,731**
KT-17B, Plagioclase Concentrate	**6,204** **6,904** **6,560** **6,544** **5,105**
B56-A, First Run **B56-A, Second Run**	**7,810** **4,864** **4,890** **5,597**

Zircon Uranium/Lead Ages Of Guyana Greenstone

These mineral samples were dated in 1982 by scientists from the Department of Geological Sciences, Cornell University, New York and the Department of Earth Sciences, University of New Hampshire. [52] According to the article the true age of the specimen is 2250 Million years old. [53] If we run the isotopic ratios [54] through Isoplot we find that there is a 43,364 million difference between the oldest and youngest date.

Table 14

Sample Number	207Pb/206Pb Million Years	206Pb/238U Million Years	207Pb/235U Million Years
1a	2,226	2,218	44,242
1b	2,217	2,021	42,199
1d	2,210	1,806	39,839
1e	2,177	1,838	39,861
3a	2,249	1,835	40,561
3b	2,236	878	27,142
4a	2,206	1,617	37,640
4c	2,155	1,327	33,447
4d	2,183	1,339	33,871
5a	2,242	1,776	39,833

References

1 Comparison of African and Canadian Diamonds, Geochimica et Cosmochimica Acta, 2000, Volume 64, Number 4, Pages 717–732
2 Reference 1, Page 725
3 Reference 1, Page 724
4 Archaean Barberton Greenstone Belt, Precambrian Research, 1998, Volume 92, Pages 129–144
5 Reference 4, Page 129
6 Reference 4, Page 140
7 Reference 4, Page 136
8 Laser argon-40-argon-39 age determinations, Meteoritics & Planetary Science, 1998, Volume 33, Pages 921-935
9 Reference 8, Page 932-935
10 Reference 8, Page 925
11 Reference 8, Page 929
12 http://en.wikipedia.org/wiki/Moon_rock
13 Northwest Africa 482, Meteoritics & Planetary Science, 2002, Volume 37, Page 1797
14 Reference 13, Page 1806
15 Reference 13, Page 1805
16 Re–Os isotopic composition in diamonds, Geochimica et Cosmochimica Acta, 2010, Volume 74, Pages 3292–3306
17 Reference 16, Page 3296
18 Reference 16, Page 3297
19 K-Ar Dating of Diamonds, Geochimica et Cosmochimica Acta, 1983, Volume 47, Pages 2217
20 Reference 19, Page 2221
21 Reference 19, Page 2220
22 Isotopic And Petrographic Evidence, Geochimica et Cosmochimica Acta, 2008, Volume 72, Pages 5819–5837
23 Reference 22, Page 5826
24 Reference 22, Page 5826-5827
25 Reference 22, Page 5832
26 Re–Os Systematics Of Diamond-Bearing Eclogites, Lithos, 2003, Volume 71, Pages 323– 336
27 Reference 26, Page 329
28 Reference 26, Page 331
29 Reference 26, Page 333
30 A Study Of Northern Canadian Cordillera Xenoliths, Geochimica et Cosmochimica Acta, 2000, Volume 64, Number 17, Pages 3061–3071
31 Reference 30, Page 3063
32 Reference 30, Page 3064
33 Reference 30, Page 3067
34 Ar-Ar Chronology Of The Martian Meteorite, Geochimica et Cosmochimica Acta, 1997, Volume 61, Number 18, Pages 3835
35 Reference 34, Page 3839
36 Reference 34, Page 3842
37 The Slave Craton, Canada, Geochimica et Cosmochimica Acta, 2010, Volume 74, Pages 5368
38 Reference 37, Page 5375
39 Reference 37, Page 5372
40 Reference 37, Page 5377
41 U-Th-Pb Systematics In Lunar Highland Samples, Earth And Planetary Science Letters, 1972, Volume 17, Pages 36-51
42 Reference 41, Page 45, 46
43 Reference 41, Page 42, 43
44 Reference 41, Page 44
45 Reference 41, Page 39, 40
46 A 40Ar/39Ar Geochronological Study, Precambrian Research, 1992, Volume 57, Pages 91-119
47 Reference 46, Page 109
52 Zircon U-Pb Ages Of Guyana Greenstone, Precambrian Research, 1982, Volume 17, Pages 199-214
53 Reference 52, Page 199
54 Reference 52, Page 207

Chapter 22

Modern Dating Methods

By Paul Nethercott

April 2014

Introduction

How reliable is radiometric dating? We are repeatedly told that it proves the Earth to be billions of years old. If radiometric dating is reliable than it should not contradict the evolutionary model. According to the Big Bang theory the age of the Universe is 10 to 15 billion years.[1] Standard evolutionist publications give the age of the universe as 13.75 Billion years. [2, 3]

Standard evolutionist geology views the Earth as being 4.5 billion years old. Here are some quotes from popular text: "The age of the Earth is 4.54 ± 0.05 billion years." [4] "The Solar System, formed between 4.53 and 4.58 billion years ago." [1] "The age of 4.54 billion years found for the Solar System and Earth." [1] "A valid age for the Earth of 4.55 billion years." [5, 6]

Evolutionists give the age of the galaxy as "11 to 13 billion years for the age of the Milky Way Galaxy." [1, 7] Let us remember this as we look at the following dating as given in secular science journals.

Isotopic and Trace Element Geochemistry

These rock samples from the Bangladesh border North east India (West Bengal, north of Kolkata) were dated in 2013 by scientist from the University of Rochester, New York using the Neodymium, Strontium, Lead age dating methods. [1] The true age of the rock formation is supposed to be 115 million years old. "40Ar/39Ar data in basalts from these drillings suggest ages of 117 Ma. More recent 40Ar/39Ar results from the Rajmahal hills and the Sylhet basalts are consistent with an 118 Ma age." [2] "This complex gives a Pb–Pb age of 134 ± 20 Ma and a more precise U–Pb perovskite age of 115 ± 5.1 Ma" [3] The article contains a table [4] that has four hundred and fifty seven ratios that have no dates beside them. Out of the 457 dates we calculated from these ratios there is a total disagreement with the so called 'true age.' Whichever date you choose for each rock as the true one is just a random guess.

Table 1	Average	Maximum	Minimum
147Sm/144Nd	106	117	99
87Rb/86Sr	112	117	102
207Pb/206Pb	5,041	5,055	5,009
206Pb/238U	9,888	10,609	8,839
207Pb/235U	6,161	6,358	6,058
208Pb/232Th	15,680	20,320	14,313

Table 2	Average	Maximum	Minimum
147Sm/144Nd	107	113	102
87Rb/86Sr	112	121	94
207Pb/206Pb	5,045	5,075	5,014
206Pb/238U	9,543	13,048	6,315
207Pb/235U	6,075	6,757	5,347
208Pb/232Th	18,054	28,756	11,610

Table 3	Average	Maximum	Minimum
147Sm/144Nd	108	119	92
87Rb/86Sr	108	119	70
207Pb/206Pb	5,039	5,053	5,017
206Pb/238U	10,844	17,441	6,877
207Pb/235U	6,343	7,468	5,495
208Pb/232Th	12,287	17,286	9,074

Table 4	Average	Maximum	Minimum
147Sm/144Nd	103	119	97
87Rb/86Sr	113	141	70
207Pb/206Pb	4,917	5,059	4,717
206Pb/238U	5,634	20,655	733
207Pb/235U	4,655	7,467	2,568
208Pb/232Th	7,077	21,557	318

Table 5	Average	Maximum	Minimum
147Sm/144Nd	107	119	97
87Rb/86Sr	115	141	106
207Pb/206Pb	4,952	5,060	4,912
206Pb/238U	7,600	19,375	1,996
207Pb/235U	5,376	7,470	3,777
208Pb/232Th	12,139	21,752	1,908

Table 6	Average	Maximum	Minimum
147Sm/144Nd	172	901	82
87Rb/86Sr	111	141	70
207Pb/206Pb	4,894	5,007	4,253
206Pb/238U	12,184	31,823	266
207Pb/235U	5,592	7,476	1,390
208Pb/232Th	18,102	61,342	261

Geochemistry of Hornblende Gabbros

These rock samples from Sönidzuoqi (Inner Mongolia, North China) were dated in 2008 by scientist from the Chinese Academy of Sciences, Beijing using the Potassium/Argon and Uranium/Lead age dating. [5] The true age of the rock formation is supposed to be 500 million years old. "Limited hornblende K–Ar and SHRIMP U–Pb zircon ages document the Late Silurian to Early Devonian gabbroic emplacement." [5] "The Siluro-Devonian hornblende gabbros, together with a pre-490 Ma ophiolitic melange of MORB-OIB affinity, 483–471 Ma arc intrusions, 498–461 Ma trondhjemite-tonalite-granodiorite plutons, and 427–423 Ma calc-alkaline granites from the same area." [5] The article contains a table [6] that has twenty eight ratios that have no dates beside them. Out of the twenty eight dates we calculated from these ratios there is a total disagreement with the so called 'true age.' The 'true ages' [7] listed in the article are completely different. Whichever date you choose for each meteorite as the true one is just a random guess.

Table 7	207Pb/206Pb	206Pb/238U	207Pb/235U	208Pb/232Th
Average	5,011	6,612	5,422	22,967
Maximum	5,014	7,297	5,648	24,397
Minimum	5,007	5,922	5,237	20,621

Post-Collisional Transition from Subduction

These rocks from south western Spain and Morocco were dated in 2003 by scientist from the Institute for Geosciences, University Of Kiel, Germany using the 40Ar/39Ar-age dating. [8] According to the article The true age of the rock formation is between 0.65 million years and 8 million years old: "Two groups of magmatic rocks can be distinguished: (1) an Upper Miocene to Lower Pliocene (8.2–4.8 Ma), Si–K-rich group including high-K (calc-alkaline) and shoshonitic series rocks; (2) an Upper Miocene to Pleistocene (6.3–0.65 Ma)." [9] The article contains tables [10] with Uranium/Thorium/Lead ratios that have no dates beside them. If we put the tables into Microsoft Excel and use the computer program Isoplot [11] we can calculate dates from the undated isotopic ratios. There is a 48,068 million year range between the youngest and oldest dates.

Table 8	207Pb/206Pb	208Pb/232Th	206Pb238U
Average	4,951	13,783	3,440
Maximum	4,986	48,962	7,519
Minimum	4,837	2,028	894

Nazca Ridge and Easter Seamount Chain

These rocks from Easter Island Sea floor were dated in 2011 by scientist from the University Of Hawaii using the 40Ar/39Ar-age dating. [12] According to the article the true age of the rock formation is between 1 million years and 33 million years old. [13] The article contains a table [14] with Uranium/Thorium/Lead ratios that have no dates beside them. If we put the tables into Microsoft Excel and use the computer program Isoplot, we can calculate dates from the undated isotopic ratios. There is a 22,684 million year range between the youngest and oldest dates.

Table 9	207Pb/206Pb	208Pb/232Th	206Pb/238U
Average	4,919	8,325	3,694
Maximum	4,971	23,850	9,645
Minimum	4,881	4,129	1,166

South African Off-Craton Mantle

These rocks from South Africa were dated in 2009 by scientist from the Arizona State University using the Rhenium/Osmium age dating. [15] According to the article the true age of the rock formation is between 600 million years and 2,600 million years old. "Rhenium depletion model ages (TRD) determined from 58 Osmium isotope compositions of peridotites span a range from 2.6 to 0.6 Ga, with an average of 1.67 Ga." [15] The article contains a table [16] with calculated dates beside them. Out of the 144 dates there is a 121.35 billion year range between the youngest [-76 billion years] and oldest [45 billion years] dates. The oldest sample is thirty billion years older than the Big Bang explosion.

Table 10

Largest (Ga)	Largest (Ga)	Smallest (Ga)	Smallest (Ga)
5.05	6.98	-0.14	-6.1
34.97	6.71	-1.54	-10.7
27.29	6.59	-1.62	-13.38
10.39	5.6	-4.44	-14.57
10.21	5.55	-4.48	-33.78
8.31	5.39	-5.91	-76.3

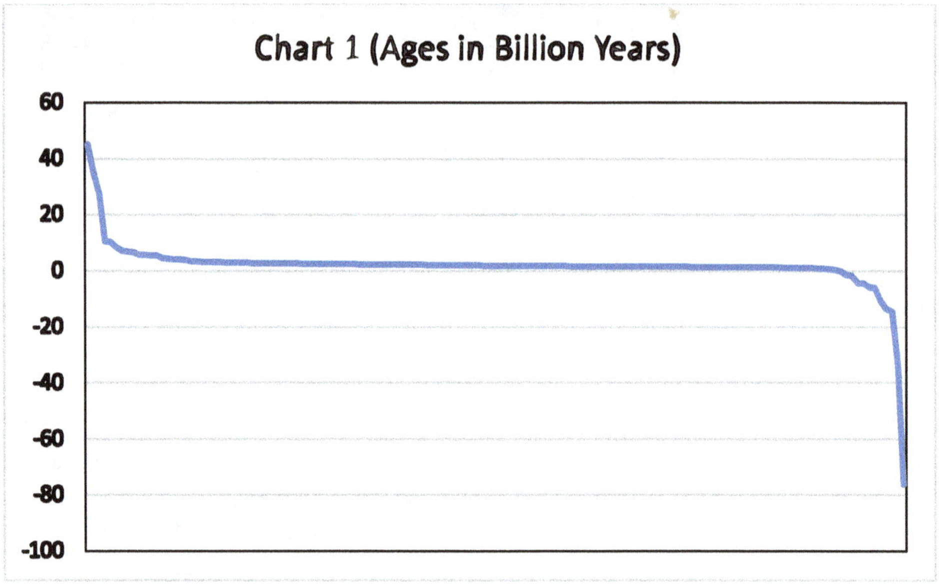

Os And Re Distribution In The Active Mound

These rocks from Mid-Atlantic Ridge were dated in 1998 by scientist from Texas AM University using the Rhenium/Osmium age dating. [17] The article contains a table [18] that has Osmium 187/186 ratios that have no dates beside them. If we put the tables into Microsoft Excel and use the formula below used in standard geology text books [19-21] we can calculate dates from the undated isotopic ratios.

(1)

$$t = \frac{1.04 - ({}^{187}Os / {}^{186}Os)}{0.050768}$$

In the above formula, t = billions of years. The same date can be calculated from the Osmium 187/188 ratios. If we use another formula [22] we can convert the Osmium 187/188 ratio to the Osmium 187/186 ratio.

(2)

$$\frac{{}^{187}Os}{{}^{186}Os} \times 0.12035 = \frac{{}^{187}Os}{{}^{188}Os}$$

(3)

$$\frac{{}^{187}Os}{{}^{186}Os} = \frac{({}^{187}Os \div {}^{188}Os)}{0.12035}$$

(4)

$$t = \frac{1.04 - \left(\frac{({}^{187}Os \div {}^{188}Os)}{0.12035} \right)}{0.050768}$$

Table 11	Million Years
Average	-123,544
Maximum	-13,394
Minimum	-154,625

The Osmium ratios yield impossible future ages. How can the rocks that formed in the past have formed 154 billion years in the future?

Osmium-Isotope Geochemistry Of Site 959

These rocks from South Africa were dated in 1998 by scientist from the Woods Hole Oceanographic Institution, Massachusetts using the Rhenium/Osmium age dating. [23] According to the article the true age of the rock formation is between 66 million years and 2 million years old. "These samples vary in age from late Neogene to Late Cretaceous." [23] The article contains a table [24] with Osmium 187/186 ratios that have no dates beside them. Out of the 19 dates there is a 246 billion year range between the youngest and oldest dates.

Table 12	Million Years
Average	-153,703
Maximum	-72,290
Minimum	-318,311

The Seve Nappe Complex of Jamtland

These rocks from Sweden were dated in 2002 by scientist from Queens College, New York using the Rhenium/Osmium and Neodymium/Samarium age dating. [25] According to the article the true age of the rock formation is 450 million years old: "Mineral isochrons from three pyroxenite layers define overlapping ages of 452.1 and 448 Ma and 451 Ma." [25] The article contains a table [26] with Osmium 187/186 ratios that have no dates beside them. Out of the forty dates there is a 41.71 billion year range between the youngest and oldest dates. Of the forty dates, twenty eight [70%] are over 5 billion years old. Fifteen [37.5%] are over 10 billion years old.

Table 13	Million Years	% Discordance	Difference
Average	-10,204	2,381	10,715
Maximum	1,205	9,201	41,406
Minimum	-40,956	170	767

The same table has calculated Rhenium/Osmium dates beside the undated ratios. Out of the 79 dates there is a 92 billion year range between the youngest and oldest dates. Of the 79 dates, twenty eight [70%] are over 5 billion years old. Fifteen [37.5%] are over 10 billion years old. Out of the 79 dates twenty [25%] are over 5 billion years old. Nine [11%] are over 11 billion years old. Forty nine [62%] are impossible future or negative ages. The oldest sample is twenty billion years older than the Big Bang explosion.

Table 14	Ga
Average	-4.25
Maximum	34
Minimum	-58

The Kaalvallei Kimberlite, South Africa

These rocks from South Africa were dated in 2004 by scientist from University Of Toronto, Canada using the Rubidium/Strontium and Neodymium/Samarium age dating. [27] According to the article the true age of the rock formation is between 990 to 1580 million years old: "All indicate Proterozoic diamond formation ages ranging from 990 to 1580 Ma, and it is, therefore, not unreasonable to assume that the

Kaalvallei Group I eclogite xenoliths are also at least Proterozoic in age." [28] The article contains a table [28] with Neodymium/Samarium dates beside them. There is a 5.4 billion year range between the youngest and oldest dates.

Table 15

Minimum Age Million Years	Maximum Age Million Years	Age Difference Million Years
-5	_	
-697	1304	2,001
-2,771	1572	4,343
-3,817	1148	4,965
-3,896	1304	5,200
-4,198	1199	5,397

Genesis of Continental Intraplate Basalts

These rocks from western Victoria were dated in 2000 by scientist from Monash University, Melbourne using the Lead/Lead, Rhenium/Osmium and Neodymium/Samarium age dating. [29] According to the article the true age of the rock formation is between 750 and 1,000 million years old: "The best fit AFC model for the group two ol-tholeiites is for assimilation with 1,000 Ma low 187Re/188Os." [30] The basalt veneer is a 10 metre deep layer. "Contamination of the Newer Volocanics Province Plains series magmas by Proterozoic crustal [>750 Ma] is considered to be more likely." [30] The article contains a table [31] with Osmium 187/188 and Lead 207/206 ratios that have no dates beside them. Out of the dates we calculated from these ratios there is a 57.45 billion year range between the youngest and oldest dates.

Table 16

Pb 207/206 Age	187Os/188Os Age	Age % Difference	Age Difference
4,979	-1,900	262	6,878
4,985	-1,484	336	6,469
4,986	-20,890	419	25,875
4,981	-23,099	464	28,081
4,984	-52,445	1,052	57,429
4,974	-39,136	787	44,109
4,975	-19,630	395	24,605
4,986	-9,132	183	14,118
5,007	-12,919	258	17,926

Xenoliths from the Colorado Plateau

These rocks from North eastern Arizona (Four corners: Utah, Colorado, Arizona, New Mexico), were dated in 2004 by scientist from Okayama University, Japan using the Uranium/Lead, Rubidium/Strontium and Neodymium/Samarium age dating. [32] The formation is supposed to have formed in the Cretaceous period: "The Late Cretaceous and Tertiary records of arc magmatism in the south western USA constrain the slab geometry and its evolution, suggesting that the migration of arc magmatism was probably caused by progressive flattening of a subducting slab." [33] The true age of the rock formation is supposed to be between 30 and 80 million years old: "Usui et al. (2003) used ion microprobe techniques to determine the U–Pb ages of zircons from the Colorado Plateau eclogite xenoliths, which yielded concordant ages from 81 to 33 Ma." [34] "The mineral isochron ages for zoisite-eclogite xenoliths are 39 Ma for the 147Sm/144Nd–143Nd/144Nd isochron diagram, and 33-20Ma for the 238U/206Pb–207Pb/206Pb isochron diagram." [35] The article contains a table [36] with Uranium/Thorium/Lead ratios that have no dates beside them. Out of the dates we calculated from these ratios there is a 39.9 billion year range between the youngest [653 million years] and oldest [40,568 million years] dates.

Table 17	Average	Maximum	Minimum
207Pb/206Pb	4,938	4,963	4,881
206Pb/238U	3,548	5,716	653
207Pb/235U	4,303	5,169	2,560
208Pb/232Th	10,765	18,206	1,806

Table 18	Average	Maximum	Minimum
207Pb/206Pb	4,956	4,961	4,949
206Pb/238U	6,799	10,481	1,894
207Pb/235U	5,303	6,231	3,732
208Pb/232Th	15,131	40,568	1,704

Table 19	Average	Maximum	Minimum
207Pb/206Pb	4,961	4,965	4,958
206Pb/238U	8,861	10,383	6,938
207Pb/235U	5,893	6,218	5,476
208Pb/232Th	14,675	16,757	11,144

Indosinian Granitoids

These rocks from The Bikou block, located along the north western margin of the Yangtze plate, were dated in 2006 by scientist from the China University of Geosciences, Wuhan, China using the Uranium/Lead, Rubidium/Strontium and Neodymium/Samarium age dating. The true age of the rock formation is supposed to be 200 to 800 million years old: "U-Pb zircon SHRIMP dating for the volcanic rocks yielded ages ranging from 840 to 776 Ma, representing formation time of the Bikou Group volcanic rocks." [38] "The magma crystallization age of the Yangba pluton was reported to be 215.4±8.3 Ma (U-Pb zircon)" [39] The article contains a table [40] with Uranium/Thorium/Lead ratios that have no dates beside them. Out of the dates we calculated from these ratios there is a 26.8 billion year range between the youngest [5,005 million years] and oldest [31,891 million years] dates.

Table 20	207Pb/206Pb	206Pb/238U	208Pb/232Th
Average	5,017	11,096	21,167
Maximum	5,028	13,173	31,891
Minimum	5,005	7,695	12,943

The Stonyford Volcanic Complex

These rocks from The San Andreas fault (San Francisco; Sacramento Valley) were dated in 2004 by scientist from the Utah State University [41] using the Uranium/Lead, Rubidium/Strontium and Neodymium/Samarium age dating. The true age of the rock formation is supposed to be 160 million years old: "Jurassic age volcanic rocks of the Stonyford volcanic complex (SFVC) comprise three distinct petrological groups." [42] "40Ar–39Ar dates on volcanic glass from the hyaloclastite breccias range from 163 to 164 Ma." [43] "Quartz diorite melange blocks that structurally underlie the SFVC yield U–Pb zircon concordia intercept ages of 163 Ma and 164 Ma." [43] The article contains a table [44] with Lead 207/206 ratios that have no dates beside them. Out of the fourteen dates we calculated from these ratios there is an agreement that the true age of the rock formation is not 160 million years but actually 5 billion years old!

Table 21	207Pb/206Pb
Average	4,952
Maximum	5,012
Minimum	4,831

Cenozoic Volcanism in Tibet

These rocks from Tibet were dated in 2002 by scientist from the University Of Arizona using the Uranium/Lead, Rubidium/Strontium and Neodymium/Samarium age dating. The rocks were also dated by two other methods (K/Ar or 40Ar/39Ar). [45] The true age of the rock formation is supposed to be 10 to 60 million years old. "Chemical data are presented for newly discovered Cenozoic volcanic rocks in the western Qiangtang and central Lhasa terranes of Tibet. Alkali basalts of 65-45Ma occur in the western Qiangtang terrane." [46] "In contrast, younger volcanic rocks in the western Qiangtang terrane (30 Ma) and the central Lhasa terrane (23, 13 and 8 Ma) are potassic to ultrapotassic and interpreted to have been derived from an enriched mantle source." [46] The article contains a table [47] 40Ar/39Ar ratios that have fifty four dates beside them. The article contains another table [48] that has thirty three Lead 207/206 ratios and fifteen Rubidium/Strontium ratios that have no dates beside them. Out of the forty eight dates we calculated from these ratios there is an agreement that the true age of the rock formation is not 60 million years but actually 5 billion years old! Whichever date you choose as the true one is just a random guess.

Table 22	207Pb/206Pb	Ar/Ar	87Rb/86Sr
Average	4,980	2.74	25
Maximum	5,014	33.50	43
Minimum	4,968	0.28	13

U-Th-Pb Analysis Of Baddeleyites

These Martian meteorites were dated in 2011 by scientist from the University Of Arizona using the Lead/Lead, Rubidium/Strontium and Neodymium/Samarium age dating. [49] The true age of the rock formation is supposed to be between 150 and 4,005 million years old. "Rb-Sr and Sm-Nd ages of basaltic shergottites consistently yield young ages (150-450 Ma). Other shergottite sub-groups also yield young ages. In contrast to these results, Pb-Pb isochron analyses yields ages on order of 4.05 Ga." [49] Such a wide age range is meaningless! The article contains a table [49] that has nine Uranium/Lead ratios from two different meteorites that have no dates beside them. Out of the nine dates we calculated from these ratios there is a total disagreement with the so called 'true age.' Whichever date you choose for each meteorite as the true one is just a random guess.

Table 23

Meteorite	206Pb/207Pb	207Pb/235U	206Pb/238U	Model Age	Error
NWA 2986	4,149	2,304	810	502	3,647
NWA 2986	4,155	3,251	1,994	1236	2,919
NWA 2986	5,199	3,644	1,501	931	4,268
NWA 2986	2,460	1,170	602	373	2,087
NWA 2986	4,022	1,368	302	187	3,835
RBT 04262	2,639	436	139	100	2,539
RBT 04262	3,956	1,485	365	263	3,693
RBT 04262	4,540	2,448	731	526	4,014
RBT 04262	4,108	1,700	429	309	3,799

Rb-Sr and Pb-Pb Geochronology

These rock samples from the alpine towns of Verbania and Locarno on the Swiss/Italian border were dated in 2007 by scientist from the University Of Milan in Italy using the Uranium/Lead, Rubidium/Strontium age dating. [50] The true age of the rock formation is supposed to between 300 and 405 million years old. "Rb-Sr whole-rock (WR) isochron (466±5 Ma) and Pb-Pb single zircon evaporation ages (458±6 Ma and 463±4 Ma) on meta-granites date the emplacement of the older intrusive series, whereas Rb-Sr muscovite ages (311-325 Ma) approach the Carboniferous metamorphism (331-340 Ma). Rb-Sr WR isochrons (277±8 Ma) and biotite ages (276-281 Ma) on granitic plutons date the emplacement of the younger intrusive series." [50] The article contains a table [51] that has sixty five Lead 207/206 ratios that have no dates beside them. Out of the sixty five dates we calculated from these ratios there is a total disagreement with the so called 'true age.' Whichever date you choose for each sample as the true one is just a random guess.

Table 24	207Pb/206Pb
Average	4,992
Maximum	5,237
Minimum	4,924

U-Th-Pb Isotope Data

These rock samples from the Marble Bar area of the Pilbara Craton (Western Australia) were dated in 2011 by scientist from the University of Wisconsin-Madison using the Uranium/Lead age dating. [52] The true age of the rock formation is supposed to be 3,400 million years old. "The first core of the Archean Biosphere Drilling Project (ABDP-1) documented hematite as alteration products in 3.4 Ga basalts from the Marble Bar area of the Pilbara Craton, NW Australia." [53] "The best-fitting isochrons for the basalts from Marble Bar at 3.4 Ga, which is the approximate formation age of these basalts. Secondary Pb growth curves were made using the Pb isotope composition of the primary Pb growth curve at 3.4 Ga as the starting point." [54] The article contains a table [55] that has thirteen Uranium/Thorium/Lead ratios that have no dates beside them. Out of the thirteen dates we calculated from these ratios there is a total disagreement with the so called 'true age.' There is a 95 billion year difference between the youngest and oldest dates. Whichever date you choose for the true one is just a random guess.

Table 25	206Pb/238U	207Pb/235U	207Pb/206Pb	208Pb/232Th
Average	15,192	7,319	5,325	56,976
Maximum	31,005	10,054	5,403	100,601
Minimum	7,138	5,795	5,222	24,980

GSA Data Repository

These rock samples from the Guyot Province and the Walvis Bay Ridge, Namibia were dated in 2013 by scientist from the Geological Society of America using the Uranium/Thorium/Lead age dating. [56] The true age of the rock formation is supposed to be 100 million years old. "The samples display an age range of ~100 Ma and are thus difficult to compare at a common age without making additional assumptions, such as parent/daughter ratios of the source." [57] The article contains a table [58] that has different isotopic ratios that have no dates beside them. Out of the one hundred and twelve dates we calculated from these ratios there is a total disagreement with the so called 'true age.' The sixty four Uranium/Lead dates totally contradict the forty eight Rb/Sr, Nd/Sm dates. Whichever date you choose for each sample as the true one is just a random guess.

Table 26	Average	Maximum	Minimum
207Pb/206Pb	4,996	5,015	4,981
207Pb/235U	4,760	5,033	4,599
208Pb/232Th	7,484	8,770	7,097
206Pb/238U	4,243	4,929	3,711

Table 27	Average	Maximum	Minimum
207Pb/206Pb	5,019	5,044	5,008
207Pb/235U	5,167	5,493	4,948
208Pb/232Th	8,727	9,496	7,516
206Pb/238U	5,514	6,675	4,782

Table 28	Average	Maximum	Minimum
207Pb/206Pb	5,012	5,022	5,005
207Pb/235U	4,726	5,038	4,340
208Pb/232Th	7,571	8,821	6,211
206Pb/238U	4,115	5,049	3,015

Table 29	Average	Maximum	Minimum
207Pb/206Pb	5,018	5,029	5,006
207Pb/235U	4,765	4,869	4,662
208Pb/232Th	10,476	10,553	10,400
206Pb/238U	4,179	4,503	3,854

Table 30	87Rb/86Sr	147Sm/144Nd	176Lu/177Hf
Average	49	49	52
Maximum	70	70	65
Minimum	30	30	31

Lead in Galena from Ore Deposits

These rock samples from the Khanka Massif range (north of Vladivostok) were dated in 2002 by scientist from the Russian Academy of Sciences in Irkutsk using the Lead 207/206 age dating. [59] The true age of the rock formation is supposed to be 100 to 245 million years old. "Lead from galena of the Taukha terrane has a wide range of model ages (245–109 Ma). The range of 109–141 Ma corresponds to the Early Cretaceous accretion of the Taukha terrane, whereas the range of 157−245 Ma corresponds to the formation of the Early Triassic–Late Jurassic oceanic fragment." [60] The article contains a table [61] that has Lead 207/206 ratios that have no dates beside them. Out of the forty three dates we calculated from these ratios there is a total disagreement with the so called 'true age.' Whichever date you choose for each sample as the true one is just a random guess.

Table 31	207Pb/206Pb	Model Age
Average	5,009	156
Maximum	5,063	736
Minimum	5,000	66

The Caribbean Large Igneous Province

These rock samples from the southern Caribbean Sea, off the Venezuelan coast were dated in 1998 by scientist from the University of California using the Lead/Lead, Rubidium/Strontium and Neodymium/Samarium age dating. [62] The true age of the rock formation is supposed to be 80 million years old. "The uniqueness of the Caribbean Large Igneous Province (CLIP, 92-74 Ma) with respect to other Cretaceous oceanic plateaus is its extensive sub-aerial exposures." [63] "Nanno fossils and 40Ar/39Ar ages suggest that the main pulse of volcanism forming the CLIP occurred primarily between 92 and 88 Ma but continued to V74 Ma." [64] The article contains a table [65] that has 147Sm/144Nd and 206Pb/207Pb ratios that have no dates beside them. Out of the thirty three dates we calculated from these ratios there is a total disagreement with the so called 'true age.' Whichever date you choose for each sample as the true one is just a random guess.

Table 32	147Sm/144Nd	206Pb/207Pb
Average	84	4,940
Maximum	91	4,973
Minimum	60	4,895

Nd-Hf-Sr-Pb isotopes

These rock samples from the Krishna River, east of Hyderabad were dated in 2006 by scientist from the University of Rochester, New York using the Neodymium, Strontium, Lead and Hafnium age dating methods. [66] The true age of the rock formation is supposed to be 1,224 million years old. "The probable sources of some of the famous Indian diamonds are the 1.2 Ga old Krishna lamproites of Southern India, a rare Proterozoic occurrence of lamproites." [67] "The initial isotopic ratios of these elements are calculated based on the ~1,224 Ma Rb–Sr age of emplacement for these lamproites." [68] The article contains a table [69] that has Rubidium/Strontium and Uranium/Lead ratios that have no dates beside them. Out of the twenty dates we calculated from these ratios there is a total disagreement between the U/Pb with the so called 'true age.' Whichever date you choose for each sample as the true one is just a random guess.

Table 33	207Pb/206Pb	208Pb/232Th	206Pb/238U	87Rb/86Sr
Average	4,953	9,685	6,472	1,221
Maximum	5,162	23,132	14,131	1,232
Minimum	4,408	4,854	3,443	1,207

Conclusion

Evolutionists Schmitz and Bowring claim that Uranium/Lead dating is 99% accurate. [70] Looking at some of the dating it is obvious that precision is much lacking. The Bible believer who accepts the creation account literally has no problem with such unreliable dating methods. Much of the data used in this dating method is selectively taken to suit and ignores data to the contrary.

Yuri Amelin states in the journal Elements that radiometric dating is extremely accurate: "However, four 238U/235U-corrected CAI dates reported recently (Amelin et al. 2010; Connelly et al. 2012) show excellent agreement, with a total range for the ages of only 0.2 million years – from 4567.18 ± 0.50 Ma to 4567.38 ± 0.31 Ma." [71-73] To come within 0.2 million years out of 4,567.18 million years means an accuracy of 99.99562%. Looking at some of the dating it is obvious that precision is much lacking. The Bible believer who accepts the creation account literally has no problem with such unreliable dating methods. Much of the data in radiometric dating is selectively taken to suit and ignores data to the contrary.

Prominent evolutionist Brent Dalrymple states: "Several events in the formation of the Solar System can be dated with considerable precision." [74] Looking at some of the dating it is obvious that precision is much lacking. He then goes on: "Biblical chronologies are historically important, but their credibility began to erode in the eighteenth and nineteenth centuries when it became apparent to some that it would be more profitable to seek a realistic age for the Earth through observation of nature than through a literal interpretation of parables." [75] The Bible believer who accepts the creation account literally has no problem with such unreliable dating methods. Much of the data in Dalrymple's book is selectively taken to suit and ignores data to the contrary.

References

1 **Geochimica et Cosmochimica Acta, 2013, Volume 115, Pages 46–72, Isotopic and trace element geochemistry**
2 **Reference 1, page 48**
3 **Reference 1, page 50**
4 **Reference 1, page 62-66**
5 **International Geology Review, 2009, Volume 51, Number 4, Pages 345, Geochemistry of hornblende gabbros**
6 **Reference 5, page 361**
7 **Reference 5, page 359**
8 **Journal Of Petrology, 2005, Volume 46, Number 6, Pages 1155–1201, Post-Collisional Transition from Subduction**
9 **Reference 8, page 1155**
10 **Reference 8, page 1181-1183**
11 http://www.creationismonline.com/Isoplot/Isoplot.html, https://www.bgc.org/isoplot
12 **Journal of Petrology, 2012, Volume 53, Number 7, Pages 1417-1448, Nazca Ridge and Easter Seamount Chain**
13 **Reference 12, page 1421**
14 **Reference 12, page 1428**
15 **Journal of Petrology, 2010, Volume 51, Number 9, Pages 1849, South African Off-Craton Mantle**
16 **Reference 15, page 1869-1870**
17 **Proceedings of the Ocean Drilling Program, Volume 158, Page 91, Os And Re Distribution In The Active Mound** **http://www-odp.tamu.edu/publications/158_SR/VOLUME/CHAP_07.PDF**
18 **Reference 17, page 95**
19 **Principles of Isotope Geology, Second Edition, By Gunter Faure, John Wiley And Sons, New York, 1986, Page 269**
20 **Introduction to Geochemistry: Principles and Applications, Page 241 By Kula C. Misra, Wiley-Blackwell Publishers, 2012** http://books.google.com.au/books?id=ukOpssF7zrIC&printsec=frontcover

21 **Radioactive and Stable Isotope Geology, Issue 3**
By H. G. Attendorn, Robert Bowen, Page 298, Chapman and Hall Publishers, London, 1997
http://books.google.com.au/books?id=-bzb_XU7OdAC&printsec=frontcover
22 **http://www.geo.cornell.edu/geology/classes/Geo656/656notes03/656%2003Lecture11.pdf**
23 **Proceedings of the Ocean Drilling Program, Volume 159, Page 181, Osmium-Isotope Geochemistry Of Site 959**
http://www-odp.tamu.edu/publications/159_SR/CHAPTERS/CHAP_18.PDF
24 **Reference 23, page 183**
25 **Journal Of Petrology, 2004, Volume 45, Number 2, Pages 415, The Seve Nappe Complex of Jamtland**
26 **Reference 25, page 432**
27 **Journal Of Petrology, 2005, Volume 46 Number 10 Pages 2059, 2078, 2079, The Kaalvallei Kimberlite, South Africa**
28 **Reference 27, page 2081**
29 **Journal Of Petrology, 2001, Volume 42 Number 6 Pages 1197–1218, Genesis of Continental Intraplate Basalts**
30 **Reference 29, page 1214**
31 **Reference 29, page 1204**
32 **Journal Of Petrology, 2006, Volume 47, Number 5, Pages 929, Xenoliths from the Colorado Plateau**
33 **Reference 32, page 931**
34 **Reference 32, page 930**
35 **Reference 32, page 953**
36 **Reference 32, page 953, 955, 956**
37 **Science in China Series D: Earth Sciences, 2007, Volume 50, Number 7, Pages 972-983, Indosinian granitoids**
38 **Reference 37, page 973**
39 **Reference 37, page 974**
40 **Reference 37, page 978**
41 **Journal Of Petrology, 2005, Volume 46, Number 10, Pages 2091–2128, The Stonyford Volcanic Complex**
42 **Reference 41, Page 2091**
43 **Reference 41, Page 2096**
44 **Reference 41, Page 2116**
45 **Journal Of Petrology, 2003, Volume 44, Number 10, Pages 1833-1865, Cenozoic Volcanism in Tibet**
46 **Reference 45, page 1833**
47 **Reference 45, page 1841, 1842**
48 **Reference 45, page 1847, 1848**
49 **http://www.lpi.usra.edu/meetings/lpsc2011/pdf/1243.pdf, U-Th-Pb Analysis Of Baddeleyites**
50 Periodico Di Mineralogia (2007), Volume 76, Pages 5, Rb-Sr and Pb-Pb Geochronology
51 **Reference 50, pages 10-11**
52 **Earth and Planetary Science Letters, Volume 319-320 (2012), Pages 197–206, U–Th–Pb Isotope Data**
53 **Reference 52, page 197**
54 **Reference 52, page 201**
55 **Reference 52, page 199**
56 **ftp://rock.geosociety.org/pub/reposit/2013/2013089.pdf, GSA Data Repository, The Geological Society of America Publication, 2013**
57 **Reference 56, page 2**
58 **Reference 56, page 4-14**
59 **Doklady Earth Sciences, 2002, Volume 387A, Number 9, Pages 1083, 1084, Lead in Galena from Ore Deposits**
60 **Reference 59, page 1086**
61 **Reference 59, page 1085**
62 **Earth and Planetary Science Letters, Volume 174, (2000) Pages 247, 251, The Caribbean Large Igneous Province**
63 **Reference 62, page 247**
64 **Reference 62, page 248**
65 **Reference 62, page 253**
66 **Chemical Geology, Volume 236, 2007, Pages 291–302, Nd–Hf–Sr–Pb isotopes**
67 **Reference 66, page 291**
68 **Reference 66, page 297**
69 **Reference 66, page 297**
70 **High-precision U-Pb geochronology. Geochimica et Cosmochimica Acta, 2001, Volume 65, Pages 2571-2587**
71 **Dating the Oldest Rocks in the Solar System, Elements, 2013, Volume 9, Pages 39-44**
72 **Amelin, Earth and Planetary Science Letters, 2010, Volume 300, Pages 343-350**
73 **Connelly, Science, 2012, Volume 338, Pages 651-655**
74 **The Age Of The Earth, By G. Brent Dalrymple, 1991, Stanford University Press, Stanford, California, Page 10.**
75 **Reference 74, Page 23**

Chapter 23

Radiometric Dating of Historic Volcanic Eruptions

Paul Nethercott
December 2015

Abstract

If evolutionary radiometric dating methods are true then historic volcanic eruptions should be a good test. Since people actually saw the eruption happening we can know its true age to the very year. When we look at dating (K/Ar, Ar/Ar) done by secular geology we find dates that are 500 thousand times in error. If we use isotope the ratio tables ($^{207}Pb/^{207}Pb$) in journal articles that have no dates beside them and run them through Isoplot **(Ludwig, 2010)** we get dates that are between 12 million to 30 billion times in error.

Introduction

Evolutionists have long claimed that radiometric dating supports an old age of the Earth and the first appearance of life on Earth. Creationists contend that such methods are wrong and that the earth is only six thousand years old. Jesus stated (Matthew 19:4-6) that the age of the Earth is the age of mankind. The Bible says (Genesis 1:9-13) says that dry land was created 6,000 years ago so such dry air eruptions as opposed to underwater pillow lava cannot be millions of years old. Many fossil deposits are deposited underneath lava flows which means that if the lava flow was millions of years ago then so was death and suffering. The Bible says (Genesis 1:1-31) that life was only created 6,000 years ago. Scientists have used a variety of methods to examine the isotopic composition of lava such as Potassium/Argon, Argon/Argon, Lead/Lead and others. If these methods that have human eye witness to test their validity are millions or billions of times in error with historic flows how can they be trusted with lava flows that have no human eye witnesses? This is a good way to test the accuracy of such dating methods against an absolute standard.

The author has compiled a list of 1003 dates form 68 volcanoes and 273 eruptions (1600 AD – 2010 AD) which were analysed between 2 and 406 years after the eruption happened. These were dated by a variety of methods: $^{207}Pb/^{206}Pb$, $^{230}Th/^{238}U$, K/Ar, Ar/Ar Dates) The dates that are calculated from $^{207}Pb/^{207}Pb$ isotope ratios using Isoplot vary between 1 million times to 30 billion times in error. The complete failure of the Uranium/Lead dating system to even come close to the true age reassures creationists that such a system offers meaningless numbers rather than real ages.

Aleutian Islands Volcanic Chain

The Aleutian Islands are a chain of 69 volcanic islands belonging to both the United States and Russia. They form part of the Aleutian Arc in the Northern Pacific Ocean, occupying an area of 17,666 km^2 and extending about 1,900 km westward from the Alaska Peninsula toward the Kamchatka Peninsula in Russia. Scientists from the University Of Wisconsin–Madison analysed these samples in 2003. Isotope ratios of Aleutian island arc lavas from Kanaga, Roundhead, Seguam, and Shishaldin volcanoes have been subjected to $^{40}Ar/^{39}Ar$ dating and returned dates of two to seven hundred thousand years old **(Jicha, 2004, P. 1852-1855)**. Two of the eruptions (Kanaga 1906 AD, and Seguam 1977 AD) are historic as well. If we examine the $^{207}Pb/^{206}Pb$ ratios **(Jicha, 2004, P. 1852-1855)** for all the eruptions and run them through Isoplot we get dates between 4,974 million years and 4,962 million years old with only 12 million years difference (99.76% accuracy). The percentage accuracy is defined by the formula below.

$$a = 1 - \left(\frac{Max - Min}{Average} \right) \quad \textbf{(1)}$$

The error ratio for $^{40}Ar/^{39}Ar$ ages versus $^{207}Pb/^{206}Pb$ isotope ages varies between 7 million to 2,485 million times in error.

The error ratio for historic real ages versus imaginary isotope ages varies between 50 million to 183 million times in error. The MSWD **(Jicha, 2004, P. 1857-1858)** tables give values between 0.04 and 2.77 which shows that the $^{40}Ar/^{39}Ar$ ages for Kanaga, Seguam, and

Shishaldin are supposed to be very accurate. The Seguam dates vary from 28 to 133 thousand years. The Kanaga dates vary from 112 to 386 thousand years. The Shishaldin dates vary from 10 to 711 thousand years. All dates use a weighted mean plateau and combined isochrons.

Table I. Aleutian Island Volcanoes.

Volcano	Historic Eruption	Ar/Ar Age (ka)	Pb/Pb Age (Ma)	Error Ratio (10^6)
Kanaga	1906 AD	0.098	4,970	50,711
		198	4,971	25
		384	4,973	13
		352	4,971	14
		199	4,971	25
Roundhead		131	4,974	38
Seguam	1977 AD	0.027	4,963	183,815
		49.2	4,962	101
		48.9	4,963	101
		93.1	4,962	53
		33.3	4,962	149
Shishaldin		2	4,970	2,485
		28	4,969	177
		30	4,972	166
		713	4,966	7

Isoplot has four different **(Ludwig, 2012)** Uranium/Lead dating formulas for Microsoft Excel.

$^{207}Pb/^{206}Pb$ Age (Ma)

$$a = Pb76\left(\frac{^{207}Pb}{^{206}Pb}\right) \quad (2)$$

$^{206}Pb/^{238}U$ Age (Ma)

$$a = Pb6U8\left(\frac{^{206}Pb}{^{238}U}\right) \quad (3)$$

$^{207}Pb/^{235}U$ Age (Ma)

$$a = Pb7U5\left(\frac{^{207}Pb}{^{235}U}\right) \quad (4)$$

$^{208}Pb/^{232}Th$ Age (Ma)

$$a = Pb8Th2\left(\frac{^{208}Pb}{^{232}Th}\right) \quad (5)$$

$^{230}Th/^{238}U$ Age (Ka)

$$a = Th230Age\left(\frac{^{230}Th}{^{238}U}, \frac{^{234}U}{^{238}U}\right) \quad (6)$$

Anatahan Volcano

Anatahan is an island in the Northern Mariana Islands in the Pacific Ocean, and has one of the most active volcanoes of the archipelago. The first historical eruption of Anatahan Volcano began suddenly on the evening of May 10, 2003. Anatahan Volcano is a small volcanic island located 120 kilometres north of Saipan Island and 320 kilometres north of Guam. The island is about 9 kilometres long and 3 kilometres wide. Anatahan is a stratovolcano that contains the largest known caldera in the Northern Mariana Islands. In 1990, when geologists from the USGS Hawaiian Volcano Observatory and the University of Hawaii examined the rock layers of Anatahan, they discovered abundant evidence of ancient explosive eruptions that sent fast-moving flows of hot ash and rocks across the island. **(HVO, 2003)**

In 2005 scientists from nine universities analysed mineral samples from the 2003 volcanic eruption. Samples of tephra from early in the eruption were analyzed for major and trace elements, and Sr, Nd, Pb, Hf, and O isotopic compositions. **(Wade, 2005, P. 139)** If take isotope data from three tables (Wade, 2005, P. 149, 150, 165) in the article and run them through Isoplot we get the ages listed below.

Table II. Anatahan Volcano (2003 AD)

Pb Age (Ma)	Pb Age (Ma)	Pb Age (Ma)
4,975	4,971	4,971
4,973	4,971	4,971
4,973	4,971	4,970
4,972	4,971	4,970
4,971	4,971	4,970
4,971	4,971	4,970
4,971	4,971	4,970

We get twenty three dates with an average age of 4,970 million years old and an accuracy of 99.94%. The minimum age is 4,969 million and the maximum age is 4,975 million years old. Since the lava was only two years old when dated the dates are 2,485 million times in error. As a distance comparison, that is one millimetres versus 2,485 kilometres (1,544 miles). That is the same distance as Denver to New York City or Sydney to Townsville. Such a fantastic error ratio shows that when the same dating technique is used to date non historic eruptions the same error ratio will be just as enormous.

Earthquake Swarm of Loihi Seamount

The Lōihi Seamount is an active submarine volcano located about 35 km off the southeast coast of the island of Hawaii. "^{210}Po/^{210}Pb dating of two fresh lava blocks from this breccia indicates that they were erupted during the first half of 1996, making this the first documented historical eruption of Loihi." **(Garcia, 1998, P. 577)** This research was done eighteen months after the eruption by scientists from the University of Hawaii. Within two months of the eruption **(Garcia, 1998, P. 584)** glass samples were recovered and dates with the ^{210}Po/^{210}Pb method. According to this method the samples were 57 days old **(Garcia, 1998, P. 586)**. If we calculate dates from the from the two ^{207}Pb/^{207}Pb isotope ratios **(Garcia, 1998, P. 585)** we get two dates of 4,997 million years old. Other researchers **(Dixon, 2001, P. 634, 635, 646)** have found identical ^{207}Pb/^{207}Pb isotope ratios from Lōihi Seamount glass samples. Since the two glass samples Garcia found were only two months old (1/6 of a year) both dates were 30 billion times in error.

Table III. Loihi Seamount Ages

Pb Age (Ma)	4,997	4,997
Historic Age	0.1666	0.1666
Age Ratio (10^6)	29,992	29,994

Excess Argon In Melt Inclusions

Mount Erebus is the second highest volcano in Antarctica (after Mount Sidley) and the southern most active volcano on earth. It overlooks the McMurdo research station on Ross Island. The 3,794 metre high Erebus is the largest of three major volcanoes forming the crudely triangular Ross Island. The summit of the dominantly phonolitic volcano has been modified by one or two generations of caldera formation. A summit plateau at about 3200 m elevation marks the rim of the youngest caldera. Two scientists from the New Mexico Bureau of Mines and Mineral Resources, analysed "Historically erupted (1984) anorthoclase phenocrysts from Mt. Erebus yield K/Ar and $^{40}Ar/^{39}Ar$ apparent ages as old as 700 thousand years indicating the presence of excess argon" **(Esser, 1997, P. 3789)**. Since the material was only thirteen years old when dated there is a major discrepancy. The article has a table with 180 dates varying from 9.419 million ago to -19.111 million years in the future. **(Esser, 1997, P. 3789)** That is a 28.53 million year range for a rock formation that is only thirteen years old. The range is 2.149 million times the age of the rock. Thirty three dates are negative or future ages.

Table IV. Excess Argon In Mount Erebus Melt Inclusions (1984 Eruption)

S.1316 Age Ka	Ratio	S.1317 Age Ka	Ratio	S.1330 Age Ka	Ratio	S.1331 Age Ka	Ratio	S. 1332 Age Ka	Ratio
273	21,000	374	28,769	4,371	336,231	5,867	451,308	-999	-76,846
97	7,462	227	17,462	507	39,000	4,716	362,769	-999	-76,846
38	2,923	275	21,154	86	6,615	1,801	138,538	-86	-6,615
28	2,154	281	21,615	166	12,769	1,120	86,154	43	3,308
25	1,923	246	18,923	41	3,154	984	75,692	23	1,769
28	2,154	240	18,462	33	2,538	1,010	77,692	33	2,538
19	1,462	263	20,231	21	1,615	950	73,077	35	2,692
10	769	225	17,308	16	1,231	834	64,154	246	18,923
12	923	209	16,077	9	692	824	63,385	64	4,923
16	1,231	78	6,000	10	769	1,278	98,308	82	6,308
33	2,538	96	7,385	5	385	628	48,308	104	8,000
38	2,923	11	846	28	2,154	732	56,308	126	9,692
111	8,538	41	3,154	15	1,154	1	77	126	9,692
77	5,923	15	1,154	142	10,923	109	8,385	153	11,769
139	10,692	78	6,000	86	6,615	-24	-1,846	168	12,923
61	4,692	48	3,692	149	11,462	154	11,846	172	13,231
139	10,692	57	4,385	82	6,308	84	6,462	263	20,231
66	5,077	128	9,846	72	5,538	90	6,923	337	25,923
		-894	-68,769	88	6,769	56	4,308	9,419	724,538
				88	6,769	71	5,462	650	50,000
				129	9,923	248	19,077		
				671	51,615				

Alkaline Magmas from Erebus Volcano

This set of fifty three mineral samples were analysed in 2008 by nine scientists from seven institutions. Thirty samples **(Sims, 2008, P. 607-608)** were taken from eruptions that happened between 1972 and 2005. Twenty three samples were taken from non-historical eruptions supposed to be between 1 thousand and 1,311 thousand years old. Erebus is an active composite volcano and is the largest of four volcanic centres forming Ross Island: Mt. Erebus (3,794 metres elevation, 2170 km^3), Mt. Terror (3,262 metres, 1700 km^3), Mt. Bird (1,800 metres, 470 km^3), and Hut Point Peninsula (100 km^3). If we calculate dates from the from the fifty three $^{207}Pb/^{207}Pb$ isotope ratios we get 53 dates between 4,875 million and 4,935 million years old with an accuracy of 98.77%.

The author claims that the samples cover an age range supposed to be between 3 years and 1,311 thousand years old. "The samples cover the entire compositional range from basanite to phonolite and trachyte, and represent all three phases of the volcanic evolution from 1.3 Ma to the present. Isotopic analyses of 7 samples from Mt. Morning and the Dry Valley Drilling Project (DVDP) are given for comparison." **(Sims, 2008, P. 606)** The $^{207}Pb/^{207}Pb$ dates calculated from the historical (1972-2005) eruptions are between 136 million and 1,224 million times in error. The $^{207}Pb/^{207}Pb$ dates calculated from the non-historical eruptions are between 0.0037 and 4.8 million times in error.

Table V. Alkaline magmas from Erebus volcano

Eruption	True Age	Pb Age (Ma)	Age Ratio 10^6
1972	36	4,894	136
1974	34	4,894	144
1974	34	4,894	144
1974	34	4,894	144
1975	33	4,894	148
1977	31	4,894	158
1977	31	4,894	158
1979	29	4,894	169
1980	28	4,894	175
1981	27	4,894	181
1982	26	4,894	188
1983	25	4,894	196
1984	24	4,896	204
1984	24	4,898	204
1984	24	4,895	204
1985	23	4,894	213
1986	22	4,894	222
1989	19	4,894	258
1991	17	4,894	288
1992	16	4,894	306
1992	16	4,895	306
1993	15	4,894	326
1993	15	4,894	326
1993	15	4,894	326
1996	12	4,894	408
1997	11	4,894	445
1997	11	4,894	445
1999	9	4,894	544
1999	9	4,894	544
1999	9	4,894	544
2000	8	4,894	612
2001	7	4,896	699
2004	4	4,894	1,224

Comparison of Th, Sr, Nd and Pb isotopes

These seventeen lava samples were examined in 2005 by two scientists from Woods Hole Oceanographic Institution in Massachusetts. Thirteen were from Samoa and the other four from historic eruptions at Mount Erebus **(Sims, 2006, P. 745)**. If we calculate dates from the from the seventeen $^{207}Pb/^{207}Pb$ isotope ratios we get 17 dates between 4,894 million and 4,976 million years old with an accuracy of 99%. The author claims that the ^{230}Th dates show the Samoan samples are young: "Note that the new Mt. Erebus samples are historic, and the new Samoan samples are demonstrably young based upon historical constraints, or measurement of ^{230}Th–^{226}Ra–^{210}Pb and ^{210}Po disequilibria." **(Sims, 2006, P. 744)**

According to $^{40}Ar/^{39}Ar$ dating the Samoan Islands are only five million years old: "Three different volcanic samples from dredge ALIA-115, on the deepest portion of the SW flank of Savai'i Island, give indistinguishable ages (2σ confidence level) ranging from 4.99 to 5.21 Ma. In addition, a sample from dredge ALIA-128, on the NE flank of Savai'i, gives an age of 4.74 Ma." **(Koppers, 2006)** The $^{207}Pb/^{207}Pb$ dates are a hundred million times older than the $^{40}Ar/^{39}Ar$ dates.

Table VI. Comparison of Th, Sr, Nd and Pb isotopes (historic eruptions)

Samoa	Samoa	Samoa	Mt. Erebus
Pb Age (Ma)	Pb Age (Ma)	Pb Age (Ma)	Pb Age (Ma)
4,935	4,935	4,944	4,894
4,935	4,932	4,937	4,896
4,932	4,944	4,937	4,894
4,932	4,935	4,945	4,894

Mount Hualalai

Mount Hualālai is a dormant volcano on the island of Hawaii in the Hawaiian Islands. It is the third-youngest and the third most active of the five shield volcanoes that form the island of Hawaii, following Kilauea and the much larger Mauna Loa, and also the westernmost. Its peak is 2,521 metres above sea level. Hualālai is estimated by evolutionists to have risen above sea level about 300,000 years ago. Funkhouser and Naughton **(Funkhouser, 1968, P. 4602-4603)** dated samples from the 1800 volcanic eruption using the K/Ar method in 1967 and obtained ages between 160 and 2,960 million years old. Since the samples were only 167 years old when dated, the ages are between million 950 thousand and 17 million times in error.

Evolutionists frequently claim the creationists misquote Funkhouser and Naughton. Five scientists from three universities **(Blichert-Toft, 2002, P. 30-33)** collected samples from Mauna Loa (27 samples) and Mauna Kea (104 samples) volcanoes. If we calculate $^{207}Pb/^{207}Pb$ dates from the from the 27 Mauna Loa samples we get dates between 5,003 million and 5,018 million years old with an accuracy of 99.69%. If we calculate $^{207}Pb/^{207}Pb$ dates from the from the 104 Mauna Kea samples we get dates between 4,974 million and 6,474 million years old with an accuracy of 70%. The tables also list the samples depth (metres below sea level). If the dating is accurate the deeper you go the older the sample. There is however no relationship between age and depth. The dates are all meaningless.

Table VII. Mount Hualalai (1800 AD)

K-Ar	K-Ar	Helium	Helium
Min Age	Max Age	Min Age	Max Age
1,030	1,030	140	350
2,480	2,480	670	670
2,040			
960			
1,500			
1,580			
791			
160			
2,470			
2,960			

Figure 1. Mauna Loa Ages (Ma) Versus Depth below Sea level

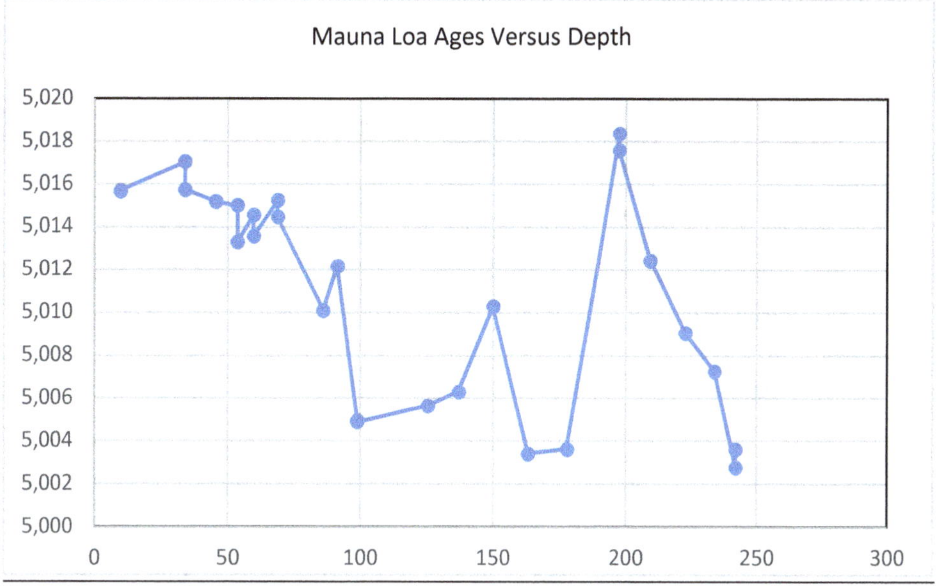

Figure 2. Mauna Kea Ages (Ma) Versus Depth below Sea level

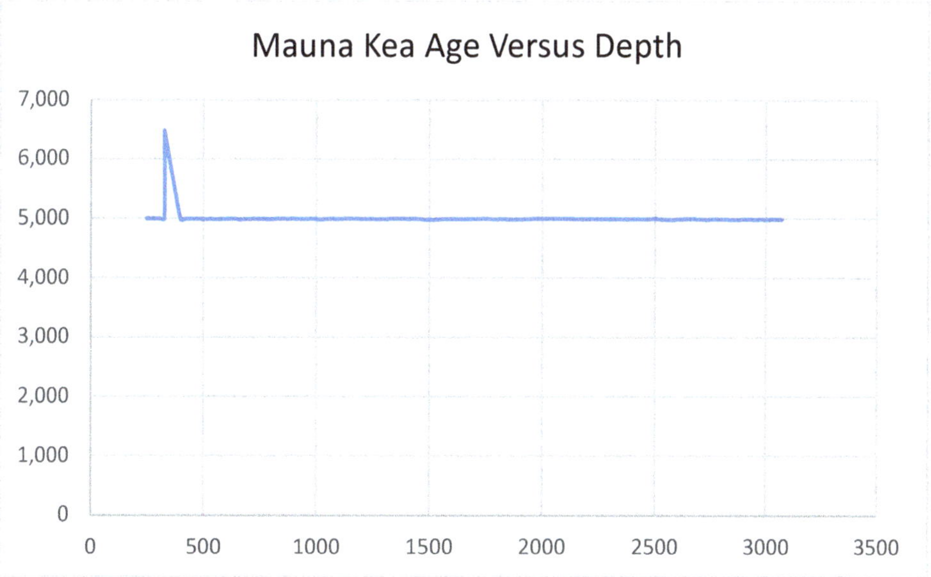

Tres Vírgenes (three virgins) is a complex of volcanoes located Mulegé Municipality in the state of Baja California Sur, on the Baja California Peninsula in north western Mexico. It is composed of three volcanoes, aligned northeast-southwest, with El Viejo, to the northeast, El Azufre in the middle, and El Virgen, to the southwest. **(Wikipedia, 2015)** "U–Pb zircon analysis of ignimbrites erupted from the adjacent Early Pleistocene La Reforma and El Aguajito calderas yielded ages of 1.38±0.03 Ma and 1.17±0.07 Ma respectively. No evidence for these ages is found among La Vírgen zircons, whereas pre-Quaternary zircon xenocrysts are common. The La Vírgen magma, therefore, evolved unrelated to Early Pleistocene magmatism in adjacent calderas, but assimilated local basement rocks." **(Schmitt, 2006, P. 281)**

The articles has two tables **(Schmitt, 2006, P. 285, 286)** with $^{230}Th/^{238}U$ ages between 25 thousand and 233 thousand years old. The articles has a third table **(Schmitt, 2006, P. 288-289)** with $^{230}Th/^{238}U$ ages between 40 thousand and 255 million years old.

If we place the $^{206}Pb/^{238}U$, $^{207}Pb/^{235}U$ and $^{207}Pb/^{206}Pb$ ratios from third table into Iosplot we get 105 dates between 315 million and 24 billion years old. Ten dates are over 4.6 billion years old and six are over 5 billion years old. Three dates are over fifteen billion years old which makes the Funkhouser and Naughton article very miniscule. If we assume that the $^{230}Th/^{238}U$ dates the author calculated are the true age then the other three isotope ratios are between 9 and 14,400 times in error.

Table VII. Crystallization ages of Las Tres Vírgenes

Volcano's Name	Statistical Data	206Pb/238U Age (Ma)	207Pb/235U Age (Ma)	207Pb/206Pb Age (Ma)	230Th/238U Age (Ma)
La Reforma	Average	2,473	3,180	3,708	1.45
	Maximum	4,195	4,623	4,841	2.05
	Minimum	1,170	1,281	1,467	0.63
	Difference	3,025	3,342	3,374	1.42
Aguajito	Average	3,323	4,090	4,553	1.15
	Maximum	5,855	5,088	4,810	2.59
	Minimum	1,541	2,802	3,905	0.64
	Difference	4,314	2,285	905	1.95
La Vírgen	Average	15,991	4,476	1,148	127.04
	Maximum	24,048	5,867	4,799	255.00
	Minimum	315	1,831	195	0.13
	Difference	23,734	4,035	4,604	254.87

Smyth **(Smyth, 2011, P. 206)** has a similar table ($^{206}Pb/^{238}U$, $^{207}Pb/^{235}U$ and $^{207}Pb/^{206}Pb$ ratios) which he claims supports an age of 20 million years for a Miocene eruption in Indonesia "Here we report an Early Miocene major eruption, the Semilir eruption, in south Java, the main phase of which occurred at 20.7±0.02 Ma." **(Smyth, 2011, P. 198)** If using isotope ratios and getting dates from those ratios by using Isoplot is true then the dates I have calculated for various eruptions must be equally true. The method is faultless by evolutionary standards and used in many geology journal articles.

The West Maui Mountains or West Maui Volcano, known to the Hawaiians as Maui Komohana and to geologists as Mauna Kahalawai, forms a much eroded shield volcano that constitutes the western one-quarter of the Hawaiian Island of Maui. Evolutionists believe that its last eruption was approximately 320,000 years ago. Tatsumoto says the K/Ar ages for the samples are very accurate at 1.16 million years old. "K-Ar ages by McDougall (1964) are 1.20± 0.02 Ma for olivine basalt from the upper part of the Wailuku Basalt, 1.16±0.01 Ma for mugearite and trachyte of the Honolua Volcanics, and 0.44-0.86 Ma for alkalic rocks of the Kula Volcanics of Haleakala. Therefore, the isotopic data obtained from these samples require no age correction." **(Tatsumoto, 1987, P. 725)** The article has two tables **(Tatsumoto, 1987, P. 726, 728)** that we can join together to get three sets ($^{206}Pb/^{238}U$, $^{208}Pb/^{232}Th$ and $^{207}Pb/^{206}Pb$) of isotopic ratios. We need to use the formulas below to rearrange the isotope ratios into a format Isoplot can use.

$$\frac{^{232}Th}{^{204}Pb} = \frac{^{232}Th}{^{238}U} \div \left(1 \div \frac{^{238}U}{^{204}Pb}\right) \quad (6)$$

$$\frac{^{208}Pb}{^{232}Th} = \frac{^{208}Pb}{^{204}Pb} \div \left[\frac{^{232}Th}{^{238}U} \div \left(1 \div \frac{^{238}U}{^{204}Pb}\right)\right] \quad (7)$$

We then get sixty six dates are between 3.1 billion and 26.7 billion years old. Fifty four dates over 4.9 billion years old. Thirty four dates over 5 billion years old. Ten dates over 10 billion years old. Since the supposed true age is just one million years old the $^{206}Pb/^{238}U$, $^{208}Pb/^{232}Th$ and $^{207}Pb/^{206}Pb$ dates are between 3,100 and 26,700 times in error.

Table IX. West Maui volcanic rocks

West Maui Rock Formation	206Pb/238U Age (Ma)	208Pb/232Th Age (Ma)	207Pb/206Pb Age (Ma)
Lahaina Volcanics	4,553	10,582	5,009
	3,147	7,319	5,008
Honolua Volcanics	7,250	11,953	4,986
	4,952	9,494	4,993
	4,541	9,238	4,992
	4,442	8,819	4,987
	3,653	8,473	4,986
	3,604	7,837	4,986
	3,493	7,472	4,986
	3,476	6,065	4,986
Wailuku Basalt	9,924	26,473	5,003
	8,129	15,405	5,002
	6,028	12,662	5,000
	5,961	12,453	4,998
	5,837	11,269	4,996
	5,275	11,139	4,995
	4,583	10,063	4,992
	4,354	9,293	4,992
Hana Volcanics	3,694	7,621	
Kula Volcanics	3,363	7,215	
Honmanu Basalt	6,104	16,457	

Recent Andesite Flows at Mount Ngauruhoe

Mount Ngauruhoe is an active stratovolcano or composite cone in New Zealand, made from layers of lava and tephra. It is the youngest vent in the Tongariro volcanic complex on the Central Plateau of the North Island, and first erupted about 2,500 years ago. Although seen by most as a volcano in its own right, it is technically a secondary cone of Mount Tongariro. Ngauruhoe erupted 45 times in the 20th century, most recently in 1974. In 1997 Dr R. A. Armstrong from the Australian National University **(Snelling, 2003, P. 8)** analysed samples and obtained $^{207}Pb/^{206}Pb$ ratios. Since the samples were from recent historic eruptions (1949, 1954 and 1975) they were only between 22 to 48 years old. The $^{207}Pb/^{206}Pb$ dates I calculated with Isoplot are between 4,974 million and 4,977 million years old with an accuracy of 99.94%. That means the $^{207}Pb/^{206}Pb$ dates are between 104 million to 226 million times to old.

Table X. Recent Andesite Flows at Mt. Ngauruhoe

Eruption Year	True Age	Pb Age (Ma)	Age Ratio 10^6
1949	48	4,974	104
	48	4,976	104
1954	43	4,975	116
	43	4,974	116
	43	4,974	116
	43	4,975	116
	43	4,977	116
	43	4,974	116
1975	22	4,974	226
	22	4,975	226

Kılauea's Puu Oo eruption (1983–2010)

Kilauea is a currently active shield volcano in the Hawaiian Islands, and the most active of the five volcanoes that together form the island of Hawaii. Located along the southern shore of the island, evolutionary science states that the volcano is between 300,000 and 600,000 years old and emerged above sea level about 100,000 years ago. There were sixty episodes in the eruptive sequence that began in January 1983 and ended in June 2011. Sixty four samples with $^{207}Pb/^{206}Pb$ ratios were collected in 2013 **(Greene, 2013, P. 4856)** from eruptions that took place between 1983 and 2010. The true age of the samples was between three and thirty years old. The $^{207}Pb/^{206}Pb$ dates I calculated with Isoplot are between 4,984 million and 4,994 million years old with an accuracy of 99.80%. That means the $^{207}Pb/^{206}Pb$ dates are between 166 million to 1,664 million times to old.

Table XI. Kılauea's Puu Oo eruption (1983–2010 AD)

Eruption Year	True Age	Pb Age (Ma)	Error Ratio (10^6)
1983	30	4,984	166
1984	29	4,988	172
1985	28	4,990	178
1986	27	4,992	185
1987	26	4,992	192
1988	25	4,993	200
1989	24	4,993	208
1990	23	4,994	217
1991	22	4,993	227
1992	21	4,992	238
1993	20	4,992	250
1994	19	4,992	263
1995	18	4,992	277
1996	17	4,992	294
1997	16	4,993	312
1998	15	4,993	333
1999	14	4,993	357
2000	13	4,992	384
2001	12	4,992	416
2002	11	4,991	454
2003	10	4,991	499
2004	9	4,991	555
2005	8	4,991	624
2006	7	4,992	713
2007	6	4,992	832
2008	5	4,992	998
2009	4	4,992	1,248
2010	3	4,993	1,664

Kilauea's Puu Oo Eruption

The Puu Oo eruption is the longest sustained (25 years) and most voluminous (3 km^3 erupted lava) historical eruption of Kilauea volcano. Jared Marske from the University Of Hawaii **(Marske, 2008, P. 1308)** collected fourteen samples in 2008 from eruptions that took place between September 1998 and August 2005. The samples were only between three and ten years old. The $^{207}Pb/^{206}Pb$ dates I calculated are between 4,991 million and 4,993 million years old with an accuracy of 99.96%. That means the $^{207}Pb/^{206}Pb$ dates are between 499 million to 1,664 million times to old.

Table XII. Kilauea's Puu Oo Eruption

Eruption Year	True Age	Pb Age (Ma)	Error Ratio (10^6)
1998	10	4,992	499
1999	9	4,992	555
1999	9	4,993	555
2000	8	4,992	624
2000	8	4,992	624
2001	7	4,992	713
2001	7	4,992	713
2002	6	4,991	832
2002	6	4,991	832
2003	5	4,991	998
2004	4	4,991	1,248
2004	4	4,991	1,248
2005	3	4,991	1,664
2005	3	4,992	1,664

Melting History of Kilauea Volcano

Eighteen samples from fifteen historic eruptions that took place over a 192 year period (1790-1982) were analysed in 1999 by two scientists from the University of Hawaii **(Pietruszka, 1999, P. 1329)**. The historical ages varied from 17 to 209 years old. The $^{207}Pb/^{206}Pb$ dates I calculated with Isoplot are between 4,974 million and 4,992 million years old with an accuracy of 99.64%. That means the $^{207}Pb/^{206}Pb$ dates are between 24 million to 293 million times to old.

Table XIII. Melting History of Kilauea Volcano

Eruption Year	True Age	Pb Ages (Ma)	Error Ratio (10^6)
1790	209	4,992	24
1820	179	4,992	28
1832	167	4,990	30
1866	133	4,981	37
1885	114	4,983	44
1894	105	4,981	47
1895	104	4,979	48
1917	82	4,975	61
1919	80	4,974	62
1929	70	4,981	71
1931	68	4,981	73
1954	45	4,981	111
1961	38	4,982	131
1971	28	4,985	178
1982	17	4,987	293

1991 Eruption Products of Mount Pinatubo

The second-largest volcanic eruption of the 20th century, and by far the largest eruption to affect a densely populated area, occurred at Mount Pinatubo in the Philippines on June 15, 1991. The eruption produced high-speed avalanches of hot ash and gas, giant mudflows, and a cloud of volcanic ash hundreds of miles across. Mount Pinatubo is an active stratovolcano in the Cabusilan Mountains on the Philippine island of Luzon. Ten scientists from seven different universities (Bernard, 1999, Castillo, 1999) examined volcanic ejecta and obtained several $^{207}Pb/^{206}Pb$ isotope ratios. This laboratory work was done in 1999 so the samples were only eight years old. The $^{207}Pb/^{206}Pb$ dates I calculated with Isoplot are between 5,000 million and 5,009 million years old with an accuracy of 99%. That means the $^{207}Pb/^{206}Pb$ dates are all 625 million times to old.

Table XIV. 1991 Eruption Products of Mount Pinatubo

Eruption	True Age	Pb Age (Ma)	Age Ratio 10^6
1991	8	5,002	625
1991	8	5,002	625
1991	8	5,001	625
1991	8	5,001	625
1991	8	5,000	625
1991	8	5,009	626
1991	8	5,001	625
1991	8	5,000	625
1991	8	5,001	625
1991	8	5,001	625

Bicol and Bataan arcs, Philippines

The Philippine island of Luzon is bounded by the Manila Trench to the west and the Philippine Trench to the east, both of which are associated with subduction of oceanic lithosphere. In contrast, the Bicol arc is associated with westward subduction of the Philippine Sea Plate with thin to modest amounts of pelagic sediments. **(DuFrane, 2006, P. 3403)** Pinatubo erupted in 1991 and 4000 years ago. Taal has been active since the early 1600's; and Mayon has erupted every 15–20 years since the late 19th century to the present Tables **(DuFrane, 2006, P. 3406-3409)**.

The $^{207}Pb/^{206}Pb$ dates I calculated with Isoplot for the 1991 Pinatubo eruption are between 4,987 million and 5,007 million years old with an accuracy of 99.60%. That means the $^{207}Pb/^{206}Pb$ dates are all 333 million times to old. Mayon Volcano's longest uninterrupted eruption occurred on June 23, 1897, which lasted for seven days of raining fire. Lava once again flowed down to civilization. Eleven kilometres eastward, the village of Bacacay was buried 15 metres beneath the lava. In Libon 100 people were killed by steam and falling debris or hot rocks. Other villages like San Roque, Misericordia and Santo Niño became death traps. Ash was carried in black clouds as far as 160 kilometres from the catastrophic event, which killed more than 400 people. The $^{207}Pb/^{206}Pb$ dates I calculated with Isoplot for the 1897 Mayon eruption are between 4,984 million and 4,987 million years old with an accuracy of 99.94%. That means the $^{207}Pb/^{206}Pb$ dates are all 46 million times to old.

There have been 33 recorded eruptions at Taal since 1572. Eruptions have also been recorded in 1634, 1635, 1641, 1645, 1790, 1825, 1842, 1873, 1885, 1903, 1966, 1967, 1968, 1969, 1970, 1976 and 1977. The $^{207}Pb/^{206}Pb$ dates I calculated with Isoplot for the 1634 Taal eruption are between 4,992 million and 4,988 million years old with an accuracy of 99.92%. That means the $^{207}Pb/^{206}Pb$ dates are all 13 million times to old. As we look at the $^{207}Pb/^{206}Pb$ dates in table 16 keep in mind how recent the eruptions really were. (Mayon 1800 AD, Bulusan 4000 BC, Taal 1600 AD, Pinatubo 1991 AD)

Table XV. Bicol and Bataan arcs Pb ages, Ma

Taal	Mayon	Bulusan	Iriga	Pinatubo
1600 AD	1800 AD	4000 BC	???	1991 AD
4,992	4,986	4,990	4,980	4,987
4,989	4,986	4,991	4,979	4,999
4,988	4,986	4,991	4,985	5,007
4,989	4,986		4,982	4,994
4,991	4,986			5,000
4,988	4,986			4,999
4,989	4,984			4,997
4,991	4,986			4,999
4,992	4,985			
4,991	4,987			
4,990				
4,992				

Table XVI. Bicol and Bataan arcs $^{230}Th/^{238}U$ ages, Ka

Taal	Mayon	Bulusan	Iriga	Pinatubo
1600 AD	1800 AD	4000 BC	???	1991 AD
276.07	0.00	332.33	281.40	505.82
287.10	0.00	345.87	184.70	298.70
286.46	0.00	470.18	204.80	324.87
251.25	0.00		187.06	0.00
278.83	0.00			323.54
177.77	0.00			293.25
275.48	0.00			326.37
310.11	0.00			278.27
317.39	0.00			
281.99				
315.43				
301.06				

Stromboli Volcano Glass-bearing Crustal Xenoliths

Stromboli is a small island in the Tyrrhenian Sea, off the north coast of Sicily, containing one of the three active volcanoes in Italy. In 2005 five scientists form Italy and Denmark examined samples from recent eruptions. "Three xenoliths erupted as ejecta during recent violent explosion of Stromboli volcano (Aeolian Islands) were investigated in this paper." **(Salvioli-Mariani, 2005, P. 255)** He goes on further and explains that the five samples were form the 1944 eruption 61 years old**.** "The three studied samples of buchite are irregularly coated with juvenile basaltic rinds; they have been sampled at the Rina Grande slope, SW of Pizzo Sopra la Fossa (RG1 and RG2), among the ejecta erupted during the recent to present-day activity, and at the bottom of Le Schicciole slope (PST107), where products of the 1944 paroxysmal eruption accumulated." **(Salvioli-Mariani, 2005, P. 260)**

The $^{207}Pb/^{206}Pb$ dates I calculated with Isoplot from an isotope table **(Salvioli-Mariani, 2005, P. 267)** are between 4,962 million and 4,965 million years old. That means the $^{207}Pb/^{206}Pb$ dates are all 81 million times to old.

Table XVII. Stromboli Volcano 1944 eruption

Historic Age	Pb Age (Ma)	Age Ratio (10^6)
61	4,965	81.39
61	4,962	81.34
61	4,965	81.39
61	4,964	81.38
61	4,963	81.36

Study of Stromboli Volcano

This research was done in 2007 by four scientists from Germany and Italy. The supposed true age varied from 85,000 years old **(Tommasini, 2007, P. 2414)** to historical or recent. Those dates were determined by $^{230}Th/^{232}Th$ dating. "U/Th disequilibria constrain the timing of the first metasomatism (Stage I) at 435 ka, whereas the second event (Stage II) occurred at 100 ka. Moreover, the ($^{230}Th/^{232}Th$) and ^{238}U and ^{230}Th excesses in the Stromboli samples provide evidence for the occurrence of dynamic melting of the metasomatized mantle wedge" **(Tommasini, 2007, P. 2426)** If we run the 29 samples through we get $^{207}Pb/^{206}Pb$ dates between 4,951 and 4,956 million years old with a 99.71% accuracy.

Table XVIII. Study of Stromboli Volcano (Pb Ages Ma)

Palaeo I (85 ka)	Palaeo II (60 ka)	Palaeo III (35 ka)	Vancori (26–12 ka)	Neo (10 ka)	S. Bartolo (2 ka)	Recent (0 ka)
4,964	4,962	4,957	4,960	4,960	4,966	4,952
4,966	4,963	4,957	4,957	4,951	4,965	4,961
4,966	4,965	4,957	4,960	4,960		4,962
			4,955	4,960		4,960
				4,960		4,960
				4,960		
				4,952		
				4,958		
				4,963		

Mount Vesuvius

Mount Vesuvius is a stratovolcano in the Gulf of Naples, Italy, about 9 kilometres east of Naples and a short distance from the shore. It is one of several volcanoes which form the Campanian volcanic arc. Mount Vesuvius is best known for its eruption in AD 79 that led to the burying and destruction of the Roman cities of Pompeii and Herculaneum

Eight scientists form the USA and Italy examined "Sanidine from pumice collected at Casti Amanti in Pompeii and Villa Poppea in Oplontis yielded a weighted-mean $^{40}Ar/^{39}Ar$ age of 1925±66 years in 2004 (1σ uncertainty) from incremental-heating experiments of eight aliquants of sanidine. This is the calendar age of the eruption. Our results together with the work of Renne and Min demonstrate the validity of the $^{40}Ar/^{39}Ar$ method to reconstruct the recent eruptive history of young, active volcanoes." (Lanphere, 2007, P. 259) If the $^{40}Ar/^{39}Ar$ is so accurate to text the dating method then the $^{207}Pb/^{206}Pb$ dates can be a test that method as completely wrong. The $^{40}Ar/^{39}Ar$ was used to date Mount Erebus rocks that were only thirteen years old and were over one million times in error. The author ends the article by saying that these dates are a sure test of its accuracy: "These results show that the $^{40}Ar/^{39}Ar$ method can be used to reconstruct the recent eruptive history of young volcanoes." (Lanphere, 2007, P. 262)

Table XIX. Mount Vesuvius 39Ar/40Ar Age

Sample	Historical Age	39Ar/40Ar Age	Accuracy
CA-1	1,928	2,061	93.55%
CA-2	1,928	1,707	88.54%
CA-4	1,928	2,099	91.85%
CA-5	1,928	1,821	94.45%
CA-6	1,928	1,938	99.48%
CA-7	1,928	2,181	88.40%
CA-8	1,928	1,667	86.46%
VP-2	1,928	1,867	96.84%

Six scientists from Germany, Italy and the USA analysed fourteen mineral samples **(Gilg, 2001, P. 166)** from two different eruption episodes (1550 BC and 79 AD). "The exact sampling site and thus eruption time of these ejecta are not known. But they most probably come from the Plinian eruptions of 'Avellino' (3550 years ago) and 'Pompeii' (79 AD), which have brought the largest amount of skarn ejecta to the surface." **(Gilg, 2001, P. 147)** The $^{207}Pb/^{206}Pb$ dates I calculated with Isoplot from the isotope table are between 4,949 million and 4,982 million years old with an accuracy of 99.34%. That means the $^{207}Pb/^{206}Pb$ dates are all 2 million times to old.

Table XX. Skarns from Vesuvius

Pb Age (Ma)	Pb Age (Ma)
4,982	4,960
4,982	4,960
4,969	4,960
4,967	4,959
4,962	4,955
4,962	4,951
4,961	4,949

Four scientists from Italy and the USA analysed fifty two mineral samples **(Somma, 2001, P. 133-134)** from three different Vesuvius eruption episodes (1550 BC to 79 AD, 79 AD to 472 AD and 472 AD to 1139 AD) **(Somma, 2001, P. 125)** The $^{207}Pb/^{206}Pb$ dates I calculated with Isoplot from the isotope table are between 4,959 million and 4,967 million years old with an accuracy of 99.84%. That means the $^{207}Pb/^{206}Pb$ dates are all 2 million times to old.

Table XXI. Pb dates of Vesuvius interplinian magmas

472 AD to 1139 AD	79 AD to 472 AD	1550 BC to 79 AD
4,963	4,960	4,967
4,965	4,959	4,965
4,963	4,960	4,966
4,965	4,960	4,966
4,964	4,959	4,965
4,964	4,959	4,965
4,964	4,960	4,966
4,963	4,959	4,966
4,964	4,960	4,966
4,964	4,961	4,966
4,963	4,960	4,967

4,963	4,960	4,966
4,962		4,966
4,961		4,966
4,963		4,965
4,964		4,967
4,964		4,967
4,959		4,967
		4,966
		4,966
		4,966

Sangeang Api Eruptions

Sangeang Api (Gunung Api or Gunung Sangeang) is an active complex volcano on the island of Sangeang in Indonesia. It consists of two volcanic cones, 1,949 metres Doro Api and 1,795 metres Doro Mantoi. Sangeang Api is one of the most active volcanoes in the Lesser Sunda Islands. Between its first recorded eruption in 1512 and 1989 it erupted 17 times. It erupted again during December 2012 and May 2014. It is an active alkaline volcano lying within an older caldera and has a known history of eruptions spanning 1512-2014. There were continuous eruptions from 1953-1958, 1964-1966, 985-1988 and finally 2012-2014. **(Turner, 2003, P. 492)** We have one sample from the 1953 eruption **(Turner, 2003, P. 495, 496)** and seven samples from the 1985 eruption. Seven scientists examined them in 2001 **(Turner, 2003, P. 491).** The eight $^{207}Pb/^{206}Pb$ dates I calculated with Isoplot from the isotope table were all over 4,962 million years old. Since the oldest sample was only 48 years old when examined, The eight $^{207}Pb/^{206}Pb$ dates I calculated were between 103 to 311 million times to old.

Table XXII. ^{207}Pb and ^{230}Th Ages of Sangeang Api

Eruption	Pb Age (Ma)	Age Ratio (10^6)	Th Ages (Ka)	Age Ratio (10^3)
1953	4,963	103	0	
1985	4,967	310	408.59	26
1985	4,968	311	0	
1985	4,970	311	0	
1985	4,969	311	247.1	15
1985	4,970	311	-	
1985	4,966	310	210.3	13
1985	4,967	310	243.73	15

The Augustine Island Volcano

Augustine Volcano is a central lava dome and lava flow complex, surrounded by pyroclastic debris. It forms Augustine Island in southwestern Cook Inlet in the Kenai Peninsula Borough of southcentral coastal Alaska, 280 kilometres southwest of Anchorage. Augustine Island has a land area of 83 square kilometres. Four scientists examined seventeen mineral samples from the volcano in 1996. These samples were taken from historic eruptions (1812-1986) and were between ten and 184 years old when examined. **(Johnson, 1996, P. 105)** The $^{207}Pb/^{206}Pb$ dates I calculated with Isoplot from the isotope table are between 4,959 million and 5,076 million years old. The dates were between 27 million and 507 million times too old.

Table XXIII. $^{207}Pb/^{206}Pb$ Ages of Augustine Volcano

Eruption	True Age	Pb Age (Ma)	Age Ratio 10^6
1812	184	4,974	27
	184	4,973	27
	184	4,984	27
	184	4,974	27
	184	5,076	28
1883	113	4,975	44
	113	4,980	44
1935	61	4,972	82
1964	32	4,969	155
	32	4,976	156
	32	4,983	156
1976	20	4,972	249
	20	4,972	249
1986	10	4,969	497
	10	5,067	507
	10	4,969	497
	10	5,067	507

Table XXIV. $^{207}Pb/^{206}Pb$ Ages of Augustine Volcano Region

Volcano	Lower Value	Upper Value
Location	Pb Age (Ma)	Pb Age (Ma)
Mount Spurr	4,964	4,976
Augustine	4,969	4,981
Okmok	4,959	4,977
Akutan	4,977	4,968
Seguam	4,966	4,966
Sediments	4,965	4,964

The Novarupta-Katmai Eruption of 1912

The explosive outburst at Novarupta (Alaska) in June 1912 was the 20th century's most voluminous volcanic eruption. During the 60-hour eruptive sequence of 6–8 June 1912, 13.5 cubic kilometres of rhyolite, dacite, and andesite magma was released at a new vent, later named Novarupta. Withdrawal of magma from beneath Mount Katmai, 10 km east of Novarupta, caused syne eruptive collapse of the 4-km-wide, kilometre-deep caldera, which has subsequently filled with a lake now 200 metres deep. The eruption occurred on the Alaska Peninsula (in Alaska), only 170 km from Kodiak and 440 km from Anchorage. If we look at isotope samples taken from the eruption **(Hildreth, 2012, P. 195)** we get $^{207}Pb/^{206}Pb$ dates of 4,966 million years old.

Table XXV. The Novarupta-Katmai Eruption of 1912

Pb Age (Ma)
4,967
4,965
4,966

Rhyolite in North American Continental Arcs

Table XXVI. Rhyolite Pb Ages (Ma)

Location	Sample	True Age	Pb Age (Ma)	Error Ratio 10^6
Medicine Lake	1545m	91	4,959	54
	1139ma	91	4,960	55
	1543m	91	4,959	54
	1331m	91	4,967	55
	733m	91	4,966	55
Novaraupta	K90Rhy	91	4,967	55
	K90Dac	91	4,965	55
Crater Lake	1533	91	4,964	55
	1532	91	4,966	55
	1534	91	4,970	55

Table XXVII. Rhyolite ^{230}Th Ages (Ka)

Location	True Age	Th Age (Ka)	Error Ratio
Novarupta	91	814	8,949
	91	350	3,843
	91	795	8,741
	91	481	5,283
	91	343	3,764
	91	319	3,502
	91	390	4,290
	91	427	4,687
Crater Lake	91	285	3,130
	91	331	3,634
	91	318	3,494
	91	273	2,998
	91	471	5,171
	91	458	5,033
Medicine Lake	91	380	4,175
	91	359	3,947
	91	333	3,663
	91	406	4,466
	91	528	5,806
	91	676	7,429
	91	562	6,178
	91	662	7,277

The 1912 Katmai-Novarupta Eruption

Five scientists form Australia, USA and UK examined chemical composition of materials **(Turner, 2010, P. 1)** that range from basalt through basaltic andesite, andesite, dacite, and rhyolite. The twenty six $^{230}Th/^{238}U$ dates I calculated with Isoplot from the isotope table

(Turner, 2010, P. 6, 7) are between 362 thousand and 610 thousand years old. Since the samples were only 98 years old when analysed the dates are between 4,000 and 6,000 times in error.

Table XXVIII. The 1912 Katmai-Novarupta Eruption

$^{230}Th/^{238}U$ Age	Historic Age	Ratio
10^3 Years	Years	Ratio
492	98	5,020
505	98	5,154
610	98	6,225
442	98	4,512
398	98	4,065
488	98	4,981
560	98	5,713
362	98	3,695
427	98	4,355
427	98	4,355
553	98	5,640
440	98	4,494
426	98	4,347
436	98	4,454
366	98	3,733
382	98	3,893
594	98	6,063
534	98	5,445
412	98	4,202
451	98	4,602
362	98	3,692
489	98	4,990
515	98	5,255
382	98	3,893
513	98	5,233
474	98	4,841

Jurassic Oceanic Crust beneath Grand Canaria

Gran Canaria is the second most populous island of the Canary Islands. Gran Canaria is located southeast of Tenerife and west of Fuerteventura. The island is of volcanic origin, mostly made of fissure vents. Gran Canaria's surface area is 1,560 km^2 and its maximum elevation is 1,949 meters (Pico de Las Nieves). It has a round shape, with a diameter of approximately 50 kilometres.

Three major volcanic structures form the 60 kilometre wide island, which has been modified by caldera collapse, gravitational edifice failure, and extensive erosion. Scoria cones and lava flows are found in the northern and eastern parts of the massive shield volcano, which is cut by a major NW-SE-trending rift zone that extends across the island and fed flows primarily to the NE. Very young basaltic cones and lava flows are situated within a NW-trending zone from Berrazales to Bandama and at Las Isletas, a peninsula on the NE coast. One cinder cone was radiocarbon dated at about 3,000 years old, and other cones and flows may be less than 1,000 years old.

If we use isotopic formulas given in standard geology text we can arrive at ages from the Rubidium/Strontium **(Attendon, 1997, P. 73)** and Neodymium/Samarium **(Attendon, 1997, P. 305)** ratios. The formula for Rb/Sr age is given as:

$$t = \frac{2.303}{\lambda}\log\left(\frac{(^{87}Sr/^{86}Sr) - (^{87}Sr/^{86}Sr)_0}{(^{87}Rb/^{86}Sr)} + 1\right) \quad \textbf{(8)}$$

Where t equals the age in years. λ equals the decay constant. ($^{87}Rb/^{86}Sr$) = the current isotopic ratio. ($^{87}Rb/^{86}Sr)_0$ = the initial isotopic ratio. ($^{87}Rb/^{86}Sr$) = the current isotopic ratio. The same is true for the formula below.

$$t = \frac{2.303}{\lambda}\log\left(\frac{(^{143}Nd/^{144}Nd) - (^{143}Nd/^{144}Nd)_0}{(^{147}Sm/^{144}Nd)} + 1\right) \quad \textbf{(9)}$$

Table XIX. Jurassic Oceanic Crust beneath Grand Canary

Sample Number	$^{207}Pb/^{206}Pb$ Age (Ma)	$^{147}Sm/^{144}Nd$ Age (Ma)	$^{87}Rb/^{86}Sr$ Age (Ma)	$^{207}Pb/^{206}Pb$ Age Ratio 10^6	$^{147}Sm/^{144}Nd$ Age Ratio 10^6	$^{87}Rb/^{86}Sr$ Age Ratio 10^6
917c	4,963	172	170	101	4	3
91249	4,974	174	170	102	4	3
DSDP397-60-4	4,971	174	170	101	4	3
DSDP397-101-1	4,975	167	171	102	3	3
DSDP397-30-1	4,972	172	170	101	4	3
DSDP397-40-2	4,972	168	166	101	3	3
DSDP397-49-1	4,971	177	165	101	4	3
303905	5,024	170	151	103	3	3
303906	5,005	172	168	102	4	3
B9117.1	5,001	166	225	102	3	5
B9117.1	4,999		204	102		4
B9117.2	4,938	173	151	101	4	3
B9117.3	4,845	174	253	99	4	5
303903	5,017	168		102	3	
B913	4,956	170	172	101	3	4
B914	4,988	166	178	102	3	4
B914	4,991	162	274	102	3	6
RNB60	4,952	153	151	101	3	3
RN1249	4,956			101		
241921B			151			3

In 1998 volcanic samples from the 1949 eruption were analysed: "Sr–Nd–Pb isotope data are presented in Table 2 for ocean crust samples from Gran Canaria and a tholeiitic gabbro xenolith in the 1949 eruption on La Palma." **(Hoernle, 1998, P. 863).** If we run the three types of ratios ($^{87}Rb/^{86}Sr$, $^{147}Sm/^{144}Nd$, $^{207}Pb/^{206}Pb$) found in table 2 in his article **(Hoernle, 1998, P. 867, 868)** through Microsoft Excel we get dates between 151 and 5,024 million years old. The dates are between 3 million and 103 million times too old.

Evolution of La Palma (Canary Islands)

The island of La Palma rises to 2,426 metres above sea level and is the westernmost and volcanically most active of the Canary Islands. "It can be divided into three major units: (1) the older basal complex (ca. 4.0 to 3.0 Ma) comprised of a Pliocene seamount complex exposed in the Caldera de

Taburiente, and a plutonic complex, uplifted and tiled by intrusions coeval with the later subaerial activity, (2) the older volcanic series (1.7 to 0.4 Ma) which include the Garafia volcano, the Taburiente shield volcano, the Bejenado edifice, and the Cumbre Nueva series, and (3) the recent Cumbre Vieja series (125 ka to present) which is confined to the southern half of the island." **(Galipp, 2005, P. 11)** Galipp has an isotope table in his article **(Galipp, 2005, P. 102, 103)** that has twelve $^{40}Ar/^{39}Ar$ dates (20 Ma - 834 Ma) and twelve $^{207}Pb/^{206}Pb$ ratios with no dates beside them. If we run the $^{207}Pb/^{206}Pb$ ratios through Isoplot we get twelve dates between 4,892 and 4,920 million years old. The twelve $^{207}Pb/^{206}Pb$ dates between six and 245 times older than the $^{40}Ar/^{39}Ar$ dates.

Table XXX. Ar/Ar Versus Pb/Pb Ages of La Palma Volcanoes

Volcano	$^{40}Ar/^{39}Ar$	$^{207}Pb/^{206}Pb$	Age Ratio
La Palma Island	Age (Ma)	Age (Ma)	
Taburiente	533	4,904	9.20
	410	4,907	11.97
	563	4,908	8.72
	585	4,902	8.38
	560	4,892	8.74
	833	4,903	5.89
Cumbre Nueva	647	4,911	7.59
	834	4,916	5.89
Cumbre Vieja	123	4,911	39.93
	90	4,917	54.63
	120	4,920	41.00
	20	4,917	245.85

The same table has sixteen $^{207}Pb/^{206}Pb$ ratios for six (1480 – 1971) historic volcanic eruptions. If we run the $^{207}Pb/^{206}Pb$ ratios through Isoplot we get sixteen dates between 4,906 and 4,932 million years old. The twelve $^{207}Pb/^{206}Pb$ dates between 9 million and 145 million times to old.

Table XXI. Pb ages of La Palma eruptions

Eruption Year	True Age	Pb Age (Ma)	Error Ratio 10^6
1480	525	4,911	9
	525	4,906	9
	525	4,911	9
1585	420	4,910	12
	420	4,910	12
	420	4,910	12
	420	4,910	12
1646	359	4,911	14
	359	4,910	14
1677	328	4,911	15
	328	4,910	15
1949	56	4,932	88
	56	4,920	88
	56	4,917	88
1971	34	4,914	145
	34	4,909	144

Volcanic rocks from Merapi Volcano

Mount Merapi, is an active stratovolcano located on the border between Central Java and Yogyakarta, Indonesia. It is the most active volcano in Indonesia and has erupted regularly since 1548. It is located approximately 28 kilometres north of the large Yogyakarta city. Merapi is the youngest in a group of volcanoes in southern Java. It is situated at a subduction zone, where the Indo-Australian Plate is subducting under the Sunda Plate. It is one of at least 129 active volcanoes in Indonesia, part of the volcano is located in the south eastern part of the Pacific Ring of Fire. Typically, small eruptions occur every two to three years, and larger ones every 10–15 years or so. Notable eruptions, often causing many deaths, have occurred in 1006, 1786, 1822, 1872, and 1930. Thirteen villages were destroyed in the later one, and 1400 people killed by pyroclastic flows. "The sample that forms the basis of this study is representative of the younger eruptive stages of Merapi. It includes (1) the lava sequences of the somma rim, (2) pyroclastic flow and tephra fall deposits of overlapping Holocene Pyroclastic Series and (3) eruptive products of selected effusive (dome forming) and pyroclastic flow forming eruptions of the recent and historical activity of Merapi, which can be traced back to the late 18th century." **(Gertisser, 2003, P. 461)**

Two Lead samples **(Gertisser, 2003, P. 467)** from two eruptions (1822, 1872) were put through Isoplot and both returned ages of 4,986 million years old. That means that the dates were 27 million and 38 million times in error respectively.

Table XXII. Volcanic rocks from Merapi Volcano

Pb Age (Ma)	Pb Age (Ma)
Holocene	**19th Century Eruption**
4,984	**4,986**
4,985	**4,986**

The two Réunion Island volcanoes

The island is 63 kilometres long and 45 kilometres wide; and covers 2,512 square kilometres. It is located above a hotspot in the Earth's crust. The Piton de la Fournaise, a shield volcano on the eastern end of Réunion Island, rises more than 2,631 metres above sea level and is sometimes called a sister to Hawaiian volcanoes because of the similarity of climate and volcanic nature. It has erupted more than 100 times since 1640 and is under constant monitoring, most recently erupting on 31 July 2015. During another eruption in April 2007, the lava flow was estimated at 3,000,000 cubic metres per day.

Twenty one samples **(Bosch, 2008, P. 752)** from twelve historic eruptions between 1700 and 1953 were analysed. The $^{207}Pb/^{206}Pb$ dates I calculated with Isoplot from the isotope table are between 4,962 million and 4,972 million years old with 99.80% accuracy. The twenty one dates were between 16 million and 90 million times too old.

Table XXXIII. The two Réunion Island Volcanoes

Eruption	Age (Years)	Pb Age (Ma)	Age Ratio (10^6)
1700	**308**	**4,964**	**16**
1802	**206**	**4,966**	**24**
1905	**103**	**4,964**	**48**
1915	**93**	**4,962**	**53**
1927	**81**	**4,968**	**61**
	81	**4,963**	**61**
	81	**4,966**	**61**
	81	**4,966**	**61**
1931	**77**	**4,971**	**65**
1934	**74**	**4,963**	**67**
1939	**69**	**4,966**	**72**

1943	65	4,965	76
	65	4,967	76
1945	63	4,972	79
	63	4,963	79
	63	4,966	79
	63	4,965	79
	63	4,968	79
1949	59	4,964	84
1953	55	4,966	90
	55	4,966	90

The Kluchevskoy volcano, Central Kamchatka

The Kamchatka Peninsula is a 1,250-kilometre-long peninsula in the Russian Far East, with an area of about 270,000 square kilometres. It lies between the Pacific Ocean to the east and the Sea of Okhotsk to the west. The volcanoes of Kamchatka are a large group of volcanoes situated on the Kamchatka Peninsula. The Kamchatka River and the surrounding central side valley are flanked by large volcanic belts containing around 160 volcanoes, 29 of them still active. Nine samples from two eruptions (1932, 1953) were analysed **(Dorendorf, 2000, P. 75)** in 2000 by three scientists. The $^{207}Pb/^{206}Pb$ dates I calculated with Isoplot from the isotope table are between 5,002 million and 5,005 million years old with 99.95% accuracy. Since the samples were only 47 and 68 years old the age ratio was 106 and 73 million times to old respectively.

Table XXXIV. The Kluchevskoy Volcano

Eruption	Historic Age	Pb Age (Ma)	Age Ratio 10^6
1932	68	5,002	73.56
1932	68	5,004	73.59
1932	68	5,004	73.59
1932	68	5,003	73.57
1932	68	5,005	73.60
1932	68	5,005	73.60
1932	68	5,004	73.58
1932	68	5,002	73.56
1953	47	5,004	106.46

In 2013 five scientists from the USA and Switzerland **(Kayzar, 2014, P. 168)** obtained sixty three samples from the Bezymianny, Klyuchevskoy, Shiveluch and Karymsky volcanoes. Theses samples were from twenty four historic eruptions (1939-2010) that were between 4 and 75 years old when analysed. The $^{207}Pb/^{206}Pb$ dates I calculated with Isoplot from the isotope table are between 4,996 million and 5,011 million years old with 99.69% accuracy. Since the samples were only between 4 and 75 years old the age ratio was between 67 and 1,253 million times too old respectively.

Table XXXV. Bezymianny and Klyuchevskoy Pb Ages

Volcano	Eruption	Age (Years)	Pb Age (Ma)	Age Ratio (10^6)
Bezymianny	1956	58	5,004	86
	1956	58	5,004	86
	1997	17	5,004	294
	2006	8	5,004	626
	2007	7	5,004	715
	2008	6	5,003	834
	2009	5	5,004	1,001

	2010	4	5,004	1,251
Klyuchevskoy	1939	75	5,002	67
	1945	69	5,002	72
	1946	68	5,002	74
	2007	7	5,001	714
Shiveluch	1964	50	4,997	100
	1980	34	4,996	147
	1993	21	4,996	238
	2001	13	4,996	384
	2007	7	4,997	714
Karymsky	1964	50	4,997	100
	1971	43	4,997	116
	1996	18	4,997	278
	1998	16	4,997	312
	2003	11	4,997	454
	2004	10	4,997	500
	2007	7	4,998	714
	2008	6	4,997	833

Element Variations in Sunda arc Lavas

This research was done in 2001 by two scientists from England and Australia **(Turner, 2001, P. 43)**. Nineteen samples from nine volcanoes and eleven historic eruptions **(Turner, 2001, P. 46, 47)** were analysed for $^{230}Th/^{238}U$ and $^{207}Pb/^{206}Pb$ ratios. The nineteen $^{207}Pb/^{206}Pb$ dates I calculated with Isoplot from the isotope table are between 4,967 million and 5,001 million years old with 99.32% accuracy. The nineteen dates were between 67 thousand and 311 million times too old. The nine $^{230}Th/^{238}U$ dates I calculated with Isoplot from the isotope table are between 160 thousand and 661 thousand years old. The nine $^{230}Th/^{238}U$ dates were between 4 and 9,600 times too old.

Table XXXVI. Sunda arc Lavas Pb/Pb Ages

Volcano	Eruption	Age (Years)	Pb Age (Ma)	Age Ratio (10^6)
Krakatau	1883	118	4,991	42
	1883	118	5,001	42
Galunggung	1982	19	4,983	262
	1982	19	4,981	262
	1982	19	4,982	262
	1918	83	4,984	60
Merapi	1006	995	4,984	5
	1006	995	4,984	5
	1006	995	4,984	5
	1006	995	4,983	5
Rindjani	1900	101	4,990	49
	1900	101	4,985	49
Tambora	1815	186	4,967	27
Sangeang Api	1985	16	4,970	311
Ayi Flores	1900	101	4,975	49

Table XXXVII. Sunda arc Lavas ^{230}Th Ages

Volcano	Eruption Year	Age (Years)	^{230}Th Age (Ka)	Age Ratio
Toba	-72,000	74,000	314.44	4.25
Sorikmarapi	-10,000	12,000	404.10	33.68
Krakatau	1883	118	661.98	5610.00
	1883	118	510.89	4329.58
Galunggung	1982	19	182.40	9600.00
	1918	83	271.89	3275.78
	-2200	2,400	265.73	110.72
Merapi	1006	995	269.13	270.48
	1006	995	160.17	160.97

(Negative dates = BC, positive = AD)

Contributions to Lesser Antilles magmas

The Lesser Antilles is the name given to a group of islands in the Caribbean Sea. Most form a long, partly volcanic island arc between the Greater Antilles to the north-west and the continent of South America. The islands form the eastern boundary of the Caribbean Sea with the Atlantic Ocean. In 2009 (DuFrane, 2009, P. 274) scientists collected nine samples from four different islands. Four of the samples were from historic eruptions (1530, 1995 and 1998). The nine ^{207}Pb/^{206}Pb dates I calculated with Isoplot from the isotope table are between 4,943 million and 5,008 million years old. The four ^{230}Th/^{238}U dates I calculated with Isoplot from the isotope table are between 300 thousand and 400 thousand years old. The dates are all between 30 thousand and 450 million times too old.

Table XXXVIII. Contributions to Lesser Antilles magmas

Island	Eruption Year	Pb Age (Ma)	Age Ratio (10^6)	Th Age (Ka)	Age Ratio
Grenada	-8,000	4,943	0.4943		
	-8,000	5,008	0.5008		
Guadeloupe	1,530	4,962	10	401.17	838
	-500,000	4,953	0.01	299.86	0.597
Montserrat	1,995	4,979	356		
	1,998	4,968	452	373.56	33,960
	1,998	4,962	451	333.79	30,345
Saba	-8,000	4,983	0.4983		
	-8,000	4,968	0.4968		

(Negative dates = BC, positive = AD)

Piton de la Fournaise historical lavas

Piton de la Fournaise is a shield volcano on the eastern side of Réunion Island in the Indian Ocean. It is currently one of the most active volcanoes in the world, along with Kīlauea in the Hawaiian Islands (Pacific Ocean), Stromboli, Etna (Italy) and Mount Erebus in Antarctica. A previous eruption began in August 2006 and ended in January 2007. The volcano erupted again in February 2007, on 21 September 2008 and on 9 December 2010, which lasted for two days. The most recent eruption began on 1 August 2015. The volcano is located within Réunion National Park, a World Heritage site.

In 2008 eight scientists from France **(Vlastélic, 2009, P. 66-68)** analysed isotopes ratios from 112 volcanic eruptions between 1927 and 2007. The one hundred and thirty one samples cover and eighty year time span. The 131 $^{207}Pb/^{206}Pb$ dates I calculated with Isoplot from the isotope table are between 4,963 million and 4,981 million years old. The dates are between 61 million and 2,482 million times too old.

Table XXXIX. Piton de la Fournaise historical lavas

Eruption	Age (Years)	Age (Ma)	Age Ratio (10^6)
1927	82	4,966	61
1931	78	4,965	64
1938	71	4,965	70
1943	66	4,964	75
1945	64	4,965	78
1948	61	4,966	81
1953	56	4,966	89
1961	48	4,965	103
1977	32	4,963	155
1957	52	4,965	95
1966	43	4,964	115
1972	37	4,965	134
1973	36	4,965	138
1975	34	4,965	146
1976	33	4,964	150
1979	30	4,964	165
1981	28	4,964	177
1983	26	4,963	191
1984	25	4,963	199
1985	24	4,963	207
1986	23	4,964	216
1987	22	4,965	226
1988	21	4,966	236
1990	19	4,965	261
1991	18	4,966	276
1992	17	4,966	292
2001	8	4,967	621
2006	3	4,965	1,655
2007	2	4,965	2,483

Two Volcanoes in the Alaska-Aleutian Arc

The Aleutian Islands are a chain of 14 large volcanic islands and 55 smaller ones belonging to both the United States and Russia. They form part of the Aleutian Arc in the Northern Pacific Ocean, occupying an area of 17,666 square kilometres and extending about 1,900 kilometres westward from the Alaska Peninsula toward the Kamchatka Peninsula in Russia, and mark a dividing line between the Bering Sea to the north and the Pacific Ocean to the south.

In 2002 seven scientists from England and the USA examined fifteen samples from Aniakchak volcano **(George, 2004, P. 207, 208)** and twelve samples Akutan volcano **(George, 2004, P. 210)**. These samples represent five historic eruptions (1931, 1870, 1910, 1929 and 1978) that were only 24 to 71 years old when studied. The $^{207}Pb/^{206}Pb$ dates I calculated with Isoplot from the isotope table are between 4,960 million and 4,972 million years old. Since the samples were only 24 to 71 years old when studied the age ratio was between 37 million and 191 million times too old respectively.

Table XL. Two Volcanoes in the Alaska-Aleutian Arc

Volcano	Eruption	Age	Pb Age (Ma)	Age Ratio (10^6)
Aniakchak	1931	73	4,972	68
	1931	73	4,966	68
	1931	73	4,967	68
	1931	73	4,966	68
	1931	73	4,965	68
	1931	73	4,966	68
	1931	73	4,967	68
	1931	73	4,965	68
Akutan	1870	134	4,960	37
	1910	94	4,960	53
	1910	94	4,961	53
	1929	75	4,961	66
	1978	26	4,962	191
	1978	26	4,961	191
	1978	26	4,960	191

Puu Oo Eruption of Kilauea Volcano

Kilauea is a currently active shield volcano in the Hawaiian Islands, and the most active of the five volcanoes that together form the island of Hawaii. Located along the southern shore of the island, evolutionary science states that the volcano is between 300,000 and 600,000 years old and emerged above sea level about 100,000 years ago. There were sixty episodes in the eruptive sequence that began in January 1983 and ended in June 2011. Two scientists from the USA examined five samples in 1999 **(Garcia, 2000, P. 967)** that were between one to seven years old. The $^{207}Pb/^{206}Pb$ dates I calculated with Isoplot from the isotope table **(Garcia, 2000, P. 984)** are between 4,992 million and 4,993 million years old with 99% accuracy. Since the samples were only between one and seven years old the age ratio was between 624 million and 2.5 billion times too old respectively.

Table XLI. Puu Oo Eruption of Kilauea Volcano

Eruption	Pb Age (Ma)	Age Ratio (10^6)
29/12/1992	4,992	624
25/04/1994	4,992	832
27/04/1995	4,992	998
10/01/1997	4,992	1,664
10/01/1998	4,993	2,497

Transport rates along the Alaska-Aleutians

The Aleutian Islands are a chain of 14 large volcanic islands and 55 smaller ones belonging to both the United States and Russia. They form part of the Aleutian Arc in the Northern Pacific Ocean, occupying an area of 17,666 square kilometres and extending about 1,900 kilometres westward from the Alaska Peninsula toward the Kamchatka Peninsula in Russia, and mark a dividing line between the Bering Sea to the north and the Pacific Ocean to the south.

In 2003 seven scientists **(George, 2003, P. 65-67)** from the USA and England examined thirty one samples from 22 volcanoes and 24 historic eruptions (1500 BC-1999 AD). The historical ages varied between 4 to 3,503 years old. The $^{207}Pb/^{206}Pb$ dates I calculated with Isoplot from the isotope table are between 4,949 million and 5,007 million years old. Since the samples were only 3,503 to 4 years old when studied the age ratio was between 1 million and 1,243 million times too old respectively.

Table XLII. Transport rates along the Alaska-Aleutians

Volcano	Eruption	Age	Pb Age (Ma)	Age Ratio (10^6)
Akutan	1850	153	4,960	32
	1978	25	4,961	198
Aniakchak	-1500	3,503	4,966	1
	-1500	3,503	4,967	1
	-1500	3,503	4,966	1
Augustine	-500	2,503	4,967	2
Bogoslof	1796	207	4,968	24
Buzzard Creek	-1000	3,003	4,949	2
Kanaga	1850	153	4,970	32
	1900	103	4,970	48
Kasatochi	1899	104	4,965	48
Katmai	1912	91	4,966	55
Kiska	1962	41	4,970	121
	Historic		5,007	
Little Sitkin	1900	103	4,994	48
Makushin	Historic		4,967	
Okmok	1946	57	4,967	87
Pavlof	1996	7	4,965	709
Redoubt	1989	14	4,972	355
	1989	14	4,973	355
Seguam	Historic		4,965	
Segula	Historic		4,979	
Shishaldin	1999	4	4,972	1,243
Spurr	1953	50	4,969	99
	1992	11	4,972	452
Trident	1953	50	4,967	99
	1953	50	4,967	99
Ukinrek	1977	26	4,973	191
Umnak	1946	57	4,974	87
Westdahl	1991	12	4,966	414
	Active		4,966	

(Negative dates = BC, positive = AD)

2010 Eyjafjallajökull explosive eruption

The 2010 eruptions of Eyjafjallajökull were volcanic events at Eyjafjallajökull in Iceland which, although relatively small for volcanic eruptions, caused enormous disruption to air travel across western and northern Europe over an initial period of six days in April 2010. Additional localised disruption continued into May 2010. The eruption was declared officially over in October 2010, when snow on the glacier did not melt. From 14–20 April, ash covered large areas of northern Europe when the volcano erupted. Two years after the eruption four scientists form France, Russia, Indonesia and Iceland **(Borisova,**

2012, P. 1) examined six samples. The six $^{207}Pb/^{206}Pb$ dates I calculated with Isoplot from the isotope table **(Borisova, 2012, P. 10)** are between 4,942 million and 4,944 million years old. Since all the samples were only 2 years old when studied the age ratio was 2.47 billion times too old.

Table XLIII. 2010 Eyjafjallajökull Pb ages

Pb Age (Ma)	Historic Age	Age Ratio 10^6
4,942	2	2,471
4,943	2	2,472
4,943	2	2,472
4,943	2	2,472
4,942	2	2,471
4,944	2	2,472

Disequilibria at Volcano Llaima, Chile

The Llaima Volcano is one of the largest and most active volcanoes in Chile. It is situated 82 km northeast of Temuco and 663 km southeast of Santiago, within the borders of Conguillío National Park. Llaima is one of Chile's most active volcanoes and has frequent but moderate eruptions. Llaima's activity has been documented since the 17th century, and consists of several separate episodes of moderate explosive eruptions with occasional lava flows. The last major eruption occurred in 1994.

An eruption on January 1, 2008 forced the evacuation of hundreds of people from nearby villages. A column of smoke approximately 3000 m high was observed. An amateur caught the early eruption phase on video. The volcanic ash expelled by Llaima travelled east over the Andes into Argentina.

In 2007 seven scientists from Switzerland, USA and Chile **(Reubi, 2011, P. 37)** analysed mineral samples for isotopic composition. Sixteen samples from seven (1640-2008) historic eruptions were analysed **(Reubi, 2011, P. 43)** and returned accurate $^{230}Th/^{238}U$ ratios. The sixteen $^{230}Th/^{238}U$ dates I calculated with Isoplot from the isotope table are between 174 thousand and 257 thousand years old. Since the samples were between 3 and 371 years old when dated the dates are all between 0.5 thousand and 60 thousand times too old.

Table XLIV. Disequilibria at Volcano Llaima, Chile

Eruption Year	$^{230}Th/^{238}U$ Age 10^3 Years	Historic Age Years	Age Ratio 10^3
2008	178.35	3	59.45
1957	201.26	54	3.73
1957	177.23	54	3.28
1780	211.51	231	0.92
1780	190.13	231	0.82
1780	183.52	231	0.79
1780	174.54	231	0.76
1751	186.49	260	0.72
1903	257.33	108	2.38
1640	213.10	371	0.57
1640	220.89	371	0.60
1640	181.19	371	0.49
1903	228.18	108	2.11
1852	185.29	159	1.17
1852	242.77	159	1.53
1852	185.89	159	1.17

Izu arc Volcanoes

The Izu-Bonin-Mariana (IBM) arc system is a tectonic-plate convergent boundary. IBM extends over 2800 kilometres south from Tokyo, Japan, to beyond Guam, and includes the Izu Islands, Bonin Islands, and Mariana Islands; much more of the IBM arc system is submerged below sea level. The IBM arc system lies along the eastern margin of the Philippine Sea Plate in the Western Pacific Ocean. It is most famous for being the site of the deepest gash in Earth's solid surface, the Challenger Deep in the Mariana Trench.

In 2011 three scientists from Japan **(Kurihara, 2011, P. 335)** examined sixteen samples from two volcanoes. These samples represent eleven historic (800 AD - 1950 AD) eruptions **(Kurihara, 2011, P. 336)**. The sixteen $^{230}Th/^{238}U$ dates I calculated with Isoplot from the isotope table are between 108 thousand and 320 thousand years old. Since the samples were between 61 and 1211 years old when dated the dates are all between 150 and 1700 times too old.

Table XLV. Izu arc volcanoes, Japan

Volcano	230Th Age 10^3	Eruption AD	True Age	Age Ratio
Fuji	191.61	800	1,211	158.22
	227.87	864	1,147	198.66
	192.43	864	1,147	167.77
	213.72	864	1,147	186.33
	193.96	864	1,147	169.10
	222.23	864	1,147	193.75
	290.82	937	1,074	270.78
	319.15	1033	978	326.33
Izu-Oshima	155.10	800	1,211	128.08
	183.97	1338	673	273.36
	151.13	1421	590	256.16
	134.08	1552	459	292.10
	119.68	1684	327	366.00
	220.25	1777	234	941.23
	240.36	1777	234	1,027.20
	108.36	1950	61	1,776.47

Volcanic Products from Izu arc volcanoes

In 2007 three scientists from Japan **(Kurihara, 2007, P. 795)** examined thirty three samples from five volcanoes. These samples represent fifteen historic (800 AD - 1986 AD) eruptions **(Kurihara, 2007, P. 800-802)**. The thirty three $^{230}Th/^{238}U$ dates I calculated with Isoplot from the isotope table are between 80 thousand and 600 thousand years old. Since the samples were between 20 and 1200 years old when dated the dates are all between 100 and 6,000 times too old.

Table XLVI. Volcanic Products from Izu arc volcanoes

Volcano	Eruption	True Age	230Th Age 10^3	Age Ratio
Younger Fuji	1707	300	164.04	547
	1707	300	150.76	503
	1707	300	164.66	549
	1707	300	142.94	476
	1707	300	164.66	549
	1707	300	164.66	549
	1707	300	164.96	550

	1707	300	142.94	476
	1707	300	169.48	565
	1707	300	150.76	503
	1707	300	164.36	548
Isu-Oshima	800	1207	155.62	129
	1338	669	186.49	279
	1421	586	151.23	258
	1552	455	134.62	296
	1684	323	118.08	366
	1777	230	222.67	968
	1950	57	108.94	1,911
	1986	21	82.11	3,910
	1986	21	125.32	5,968
	1986	21	102.40	4,876
	1986	21	105.23	5,011
	1874	133	186.51	1,402
Miyake-Jima	1940	67	213.82	3,191
	1962	45	123.92	2,754
	1983	24	147.01	6,125
Nijima	886	1121	116.46	104
	886	1121	447.11	399

Northern Izu arc Volcanoes

In 2008 seven scientists from Japan **(Fukuda, 2008, P. 461)** examined fifteen samples from four volcanoes. These samples represent thirteen historic (864 AD - 2000 AD) eruptions **(Fukuda, 2008, P. 463-465)**. The fifteen $^{230}Th/^{238}U$ dates I calculated with Isoplot from the isotope table are between 100 thousand and 400 thousand years old. Since the samples were between 8 and 1100 years old when dated the dates are all between 230 and 19,000 times too old.

Table XLVII. Northern Izu arc Volcanoes

Volcano	Eruption AD	True Age	230Th Age (Ka)	Age Ratio
Fuji	864	1,144	319.05	279
	1033	975	398.93	409
	1707	301	271	900
Miyakejima	1469	539	167.88	311
	1643	365	154.41	423
	1874	134	157.21	1,173
	1962	46	164.08	3,567
	1983	25	154.81	6,192
	2000	8	142.66	17,833
	2000	8	152.66	19,083
	2000	8	157.68	19,710
Oshima	1950	58	111.53	1,923
	1550	458	106.6	233
	1778	230	134.91	587
Teishi Knoll	1989	19	262.95	13,839

The Great Tambora Eruption in 1815

In 2010 six scientists form Australia, England and the USA **(Gertisser, 2012, P. 271)** examined volcanic material that was 195 years old. The cataclysmic eruption of Tambora volcano (Sumbawa, Indonesia) in 1815 has long been recognized as one of the largest explosive eruptions in historical time. The eight $^{230}Th/^{238}U$ dates I calculated with Isoplot from the isotope table **(Gertisser, 2012, P. 277)** are between 220 thousand and 520 thousand years old. Since the samples were all 203 years old when dated the dates are all between 1000 and 2600 times too old.

Table XLVIII. The Great Tambora Eruption in 1815

Erupted	True Age	230Th Age 10^3	Age Ratio
1815	203	451.58	2,225
1815	203	301.81	1,487
1815	203	504.04	2,483
1815	203	219.64	1,082
1815	203	233.17	1,149
1815	203	336.91	1,660
1815	203	528.61	2,604
1815	203	230.08	1,133

1585 Eruption on La Palma, Canary Islands

Johansen, T. S., 2005, 1585 eruption on La Palma, Canary Islands, Geology, 33:897-900

Table XLIX. 1585 eruption on La Palma, Canary Islands

Pb Age (Ma)	Eruption	True Age	Age Ratio 10^6
4,910	1585	420	11.69
4,910	1585	420	11.69
4,910	1585	420	11.69
4,910	1585	420	11.69

Active Caldera, Rabaul, Papua New Guinea

Cunningham, H. S., 2009, Active Caldera, Rabaul, Papua New Guinea, Journal of Petrology, 50(3):507-529

Table L. Active Caldera, Rabaul, Papua New Guinea

Eruption	Historic Age	$^{230}Th/^{238}U$ Age	Age Ratio
Year AD	Years	Thousand Years	
640	1369	311	227
1250	759	315	415
	759	0	759
1850	159	297	1,865
1878	131	261	1,993
1937	72	173	2,402
1994	15	278	18,511
	15	319	21,293
	15	315	21,030

1996	**13**	**289**	**22,213**
	13	**312**	**24,023**
	13	**343**	**26,382**
1997	**12**	**286**	**23,844**
	12	**333**	**27,760**
	12	**272**	**22,641**
	12	**274**	**22,847**
1998	**11**	**349**	**31,723**
	11	**444**	**40,348**
	11	**295**	**26,805**
	11	**325**	**29,518**
	11	**315**	**28,612**
1999	**10**	**297**	**29,656**
2000	**9**	**169**	**18,799**
2001	**8**	**309**	**38,660**

References

Attendon, H. G., 1997, Radioactive and Stable Isotope Geology, P. 1-305, Chapman And Hall Publishers

Bernard, Alain, 1999, 1991 Eruption Products of Mount Pinatubo, http://pubs.usgs.gov/pinatubo/bernard/table9.html

Blichert-Toft, Janne, 2002, Isotope evolution of Mauna Kea volcano, http://perso.ens-lyon.fr/francis.albarede/JBT_Rev_text.pdf

Borisova, Anastassia Y., 2012, 2010 Eyjafjallajökull explosive eruption, Journal Of Geophysical Research, 117(B05202):1-18

Bosch, Delphine, 2008, The two Réunion Island volcanoes, Earth and Planetary Science Letters 265:748-765

Bühler, Alena, 2011, Lavas from the islands of Upolu and Savaii, Ph. D Thesis, San Diego State University

Castillo, Paterno R., 1999, Geochemistry of Mount Pinatubo Volcanic Rocks, http://pubs.usgs.gov/pinatubo/castillo/index.html

Cooper, Kari M., 2008, Eruption of Mount St. Helens, 2004–2006, http://pubs.usgs.gov/pp/1750/chapters/pp2008-1750_chapter36.pdf

Cunningham, H. S., 2009, Active Caldera, Rabaul, Papua New Guinea, Journal of Petrology, 50(3):507-529

Dorendorf, Frank, 2000, The Kluchevskoy volcano, Kamchatka/Russia, Earth and Planetary Science Letters, 175:69-86

DuFrane, S. Andrew, 2006, Bicol and Bataan arcs, Philippines, Geochimica et Cosmochimica Acta 70:3401-3420

DuFrane, S. Andy, 2009, Contributions to Lesser Antilles magmas, Chemical Geology 265:272–278

Esser, R. P., 1997, Excess Argon In Melt Inclusions, Geochemica Et Cosmochemica Acta, 61(18):3789-3801

Fukuda, Satoru, 2008, 238U–230Th radioactive disequilibrium, Geochemical Journal, 42:461-479

Funkhouser, John G., 1968, Ultramafic Inclusions from Hawaii, Journal Of Geophysical Research, 73(14):4601-4607

Galipp, Karsten, 2005, Evolution of La Palma (Canary Islands), PhD Thesis, University of Bremen, http://d-nb.info/1072301873/34

Garcia, M. O., 1998, Earthquake swarm of Loihi seamount, Bulletin Volcanology, 59:577–592

Garcia, M. O., 2000, Puu Oo Eruption of Kilauea Volcano, Journal Of Petrology, 41(7):961-990

George, Rhiannon, 2003, Transport rates along the Alaska-Aleutian, Journal Of Geophysical Research, 108:(B5-2252):1-25

George, Rhiannon, 2004, Two Volcanoes in the Alaska-Aleutian Arc, Journal Of Petrology, 45(1):203-219

Gertisser, R., 2003, Volcanic rocks from Merapi Volcano, Journal Of Petrology, 44(3):457-489

Gertisser, R., 2012, The Great Tambora Eruption in 1815, Journal of Petrology, 53(2):271-297

Gilg, H. A., 2001, Skarns from Vesuvius, Mineralogy and Petrology, 73:145-176

Greene, Andrew R., 2013, Kılauea's Puu Oo eruption (1983–2010), Geochemistry, Geophysics, Geosystems, 14(1):4849-4873

Hildreth, Wes, 2012, The Novarupta-Katmai Eruption of 1912, P. 1-259, http://pubs.usgs.gov/pp/1791/

Hoernle, Kaj, 1998, Jurassic Oceanic Crust beneath Grand Canaria, Journal Of Petrology, 39(5):859-880

Jicha, B. R., 2004, Variable Impact of the Subducted Slab, Journal Of Petrology, 45(9)1845–1875

Johansen, T. S., 2005, 1585 eruption on La Palma, Canary Islands , Geology, 33:897-900

Johnson, K. E., 1996, Geochemistry of Augustine Volcano, Journal Of Petrology, 37(1):95-115

Kayzar, Theresa M., 2014, Bezymianny and Klyuchevskoy Volcanoes, Central Kamchatka, Contributions Mineralogy Petrology, 168(1067):1-28

Koppers, A. A, 2006, New 40Ar/39Ar Ages for Savai'i Island, http://adsabs.harvard.edu/abs/2006AGUFM.V34B..02K

Kurihara, Yuichi, 2007, Volcanic Products from Izu arc volcanoes, Radio Isotopes, 56:795-809

Kurihara, Yuichi, 2011, Izu arc volcanoes, Japan, Proc. Radiochim. Acta 1:335-338

Lanphere, Marvin, 2007, 40Ar/39Ar ages of the AD 79 eruption of Vesuvius, Bulletin Volcanology, 69:259-263

Ludwig, Kenneth, 2010, Isoplot User Manual, http://www.creationismonline.com/Isoplot/Isoplot.html, https://www.bgc.org/isoplot 3_75-4_15manual.pdf

Ludwig, Kenneth, 2012, Isoplot Excel Add In, http://www.creationismonline.com/Isoplot/Isoplot.html, https://www.bgc.org/isoplot 4.13%20files.zip

Marske, Jared P., 2008, Kilauea's Puu Oo Eruption, Journal Of Petrology, 49(7):1297-1318

Millet, Marc-Alban, 2009, Isotopic variations in Ocean Island Basalts, Chemical Geology 265:289-302

Pietruszka, Aaron J., 1999, Melting History of Kilauea Volcano, Journal Of Petrology, 40(8)1321–1342

Reagan, M. K., 2003, Rhyolite in North American Continental Arcs, Journal Of Petrology, 44(9):1703-1726

Regelous, Marcel, 2008, Melting beneath Niuafo'ou Island, Contributions Mineralogy Petrology, 156:103-118

Reubi, O., 2011, Disequilibria at Volcan Llaima, Chile, Earth and Planetary Science Letters, 303:37–47

Salvioli-Mariani, E., 2005, Glass-bearing crustal xenoliths, Lithos, 81:255-277

Schmitt, Axel K., 2006, Crystallization ages of Las Tres Vírgenes, Journal of Volcanology and Geothermal Research 158:281-295

Sims, Kenneth W.W., 2006, Comparison of Th, Sr, Nd and Pb isotopes, Earth and Planetary Science Letters, 245:743-761

Sims, Kenneth W.W., 2008, Alkaline magmas from Erebus volcano, Journal of Volcanology and Geothermal Research, 177:606-618

Smyth, Helen R., 2011, A Toba-scale eruption in the Early Miocene, Lithos 126 (2011) 198–211

Snelling, Andrew, 2003, Recent Andesite Flows at Mt. Ngauruhoe, Fifth International Conference on Creationism

Somma, R., 2001, (Sr-Nd-Pb) of interplinian magmas, Mineralogy and Petrology, 73:121-143

Tatsumoto, Mitsunobu, 1987, Origin of the West Maui volcanic rocks, Volcanism in Hawaii, 26:723-744

Tommasini, Simone, 2007, Study of Stromboli Volcano, Journal Of Petrology, 48(12): 2407-2430

Turner, S., 2001, Element Variations in Sunda arc Lavas, Contributions Mineralogy Petrology, 142:43-57

Turner, Simon, 2003, Evolution Beneath the Sangeang Api Volcano, Journal Of Petrology, 44(3)491–515

Turner, Simon, 2010, The 1912 Katmai-Novarupta Eruption, Journal Of Geophysical Research, 115(B12201):1-22

Vlastélic, Ivan, 2009, Piton de la Fournaise historical lavas, Journal of Volcanology 184:63-78

Wade, Jennifer A., 2005, The May 2003 eruption of Anatahan volcano, Journal of Volcanology and Geothermal Research 146:139-170

Wikipedia, 2015, Volcanoes of east-central Baja California, https://en.wikipedia.org/wiki/Volcanoes_of_east-central_Baja_California

Paul Nethercott is available for interviews and personal appearances. For more information contact us at info@advbooks.com

To purchase additional copies of these books, visit our bookstore at: www.advbookstore.com

Advantage BOOKS

Longwood, Florida, USA
"we bring dreams to life"™
www.advbookstore.com

www.ingramcontent.com/pod-product-compliance
Lightning Source LLC
Chambersburg PA
CBHW061342210326
41598CB00035B/5859
* 9 7 8 1 5 9 7 5 5 6 6 7 5 *